Signals and Communication Technology

More information about this series at http://www.springer.com/series/4748

Ilia V. Safonov · Ilya V. Kurilin
Michael N. Rychagov · Ekaterina V. Tolstaya

Adaptive Image Processing Algorithms for Printing

Springer

Ilia V. Safonov
Moscow
Russian Federation

Ilya V. Kurilin
Moscow
Russian Federation

Michael N. Rychagov
Moscow
Russian Federation

Ekaterina V. Tolstaya
Moscow
Russian Federation

ISSN 1860-4862 ISSN 1860-4870 (electronic)
Signals and Communication Technology
ISBN 978-981-10-6930-7 ISBN 978-981-10-6931-4 (eBook)
https://doi.org/10.1007/978-981-10-6931-4

Library of Congress Control Number: 2017956337

Printed on acid-free paper

This Springer imprint is published by Springer Nature
The registered company is Springer Nature Singapore Pte Ltd.
The registered company address is: 152 Beach Road, #21-01/04 Gateway East, Singapore 189721, Singapore

Preface

Every day, people shoot millions of photographs. The absolute majority of the images stay in digital form only. A small proportion of them are still assigned for printing, but even this amount represents a huge number of pictures. Frequently, the quality of the images is not very high. Usually, users do not have time for substantial manual editing of the captured images. How can the best visual quality of a hard copy be provided in automatic mode? This book describes in detail several algorithms for the image processing pipeline of photo-printer and photo-editing software tools.

The material is based on the experience of the authors in industrial research and development. We have worked on the development of algorithms for the Printing Division of Samsung Electronics for more than 10 years. In the book, we consider several techniques that we have developed at different times for processing images in documents and photographs. The algorithms presented in the book were disclosed earlier in dozens of patents worldwide, presented at international conferences, and realized in devices and software manufactured and utilized by the company.

We should especially note that this book in no way pretends to present an in-depth review of the achievements accumulated to date in the field of applied image processing and digital printing. Instead, in the book, the main results of our own studies carried out during more than ten years, that is, 2005–2016, are summarized. We hope that the main approaches, optimization procedures, and heuristic findings are still up to date and can be used as a basis for new intelligent solutions for the processing of photographs.

There are three groups of topics covered by the book: image enhancement algorithms, smart techniques for changing the aspect ratio for borderless printing, and approaches for peculiar printing modes. In the first group, we consider an image enhancement pipeline based, in part, on Exchangeable Image File Format (EXIF) information, adaptive local and global contrast improvement, the fusion of photographs captured with exposure bracketing, red-eye correction, reduction of JPEG artefacts, inverse halftoning, and sharpening. Smart cropping, image complementing, and retargeting are employed for changing the aspect ratio in the second group. The third group includes algorithms for printing anaglyph images, the generation of colour sketches for eco-draft-printing, detection of photo-orientation, and generation of collage from photographs. Almost all enumerated algorithms can be adjusted for any image enhancement applications, and only some of the algorithms have specifics intended for printing only. We do not describe methods related to specific printing devices or technologies, for example, laser or inkjet printing. Besides, the subjects of colour management and halftoning are also out of the scope of this book.

How can algorithms capable of being adaptive to image content be developed? In many cases, inductive or deductive inference can help. Many of our algorithms include lightweight classifiers or other machine-learning-based techniques, which have low computational complexity and model size. This allows them to be implemented on embedded platforms. The quality metrics is a key component in the development of the algorithm. In this book, we demonstrate various approaches for quality assessment.

As mentioned, the majority of the described techniques were implemented in the firmware of system on chip or in computer software products. This was a challenge because for each industrial task, there are always strong specification requirements and subsequent limitations on computational complexity and memory consumption. In the book, as a rule, we do not describe any device-dependent optimization tricks, but we provide ideas for effective methods from an algorithmic point of view.

This book is intended for all those who are interested in image processing and want to know how it works for photo-printing and how to apply machine learning techniques for the creation of effective adaptive algorithms. We hope the book will serve as a useful guide for students, researchers, and practitioners.

It is the intention of the authors that each chapter can be used as an independent text. In this regard, at the beginning of a large fragment, the main provisions considered in the preceding text are briefly repeated with reference to the appropriate chapter or section. References to the works of other authors and discussion of their results are given in the course of presentation of the material.

We would like to thank our colleagues who worked with us both in Korea and at the Samsung R&D Institute Russia, Moscow, on the development and implementation of the technologies mentioned in the book: Don Chul Choi, Dr. Sangho Kim, Dr. Kimin Kang, Dr. Hokeun Lee, Ilmir R. Alikberov, Dr. Victor V. Bucha, Dr. Sergey V. Efimov, Marta A. Egorova, Konstantin A. Kryzhanovsky, Dr. Alexey V. Nefyodov, Dr. Arkady L. Panfilov, Dr. Elena I. Patana, Dr. Xenia Yu. Petrova, Dr. Adilbek K. Shakenov, Dmitry E. Smirnov, Dr. Sergey S. Zavalishin, Alexey M. Vil'kin, Sergey Yu. Yakovlev, and numerous volunteers who took part in the collection of test databases and evaluation of the quality of our algorithms.

Contributions from our former partners at academic and institutional organizations with whom we are associated through joint publications, patents, and collaborative work are also deeply appreciated, especially those of Prof. Andrey S. Krylov, Dr. Andrey V. Nasonov, Dr. Vladimir P. Vezhnevets, and Dr. Alexander P. Vezhnevets of Moscow State University and Prof. Igor S. Gruzman of Novosibirsk State Technical University.

Moscow, Russian Federation

Ilia V. Safonov
Ilya V. Kurilin
Michael N. Rychagov
Ekaterina V. Tolstaya

Acknowledgement

Proof reading of all pages of the manuscript has been performed by PRS agency (http://www.proof-reading-service.com).

Contents

About the Authors

Ilia V. Safonov graduated from Moscow Engineering Physics Institute (at present time, National Research Nuclear University MEPhI) in 1993 as engineer-physicist. He obtained PhD degree in Computer Science in 1997. Since 2000, he is Associated Professor in the Department of Computer Science and Control Systems at MEPhI. At last decade, he had senior researcher position in R&D of Samsung, Nokia and Intel. At present time, he is Principal Research Scientist at Schlumberger Moscow Research. His interests include image and signal processing, machine learning, measurement systems, computer graphics and vision.

Ilya V. Kurilin received his MS degree in Radio Engineering from Novosibirsk State Technical University (NSTU), Russia, in 1999 and his PhD degree in Theoretical Bases of Informatics from NSTU in 2006. In 2007, he joined Image Processing Group, Samsung R&D Institute in Moscow, Russia, where he is engaged in photo and document image processing projects. Since 2015, he leads video and image processing laboratory specialized in real-time semantic processing of visual data for mobile devices.

Michael N. Rychagov received MS degree in Acoustical Imaging and PhD degree from the Moscow State University (MSU) in 1986 and 1989, respectively. In 2000, he received a Dr.Sc. degree (habilitation) from the same university. From 1991, he is involved in teaching and research at the Moscow Institute of Electronic Technology (MIET) as an Associate Professor in the Department of Theoretical and Experimental Physics (1998), Professor in the Department of Biomedical Systems (2008), Professor in the Department of Informatics and SW for Computer Systems (2014). Since 2004, he joined Samsung R&D Institute in Moscow, Russia (SRR), working on imaging algorithms for printing, scanning and copying, TV and display technologies, multimedia and tomographic areas. Currently, he is Director of the Advanced Mobile Solution Division at SRR. His technical and scientific interests are image and video signal processing, biomedical visualization, engineering applications of machine learning and artificial intelligence.

Ekaterina V. Tolstaya received her MS degree in Applied Mathematics from Moscow State University, in 2000. In 2004, she completed her MS degree in geophysics from University of Utah, USA, where she worked on inverse scattering in electromagnetics. Since 2004, she worked on problems of image processing and reconstruction in Samsung R&D Institute in Moscow, Russia. Based on these investigations, she obtained in 2011 her PhD degree in Research on Image Processing Algorithms for Printing. In 2014, she continued her career with Align Technology on problems involving computer vision, 3D geometry and machine learning.

Chapter 1
Adaptive Global and Local Contrast Enhancement

1.1 Introduction

People shoots millions of photographs every day. The majority of images are not intended for printing. Nevertheless, users print parts of photos and the market for photo printing is growing due to the increasing availability of cameras. The quality of photos to be printed has become a very important issue. Many non-professional photos are damaged by various defects and need to be enhanced automatically to be more pleasing to observers. We have collected a huge set of typical photos taken by various cameras and deteriorated by different defects. The following seven items were selected and ranked as the most noticeable defects of photographs:

- exposure issues;
- colour misbalance;
- blurring;
- JPEG compression artefacts;
- high noise;
- red eyes and other artefacts due to flash;
- colour fringing.

Figure 1.1 shows the percentage of each defect in the set of damaged photos. Some photos are damaged by several defects.

In this chapter, we discuss methods for the correction of images with exposure issues that have some problems with brightness and contrast. Modern cameras have smart algorithms for adjustment of the aperture, shutter speed, and ISO sensitivity, which define the exposure of a photo. As a rule, cameras can take a well-exposed image in normal lighting conditions. A well-exposed photo is one that appears bright and dark enough such that both shadows and highlights are as natural as possible to look at. There is no loss of shadows or highlights, and all details are clearly distinguishable. However, sometimes lighting conditions in a scene are far from ideal, an automatic camera causes a fault, or a user sets the wrong shooting

I.V. Safonov et al., *Adaptive Image Processing Algorithms for Printing*,
Signals and Communication Technology,
https://doi.org/10.1007/978-981-10-6931-4_1

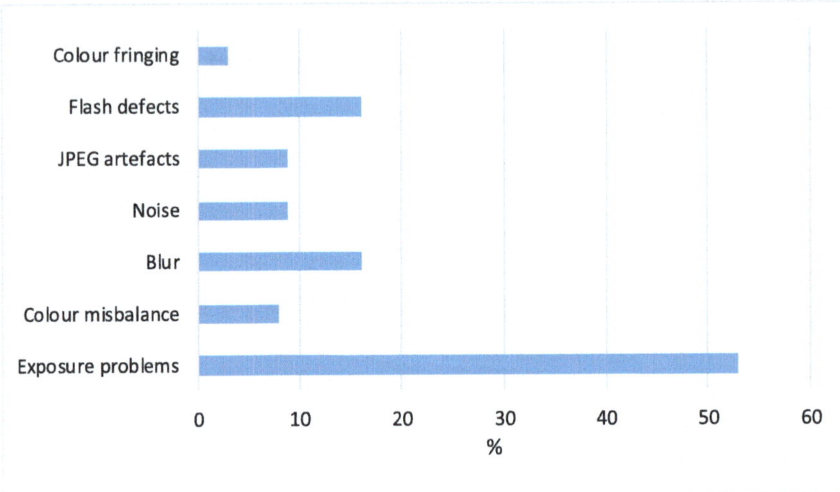

Fig. 1.1 Percentages of various defects in photos

parameters. Collectively or separately, these factors can lead to various issues with brightness and contrast.

First, let us give a definition of contrast. There are several mathematical definitions of contrast. For example, Michelson (1927) formulated contrast, also known as a visibility metrics, as:

$$Cm = \frac{I_{\max} - I_{\min}}{I_{\max} + I_{\min}},$$

where $I_{\max}$ and $I_{\min}$ are the highest and lowest brightnesses, respectively. Fechner (1860) described Weber's contrast as a rule of thumb regarding human perception. There are several mathematical formulations of Weber's contrast. For example, the following expression is used:

$$Cw = \frac{I - I_b}{I_b},$$

where I is an original image with fine-grained details, and I_b is an estimation of background via smoothing of I by Gaussian blurring. One can see that any definition of contrast is the difference or ratio between the maximal and minimal intensities in some region. When we consider local regions, we are talking about a local contrast. If we consider the entire image, then we are talking about a global contrast. Both should to be high enough to allow a good impression and perception of an image.

Figure 1.2 demonstrates an image with low global contrast as well as its brightness histogram. The histogram is narrow. Intensities of pixels lie in midtones

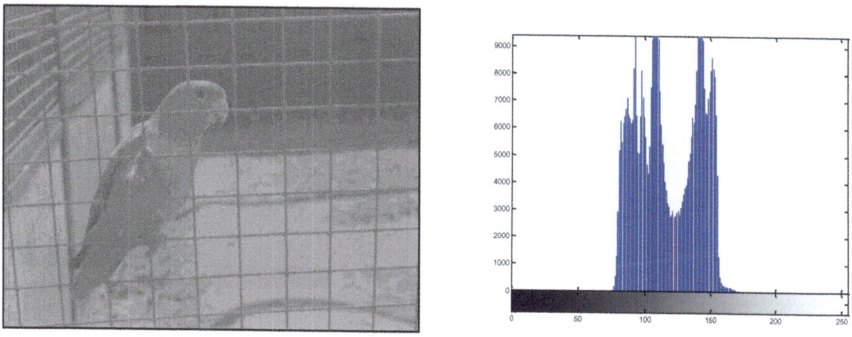

Fig. 1.2 Example of image with low global contrast as well as its brightness histogram

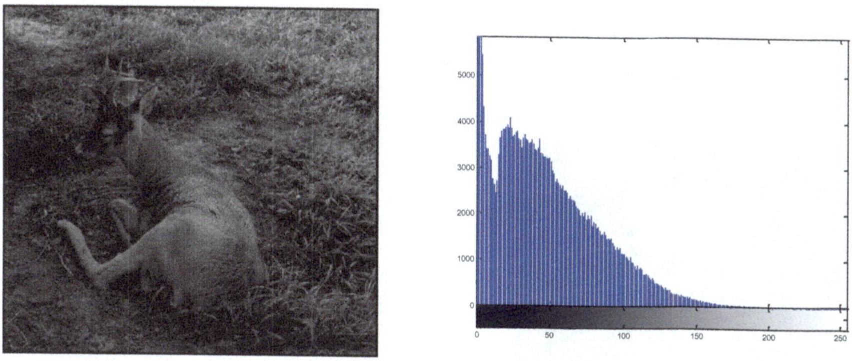

Fig. 1.3 Example of underexposed photo and its brightness histogram

only. There are no bright or dark pixels. It is hard to distinguish details in the image. The available dynamic range is employed ineffectively. Stretching of the histogram in a dynamic range could improve the quality of the photo.

An underexposed image is the sort of photograph that one might consider to be too dark. Figure 1.3 shows an example of an underexposed image as well as its histogram. The histogram is shifted to the left. Some of the intensities are cut off at zero. It is impossible to restore the original intensities of clipped pixels precisely. An image that is brighter than it should be and has no dark pixels can be considered overexposed. Figure 1.4 shows an example of an overexposed image as well as its histogram. The histogram is shifted to the right. Parts of the intensities are clipped by the maximum of the dynamic range. Identically to the case of underexposure, pixels with clipped intensities are corrupted irrevocably.

Sometimes an image has a high global contrast but local contrast in some areas is low. As a rule, this happens in shadow tones due to backlighting or shooting with a flash in a dark room. Figure 1.5 demonstrates a photo damaged by backlighting as well as its brightness histogram. One can see poorly distinguishable details in both

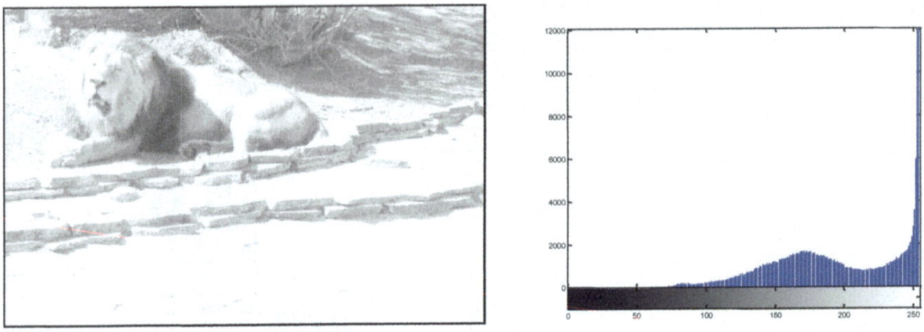

Fig. 1.4 Example of overexposed photo and brightness histogram

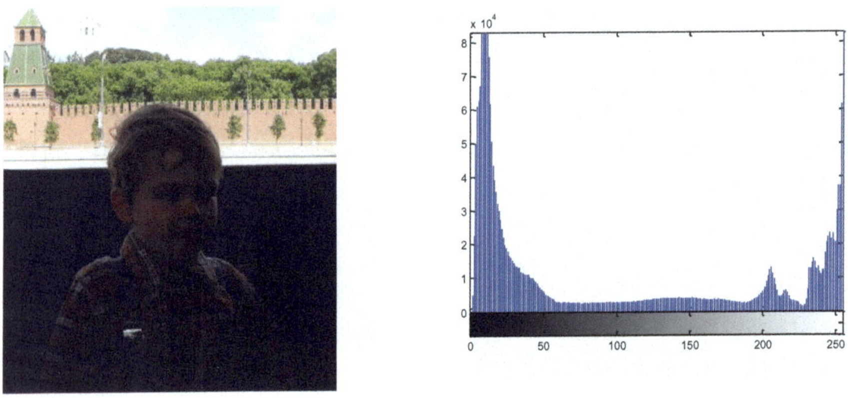

Fig. 1.5 Example of photo damaged by backlighting and its brightness histogram

shadows and highlights. As a rule, underwater photos as well as scenes with fog and smoke have low local contrast.

1.2 Global Contrast Adjustment

To improve the contrast, it is necessary to stretch a narrow histogram of an image from its lower and upper limits to the whole of the available dynamic range. There are two key problems in adaptive stretching of tones.

- How can the lower and upper limits of stretching be found automatically?
- Which function should be used for tone mapping?

These limits can be extracted from the brightness histogram. First, let us discuss how to calculate brightness channel I from an image in RGB colour space. The following four approaches are more popular.

1. Averaging of R, G, and B values:

$$I(r,c) = (R(r,c) + G(r,c) + B(r,c))/3,$$

where $(r,\ c)$ are coordinates of a pixel in an image.

2. Summation of R, G, and B with weights according to the relative sensitivity of human vision to different colours:

$$I(r,c) = 0.1R(r,c) + 0.6G(r,c) + 0.3B(r,c).$$

This is identical to obtaining the Y channel from YCbCr or YIQ colour spaces (Foley et al. 1990).

3. Obtaining the maximal from R, G, and B for a pixel:

$$I(r,c) = \max(R(r,c),\ G(r,c),\ B(r,c)).$$

This is identical to getting the V channel from the HSV colour space.

4. Getting the L channel from Lab colour space.

In general, there are two approaches to find a lower limit from a histogram H: by looking for minimal brightness that has a histogram value greater than a predefined threshold H_0 and by searching for the minimum brightness such that the sum of the histogram values to the left of this brightness is greater than the threshold C_0. An analogous method exists for calculation of an upper limit. We combine both approaches:

$$l = \min(T, \min\{i|H[i] \geq H_0\}, \min\{i|\sum_{k=0}^{i} H[k] \geq C_0\}),$$

where H is the brightness histogram; H_0, C_0, and T are thresholds for the histogram level, histogram area, and brightness, respectively. We include threshold T in the formula to avoid excessive darkening of the image.

$$u = \max(\max\{i|H[i] \geq H_1\}, \max\{i|\sum_{k=i}^{2^n-1} H[k] \geq C_1\}),$$

where H_1 and C_1 are the thresholds for the histogram level and area, and n is the bit depth of brightness channel. Figure 1.6 illustrates the thresholds for calculation of

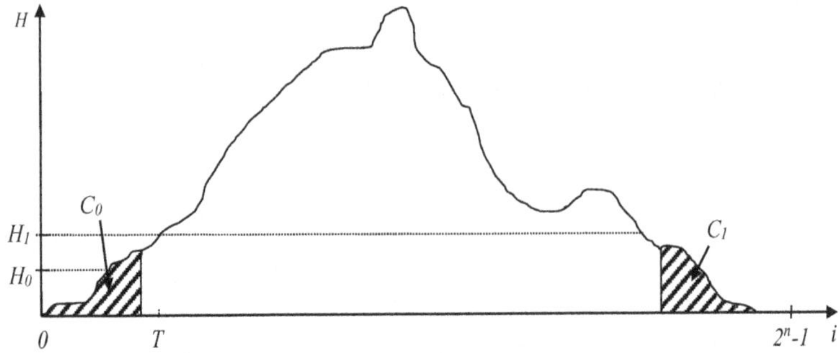

Fig. 1.6 Thresholds for adaptive contrast stretching

Fig. 1.7 Plot of function for proportional stretching

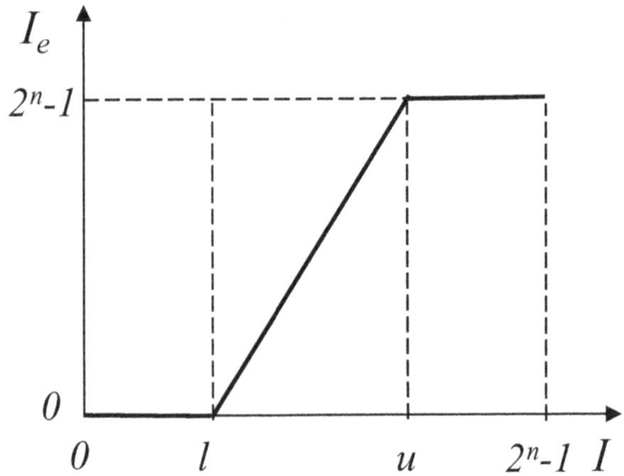

the lower and upper limits. Also, it is possible to calculate the limits from the *R*, *G*, and *B* channels instead of the brightness channel.

The simplest tone mapping function is proportional stretching by a line:

$$I_e(r,c) = \begin{cases} 0 : I(r,c) \le l \\ (2^n - 1)(I(r,c) - l)/(u - l) : l < I(r,c) < u, \\ 2^n - 1 : u \le I(r,c) \end{cases}$$

where n is the bit depth of a greyscale channel.

The same transformation can be applied for *R*, *G*, and *B* channels of the colour image. Figure 1.7 demonstrates a plot of the function for proportional stretching. An example of tone mapping via proportional stretching is shown in Fig. 1.8.

Proportional stretching provides nice outcomes in most cases. However, sometimes the application of nonlinear mapping via a curve is desirable. Gamma-correction is widely used in image processing and computer graphics for

(a) (b)

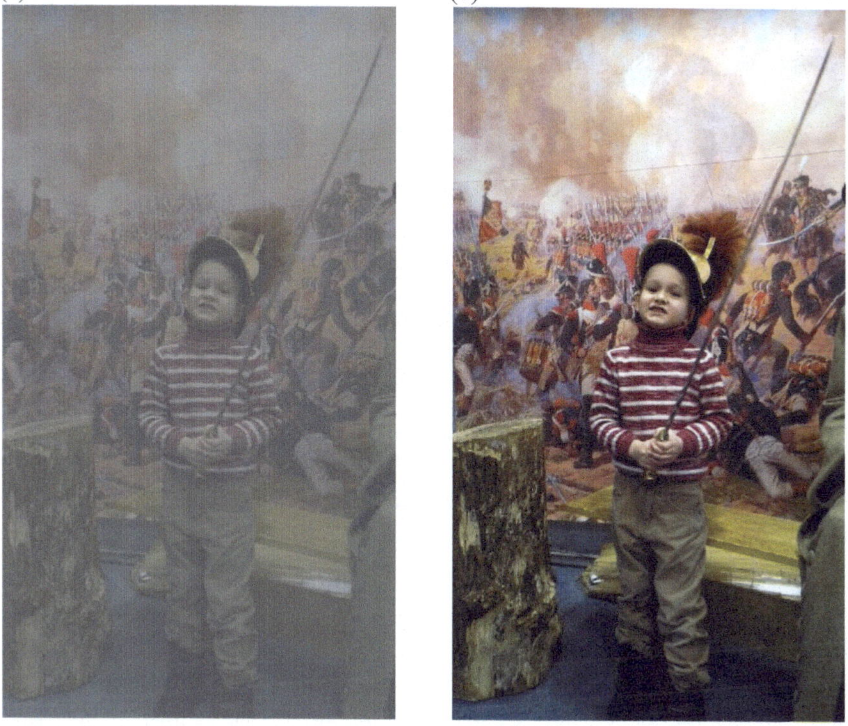

Fig. 1.8 Example of tone mapping via proportional stretching: **a** Original image; **b** outcome of correction

nonlinear reproduction of tones. The function was originally developed to compensate for the input–output characteristic of Cathode Ray Tube (CRT) displays. Poynton (1998) considers gamma-correction concepts in details. Nowadays the nonlinear function is used directly and indirectly in many imaging devices, coding standards, file formats, and so on. In particular, gamma correction or a similar function is employed in printing devices. So, there is no need to carry out gamma-correction each time before sending the image to a printing device, but it may be done to make an image darker or lighter. Stretching of tones with gamma correction can be done by means of a small modification of proportional stretching:

$$I_e(r,c) = \begin{cases} 0 : I(r,c) \le l \\ (2^n - 1)[(I(r,c) - l)/(u - l)]^{\gamma} : l < I(r,c) < u, \\ 2^n - 1 : u \le I(r,c) \end{cases}$$

A gamma value below one produces a convex curve (see Fig. 1.9) and makes an image lighter. A gamma value greater than one produces a concave curve and leads to image darkening.

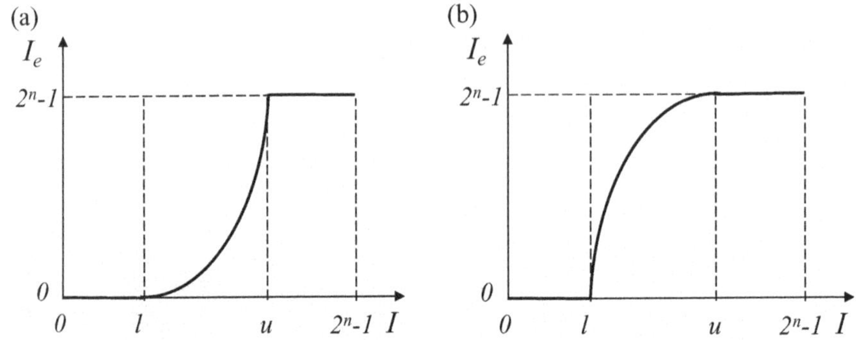

Fig. 1.9 Plots of functions for stretching with gamma-correction: **a** $\gamma > 1$; **b** $\gamma < 1$

An S-shaped curve or sigmoid is one more well-known nonlinear tone reproduction function. In Sect. 1.5 we describe the formation of a flexible S-shaped curve for joint global and local contrast improvement. The method provides fine and vivid photos.

1.3 Saturation Preservation

Most publications ignore the fact that desaturation is a side effect of increasing contrast for colour images. Sometimes noticeable desaturation occurs, while sometimes the decrease in saturation is not obvious. Let us demonstrate it analytically. The following expression is used for the calculation of saturation S in HSV colour space from R, G, and B components (Foley et al. 1990):

$$S = \frac{\max(R, G, B) - \min(R, G, B)}{\max(R, G, B)}.$$

Assume that a pixel R has maximal intensity and G has minimal intensity. Supposing we increase the contrast so that the new value of the red component is equal to $1.1R$ and the new value of the green component is equal to $0.9G$, where R and G are the initial red and green components respectively. Let us calculate the difference between the saturations of the initial and processed pixels:

$$\frac{R - G}{R} - \frac{1.1R - 0.9G}{1.1R} = -\frac{0.2G}{1.1R}.$$

The difference is negative, which means the saturation decreases.

After contrast changing, the following expressions for modification of colour channels should be used:

$$R_e(r,c) = R(r,c)I_e(r,c)/I(r,c),$$
$$G_e(r,c) = G(r,c)I_e(r,c)/I(r,c),$$
$$B_e(r,c) = B(r,c)I_e(r,c)/I(r,c),$$

where I, R, G, and B are the brightness and red, green, and blue channels of an initial image; I_e, R_e, G_e, and B_e are the brightness and red, green, and blue channels of an enhanced image.

The given expressions preserve constant ratios between RGB channels, and accordingly the hue and saturation of a colour image are kept constant. Fattal et al. (2002) depict a similar approach for the modification of colour channels after contrasting: the initial colour value is divided by the initial brightness and this ratio is raised to a power of 0.4–0.6 and then multiplied by the corrected brightness. We consider that raising to the power is not an appropriate action, because it does not preserve the saturation of a photo.

It is desirable to carry out identical modifications of the chromaticity channels of YCbCr colour space when the brightness channel Y is processed:

$$Cr_e(r,c) = I_e(r,c) \times (Cr(r,c) - 2^{n-1})/Y(r,c) + 2^{n-1},$$
$$Cb_e(r,c) = I_e(r,c) \times (Cb(r,c) - 2^{n-1})/Y(r,c) + 2^{n-1},$$

where Y, Cr, and Cb are the brightness and chromaticity channels of the initial image; I_e, Cr_e, and Cb_e are the brightness and chromaticity channels of the processed image; and n is the bit depth per chromaticity channel.

1.4 Shadow Areas Correction

1.4.1 Existing Approaches for Enhancement of Shadow Areas

One of the common approaches for local contrast correction is the application of the simplest physical model of forming a greyscale image I as an element-wise product of reflectance and luminance (Horn 1986):

$$I = R \times L,$$

where L is the luminance of the scene, and R is the reflectance (or albedo) of a fine structure. The luminance varies relatively slowly in the spatial domain. On the contrary, the reflectance contains high frequencies. The given model does not take into account several important matters such as the presence of light sources in a scene, specular reflection, and shadows from objects in a scene. Strictly speaking, the model corresponds to diffusive Lambert reflection only. In spite of the disadvantages, the model is quite applicable to a wide class of photos and is used in plenty algorithms for local contrast enhancement.

Many papers (Orsini et al. 2005; Tao and Asari 2003; Jobson 1997; Rahman et al. 2004; Watanabe et al. 2005) depict a method of correction of dark areas via various modifications of the MultiScale Retinex (MSR) algorithm. Classical MSR is determined as:

$$I_e(r,c) = \sum_{m=1}^{M} \omega_m \log(\frac{I(r,c)}{LPF_m(I)(r,c)}),$$

where M is the number of scales, and it is recommended that at least three scales should be applied; ω_m are weights; LPF_m are Gaussian blur filters with different variances for scales.

Figure 1.10 demonstrates the outcome of MSR. Certainly, MSR is able to make details distinguishable. However, it leads to colour distortions and a cartoon effect. An image processed by MSR looks unnatural. In addition, methods based on classical MSR have a high computational complexity.

Sobol (2004) describes an optimized algorithm for fast calculation of Low-Pass Filters (LPFs) for MSR and an application of a set of thresholds for prevention of the formation of a halo. Briefly, the idea of the approach is as follows. The modified

(a) (b)

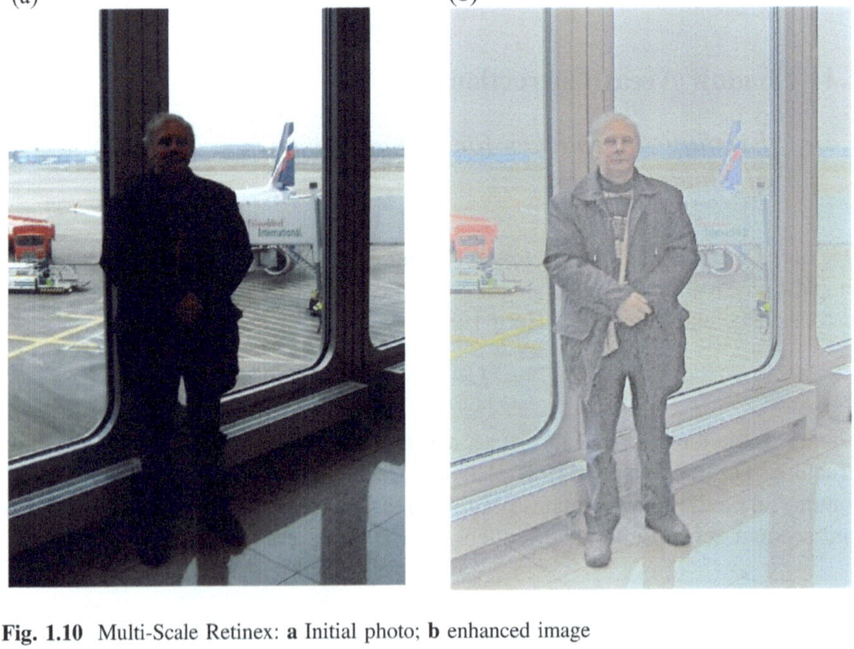

Fig. 1.10 Multi-Scale Retinex: **a** Initial photo; **b** enhanced image

(a) (b)

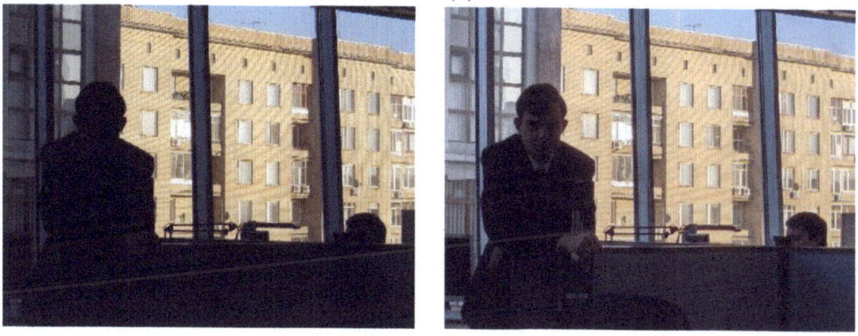

Fig. 1.11 Sobol's approach: **a** Initial photo; **b** enhanced image

MSR handles the brightness channel. The channel is calculated as maximal from *R*, *G*, and *B* channels. A mask is an outcome of the modified MSR. The mask is added to the *R*, *G*, and *B* channels. Figure 1.11 shows an image processed by Sobol's technique.

Moroney (2000) presents an intuitively clear and simple method of correction of areas of shadow and highlights. The first step is mask generation. The mask is a result of strong blurring of the brightness channel by means of a linear low-pass filter. The next step is transformation of the *R*, *G*, and *B* channels via a family of tone reproduction curves, where the curve shape depends on the mask value. For an image normalized in the range of [0, 1], the following expressions are employed:

$$R(r,c) = (R(r,c))^{2^{(1-2Mask)}},$$
$$G(r,c) = (G(r,c))^{2^{(1-2Mask)}},$$
$$B(r,c) = (B(r,c))^{2^{(1-2Mask)}}.$$

Figure 1.12 illustrates Moroney's technique. The dark areas of the image look better, but one can see a halo along the edges.

Chesnokov (2007) describes a method of image enhancement based on Orthogonal Retino-Morphic Image Transform:

$$I_e = \sum_{i=0}^{N} a_i(I) \times LPF(P_i(F(I))) \times Q_i(F(I)) + b(I),$$

where $P_i(x)$ is the orthogonal basis of functions of x defined in the range $0 < x < 1$; Q_i are anti-derivatives of P_i; *LPF* is the operator of low-pass spatial filtering; *F* is a weight function, which looks like gamma correction; *N* is the number of bands, $N \ll n$, where *n* is the bit depth per colour channel, and *a* and *b* are specific

(a) (b)

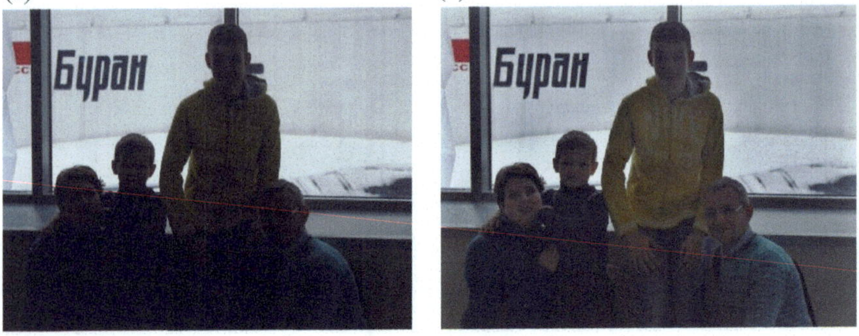

Fig. 1.12 Moroney's method: **a** Initial photo; **b** enhanced image

constants for each band. A choice of the appropriate P_i leads to a local contrast increase.

Brajovic (2004) depicts a method of dark area enhancement. The algorithm utilizes the following statement:

$$I_e = S \times (R + a \times L),$$

where a is the attenuation constant; S is a linear scaling operator that scales to the whole dynamic range of a signal; L is the outcome of edge-preserving filtering; R is a function of the ratio of brightness and L. The one-dimensional edge-preserving filter processes scanlines of the brightness channel. The parameters of the filter depend on estimation of the noise level. Noise is suppressed in a brightness channel but remains in chromaticity channels. The method does not form a distinguishable halo along edges.

Several High Dynamic Range (HDR) contrast reduction techniques can be used for correction of photos affected by backlighting (Tumblin et al. 1999; Fattal et al. 2002). However, these algorithms modify midtones and highlights as well. Frequently this leads to unwanted colour distortions, especially in highlights.

So, the mentioned techniques for local correction of dark areas have a set of essential drawbacks:

- the formation of a halo near edges;
- amplification of noise and compression artefacts;
- a cartoon effect;
- colour distortion;
- high computational complexity.

1.4.2 General Idea of Shadow Correction Technique

First, we describe the basic idea of our algorithm for a greyscale image with brightness normalized in the range from 0 to 1. Photos affected by backlighting have good global contrast but low local contrast in shadows. For an observer, it is difficult to distinguish details, especially in shadows, where tones are darker than in well-exposed areas. Therefore, it is necessary to increase the local contrast and brightness in shadow tones without any modification of other tones.

Assume that we have "an image of details" or "a fine structure image" D_s with high local contrast in shadow areas, that is, an image with easily distinguishable details in dark areas. Then it is possible to increase the local contrast and brightness of the source greyscale image I using the following alpha-blending of I and D_s:

$$I_e(r,c) = I(r,c) \times \alpha_s(r,c) + (1 - \alpha_s(r,c)) \times (k_s \times D_s(r,c) + (1 - k_s) \times I(r,c)),$$

where (r, c) are coordinates of a pixel; α_s is an alpha-channel that is treated as a transparency, pixels of α_s are close to 1 in bright areas and close to 0 in dark areas; and factor k_s controls the level of amplification of local contrast in shadows, where $k_s \in [0, 1]$.

After opening the brackets, we have the following formula:

$$I_e(r,c) = I(r,c) + k_s \times (1 - \alpha_s(r,c)) \times (D_s(r,c) - I(r,c)).$$

How can we obtain "a fine structure image" D_s? Humans distinguish details in an image because their brightness differs from that of the background in some local area. The statement inspires the application of the simplest physical model of forming a greyscale image, where "a fine structure image" is an estimate of the reflectance.

1.4.3 Reflectance Estimation and Correction of Brightness

The Single Scale Retinex algorithm estimates the local background or luminance of a scene by means of a low-pass filter with a Finite Impulse Response (FIR), for example, Gaussian blur. However, such an approach is not good, as sharp edges are degraded, producing smooth variation of L between regions with different levels of illumination. Excessive smoothing leads to image blurring and the production of a halo artefact along the edges.

To overcome the disadvantage of the FIR filter, it is possible to apply filtration that preserves the sharp edges. There are several edge-preserving filters. They are mainly considered as noise-suppressing filters. Tomasi and Manduchi (1998) describe a bilateral filter (BF) that is a family of effective and relatively simple techniques:

$$I_f(r,c) = \frac{\sum_{i=-S/2}^{S/2}\sum_{j=-S/2}^{S/2} I(r+i,c+j) \times v(i,j) \times w(I(r+i,c+j) - I(r,c))}{\sum_{i=-S/2}^{S/2}\sum_{j=-S/2}^{S/2} v(i,j) \times w(I(r+i,c+j) - I(r,c))},$$

where S is the size of the filter aperture or kernel in the spatial domain; a classical description of the BF uses Gaussian functions for the spatial domain and for the brightness range, the so-called edge-stop function:

$$v(i,j) = \exp(-\frac{i^2+j^2}{2\sigma_D^2}),$$

$$w(x) = \exp(-\frac{x^2}{2\sigma_R^2}),$$

where σ_D^2 and σ_R^2 are the variances for the spatial domain and brightness range respectively.

Computation of reflectance R for the model of the image formation is an ill-posed problem (Choudhury and Tumblin 2005). It is clear that the estimate of the reflectance is the ratio between the brightness of the initial image and the estimate of lightness and it is necessary to perform regularization. During experiments with a huge number of photos damaged by backlighting, we found a heuristic formula for the estimation of reflectance D_s:

$$D_s(r,c) = \frac{1.3 \times I(r,c)}{I_f(r,c) + 3 \times \sigma_R},$$

where I is the brightness channel of the initial image; I_f is the estimate of lightness by bilateral filtering of I with $\sigma_R = 0.1$.

Application of the expression makes details in shadow areas easily distinguishable. Correction of the brightness channel of a photo is performed using the expression from Sect. 1.4.2. The estimate of lightness I_f can be used as an alpha channel α_s. It is expedient to have an opportunity to control the tone width of shadow areas. The necessary effect is achieved by raising α_s to the power t_s, where $t_s \in [1, 6]$. In addition, we add clipping functions to the formula from Sect. 1.4.2 to prevent outcomes from being out of the allowable range:

$$I_e(r,c) = f_w(I(r,c) + k_s \times (1 - I_f(r,c))^{t_s} \times f_b(\frac{1.3 \times I(r,c)}{I_f(r,c) + 3 \times \sigma_R} - I(r,c))),$$

$$f_b(x) = \begin{cases} x : x \geq 0 \\ 0 : x < 0 \end{cases},$$

$$f_w(x) = \begin{cases} x : x \leq (2^n - 1) \\ 2^n - 1 : x > (2^n - 1) \end{cases},$$

where I is the brightness channel of the initial image; n is bit depth; I_f is the outcome of bilateral filtering; k_s is the correction factor for shadows; t_s is the tone width for shadows; σ_R is the variance of the edge-stop function of the BF.

Notice that I_e depends only on I and I_f, and thus the computation time of I_e can be reduced by constructing a 2D Look-Up Table (LUT) with indexes I and I_f and fetching the corrected values from the table. In case of limited memory resources, when it is impossible to store a $(2^n - 1)$ by $(2^n - 1)$ matrix, a decimated LUT can be constructed by considering subsequent bilinear interpolation between the decimated values. For example, when the bit depth n is equal to 8, outcomes of usage of LUT 64 by 64 or 86 by 86 looks like result of usage of LUT 256 by 256.

We noticed that depending on the parameters of the BF it is possible to achieve some increase in sharpness. Actually, if the Gaussian is an edge-stop function then some sharpening occurs for edges of less than $3\sigma_R$. The edges that have a greater local contrast do not change because the Gaussian tends rapidly to zero outside the range of $[-\sigma_R, 3\sigma_R]$. We propose to use the edge-stop function, which on the one hand is similar to the Gaussian and on the other hand does not tend to zero so rapidly. The following function is preferable to the Gaussian:

$$w(x) = \frac{1}{\sqrt{1 + \left|\frac{x}{\sigma_R^2}\right|^{\mu}}}.$$

El-Fallah and Ford (1993) described the function with $\mu = 2$. We name the function as E_μ, where μ is a concrete value. Function E_6 is close to Gaussian. Variation of μ in a range from 1 to 6 allows the sharpening to be controlled. Figure 1.13 shows a plot of $w(x)$ for $E2$ and $E6$ and also the Gaussian with the same σ_R.

Figure 1.14 demonstrates an example of a correction: part (a) shows the initial photo, (b) shows the corrected image, (c) shows the outcome I_f of bilateral filtering, and (d) shows the estimation D_s of reflectance.

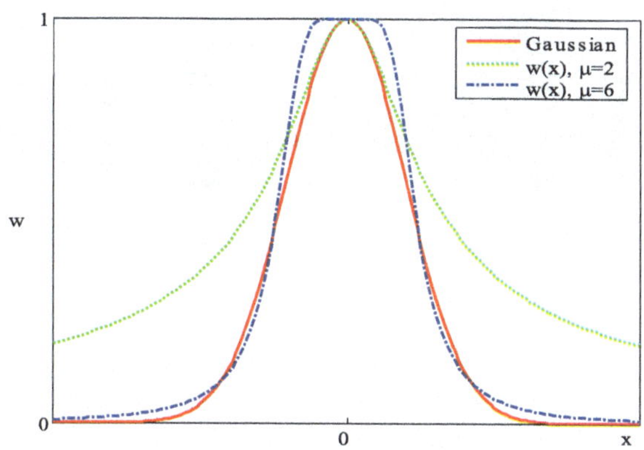

Fig. 1.13 Plots of various edge-stop functions

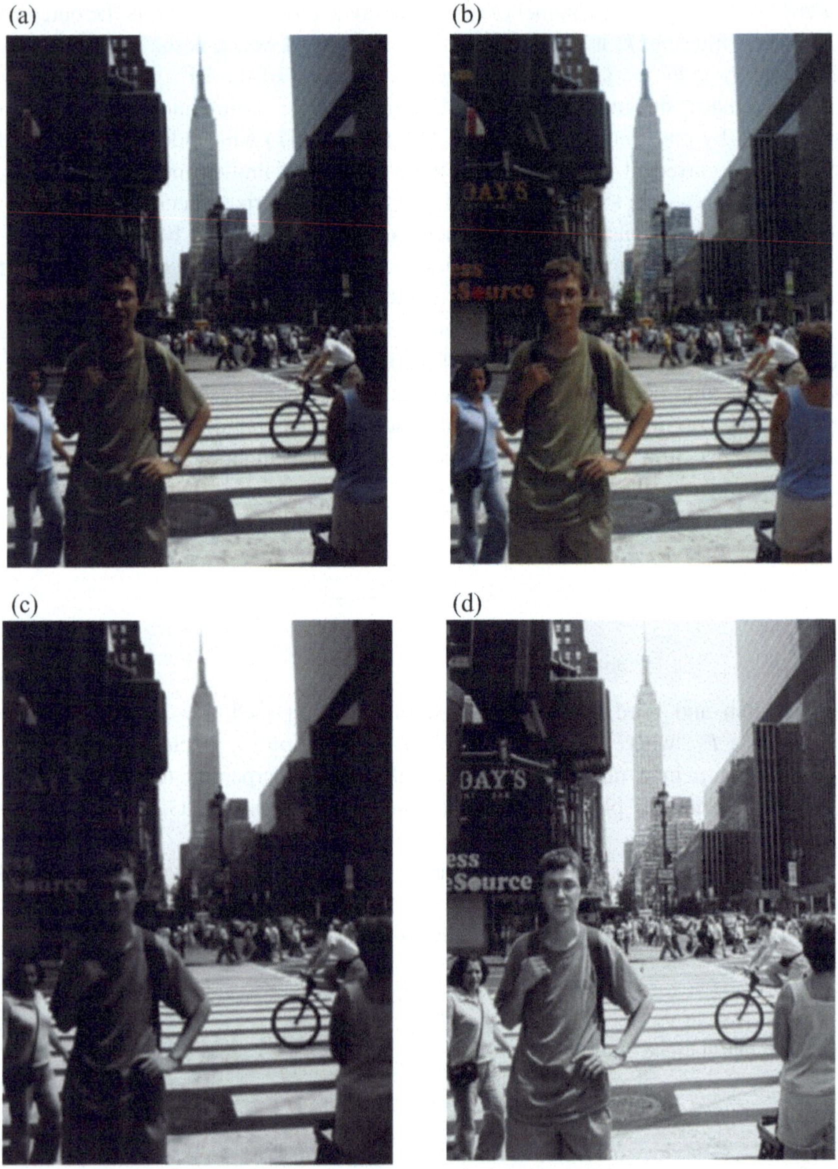

Fig. 1.14 Example of correction: **a** Initial colour photo; **b** corrected photo; **c** outcome of bilateral filtering; **d** estimation of reflectance

Images with low local contrast in the highlights are rare. Nevertheless, this sometimes happens. For example, see the tower in the top part of the photo in Fig. 1.15a. It is possible to perform a similar correction in the highlights as was

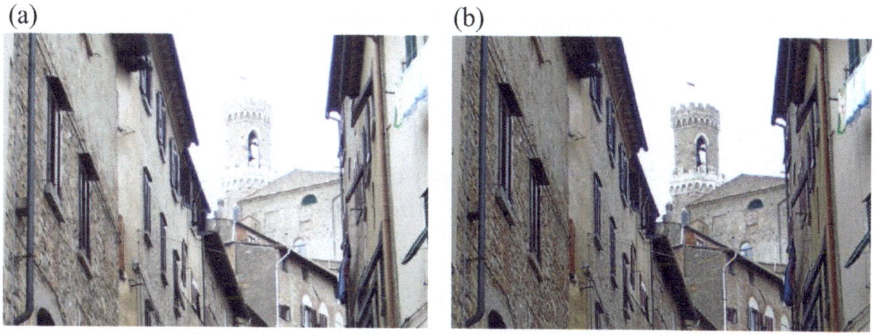

Fig. 1.15 Example of correction of highlights: **a** Initial photo; **b** enhanced image

done in the shadows and to combine both corrections into a single procedure. First, let us rewrite the formula for correction of tones in shadows:

$$I_{es}(r,c) = I(r,c) + k_s' \times (1 - \alpha_s(r,c)) \times (D_s(r,c) - I(r,c)).$$

By analogy, assume that we have "a fine structure image" D_h with high local contrast in highlights. Then the following expression increases the local contrast and decreases the brightness of I using alpha-blending of I and D_h:

$$I_{eh}(r,c) = I(r,c) - k_h' \times (1 - \alpha_h(r,c)) \times (I(r,c) - D_h(r,c)).$$

where α_h is an alpha-channel in the sense of transparency, which has values close to 1 in dark areas and values close to 0 in bright areas; factor k_h' controls the level of correction in highlights. Figure 1.15 shows an example of image correction in highlights.

The outcomes of correction of shadows and highlights can be combined:

$$I_e(r,c) = (I_{es}(r,c) + I_{eh}(r,c))/2 = I(r,c) + k_s \times (1 - \alpha_s(r,c)) \times (D_s(r,c) - I(r,c)) - k_h \times (1 - \alpha_h(r,c)) \times (I(r,c) - D_h(r,c)),$$

where k_s is the correction factor for shadows, $k_s = k_s'/2$; k_h is the correction factor for highlights, $k_h = k_h'/2$; k_s and $k_h \in [0, 1]$.

An image filtered by the bilateral filter I_f can be used as an alpha channel α_s. An inverted image I_f can be used as an alpha channel α_h. It is expedient to have an opportunity to control the tone width for shadows and highlights. The necessary effect is achieved by raising α_s to the power of t_s and α_h to the power of t_h, where t_s and $t_h \in [1, 6]$:

$$I_e(r,c) = I(r,c) + k_s(1 - I_f(r,c))^{t_s} \times (D_s(r,c) - I(r,c)) - k_h(I_f(r,c))^{t_h} \times (I(r,c) - D_h(r,c)).$$

One of the most preferable approach for calculation of D_h is:

$$D_h(r,c) = \frac{3 \times \sigma_R \times I_f(r,c)}{1.3 - I(r,c)}.$$

It is necessary to consider the bit depth of the brightness channel and to add clipping functions to prevent the outcomes from being out of the allowable range:

$$I_e(r,c) = f_w(I(r,c) + k_s \times (2^n - 1 - I_f(r,c))^{t_s} \times f_b(\frac{1.3 \times I(r,c)}{I_f(r,c) + 3 \times \sigma_R} - \frac{I(r,c)}{2^n - 1})$$
$$- k_h \times (I_f(r,c))^{t_h} \times f_b(\frac{I(r,c)}{2^n - 1} - \frac{3 \times \sigma_R \times I_f(r,c)}{1.3 - I(r,c)})),$$

where I is the brightness channel of the initial image having a bit depth of n; I_f is the outcome of bilateral filtering; k_s is the correction factor for shadows; t_s is the tone width for shadows; k_h is the correction factor for highlights; t_h is the tone width for highlights; σ_R is the variance of the edge-stop function of the BF.

1.4.4 Regression Model for the Choice of Shadow Correction Factor

Photos damaged by backlighting have high peaks in shadows and/or highlights and a gap in the midtones. Usually a histogram of photos with low local contrast in shadows is characterized by asymmetry in the shadows: its centre in the shadows is dislodged nearer to the origin of the brightness range. Figure 1.16 demonstrates a typical brightness histogram for an image affected by backlighting.

Fig. 1.16 Typical brightness histogram for photo damaged by backlighting

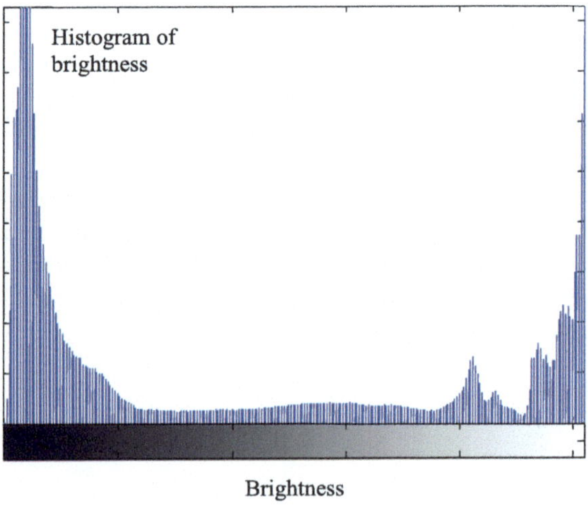

To characterize such histograms, we propose to use the following features. Parts of the tones in the shadow and middle tones are:

$$S_1 = \sum_{[0,1/3]} H(i)/(M \times N),$$

$$S_2 = \sum_{(1/3,2/3]} H(i)/(M \times N),$$

where H is the brightness histogram for an image with size $M \times N$ and its brightness normalized to the range [0, 1].

Parts of tones in the first and second halves of dark tones are:

$$S_{11} = \sum_{[0,1/6]} H(i)/(M \times N),$$

$$S_{12} = \sum_{(1/6,1/3]} H(i)/(M \times N).$$

The ratios of the histogram maxima in shadows, middle, and highlight tones per the global histogram maximum are:

$$M_1 = \max_{[0,1/3]}(H(i)) \,/\, \max_{[0,1]}(H(i)),$$

$$M_2 = \max_{(1/3,2/3]}(H(i)) \,/\, \max_{[0,1]}(H(i)),$$

$$M_3 = \max_{(2/3,\,1]}(H(i)) \,/\, \max_{[0,1]}(H(i)) .$$

Locations of the histogram maxima in shadows and highlights are:

$$P_1 = l | H(l) = \max_{[0,1/3]}(H(i)),$$

$$P_3 = l | H(l) = \max_{(2/3,1]}(H(i)) .$$

Figure 1.17 demonstrates a scatter plot for features P_1 and P_3 for 50 properly exposed photos and 50 photos affected by backlighting. Plots for other pairs of features look similar. One can see that it is possible to construct a classifier for detection of photos with backlighting. The automatic choice of correction factor k_s is a more complex task. Safonov (2006) describes a decision tree for calculation *of* k_s. The solution is based on deductive reasoning and is capable of providing only five discrete values of k_s.

It is preferable to apply an inductive inference, that is, the creation of a regression model by supervised machine learning. To create a good model, a huge number of images are required for training. Moreover, the training set should be

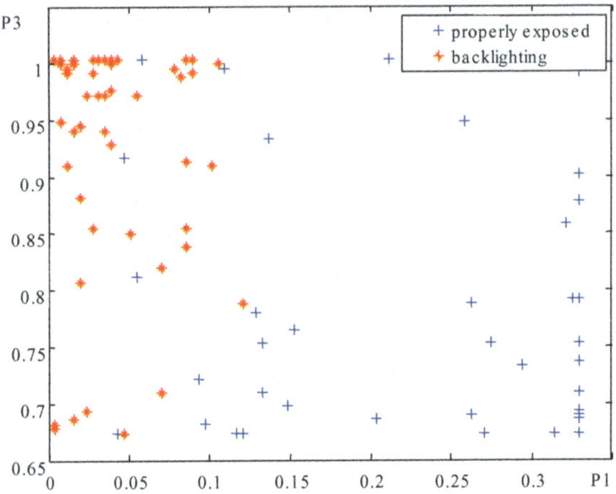

Fig. 1.17 Features P_1 and P_3 for well-exposed photos and images with backlighting

representative and values of k_s in the set should be evenly distributed. The collection of such a dataset is a tiresome task.

We employ a concept of semi-supervised learning (see the scheme in Fig. 1.18). An initial dataset containing about 200 photos was collected and labelled manually. The labelling procedure consisted of an interactive selection of correction factor of shadows k_s' providing a better visual appearance, where k_s' lies in the range from 0 to 100, with $k_s = k_s'/100$. We used the initial dataset for training of an initial regression model. The random forest (Breiman 2001) technique was selected for building the model, because, as a rule, the method makes it possible to obtain an appropriate outcome even without fine-tuning parameters and for a relatively small number of features; the size of its model is not big. Nowadays, many social networks provide an Application Interface (API) for downloading of photos, for example, flickr.com and vk.com. We made a software tool for downloading photos from several social networks in automatic mode. The initial model estimates the downloaded photos and collects photos having the required k_s' to provide a uniform distribution of images using a correction factor for shadows. Obviously, the initial model is far from ideal, and checking and sometimes relabelling of its outcome is necessary. We add checked photos to the main dataset. Photos from the initial dataset are added to the main dataset as well. We train and adjust the final regression model for images from the main dataset.

We used a five-fold cross-validation procedure for adjustment of parameters and estimation of the quality of the model. The following measures were used for quality assessment of the regression models: the Mean Absolute Error (MAE), Mean Squared Error (MSE), and Median Absolute Error (MedAE):

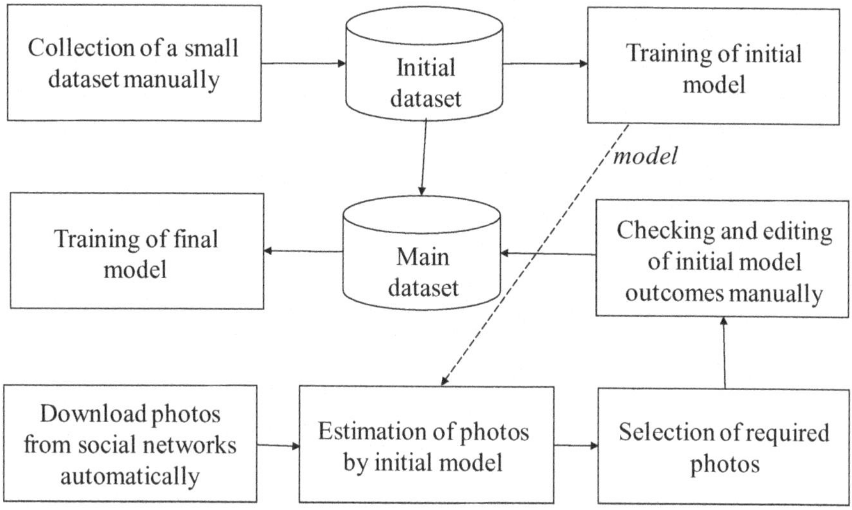

Fig. 1.18 Scheme of regression model creation

$$MAE = \frac{1}{N}\sum_{i=1}^{N}|f_i - y_i|,$$

$$MSE = \frac{1}{N}\sum_{i=1}^{N}(f_i - y_i)^2,$$

$$MedAE = median(|f_1 - y_1|, \ldots, |f_N - y_N|),$$

where N is the number of elements in a test set, f_i is the predicted value, and y_i is the true value.

The parameters of the final model were adjusted by minimizing MSE. Figure 1.19 demonstrates the dependence of MSE on the number of trees. In each cross-validation procedure, the plot changes a little due to the random nature of the method. We concluded that 40 trees should be used in our model. Other parameters are as follows: the maximal number of features to be split is equal to three, which corresponds to the value recommended for regression: the number of features divided by three; the minimal number of samples in a leaf is equal to five; the maximal depth of a tree is equal to 12; we do not use deep trees, because our training set contains outliers and noisy samples.

Table 1.1 shows a comparison of our regression model with the decision tree built on the same features (Safonov 2006) and the well-known least squares method. The random forest model outperforms the other considered approaches significantly. One more quality metrics for a regression is a distribution of residuals, which are the differences between predicted and true values. For a good regression model, the distribution should be close to normal. Figure 1.20 shows the distribution of residuals for the final model created. It looks good enough.

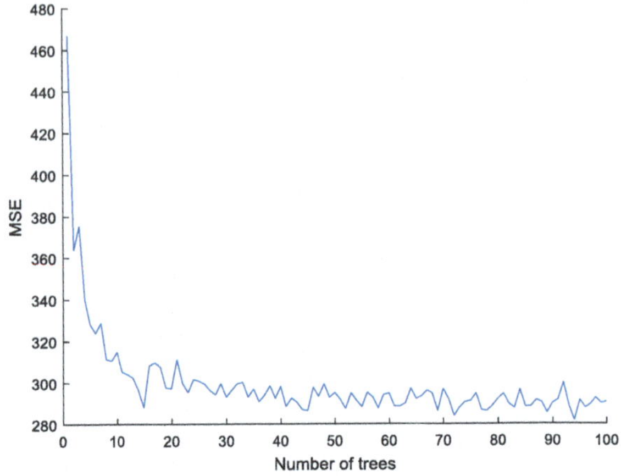

Fig. 1.19 MSE depending on the number of trees in the regression model obtained by random forest

Table 1.1 Comparison of regression models

Method	MAE	MSE	MedAE
Decision tree (Safonov 2006)	21.5	800.0	17.0
Least squares method	15.1	350.0	13.3
Random forest	**14.8**	**333.8**	**13.1**

Bold is used in order to emphasize the highest values.

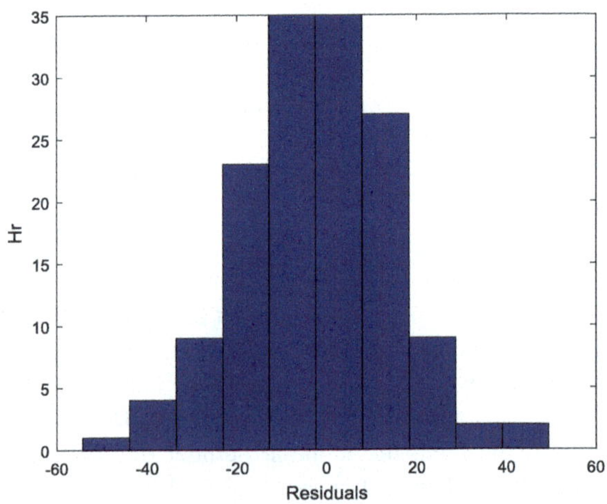

Fig. 1.20 Distribution of residuals

An important property of the random forest method is the ability to perform estimation of the relative importance of features. Figure 1.21 shows the relative importance for our features: S_{11}, S_{12}, and P_1 characterize photos affected by backlighting better than other features. Indeed, a big peak in shadows is a good indicator of backlighting. On the other hand, excluding some of the features (even M_3) leads to a decrease of the regression quality.

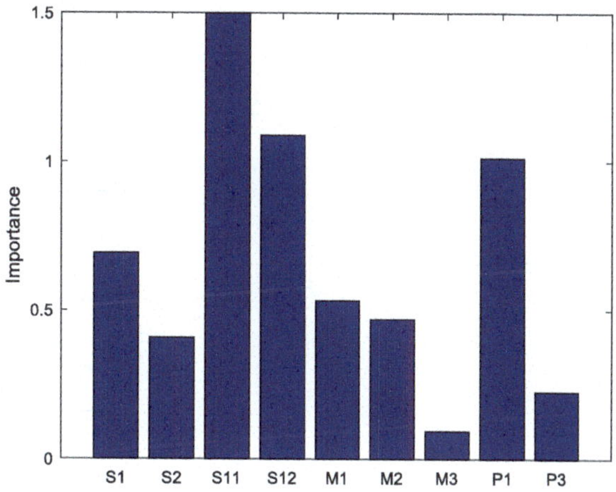

Fig. 1.21 Relative importance of features according to the random forest model

There is much room for improvement of our regression model due to smart feature engineering. The usage of features extracted from a co-occurrence matrix instead of features describing the global brightness histogram looks promising. One more direction of investigation is the partitioning of an image and calculation of features for local blocks. In addition, new important features can be obtained as combinations of existing ones.

1.4.5 Processing Workflow

Figure 1.22 shows a flow chart of our processing workflow. During the correction of shadows, the noise level and JPEG artefacts in these tones are amplified. For this reason, the suppression of noise and compression artefacts is required beforehand. The next step is automatic estimation of parameters in order to perform global and local correction. For proper correction of shadows, normal global contrast is necessary. Thus, a global contrast adjustment is required for colour channels before local contrast enhancement is applied. Further, we calculated the luminance by applying a BF and estimating the reflectance. The final stage is the modification of colour channels to preserve the saturation.

How can filtering of noise be done? First, a BF is intended for noise suppression. Tomasi and Manduchi (1998) proposed to calculate the edge-stop function in Lab colour space, because filtration of *R*, *G*, and *B* colour channels leads to the formation of colour fringes. However, for a small size of spatial kernel *S*, these halos are almost unnoticeable, so channel-wise filtering for R, *G*, and *B* is allowable.

Second, we propose to apply a Generalized Bilateral Filter (GBF) for denoising. GBF is an edge-preserving and texture-preserving noise reduction filter. It exploits the similarity between blocks as the Non-Local-means (NL-means) filter (Buades et al. 2005) instead of the similarity between pixels as in traditional BF.

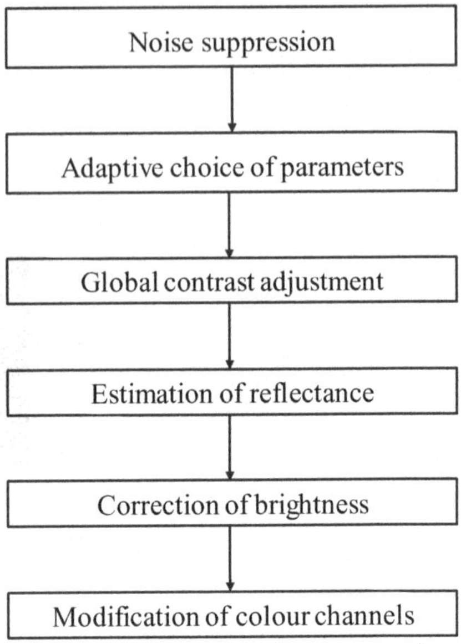

Fig. 1.22 Flow chart of the correction procedure

Traditional BF smooths textures having a low local contrast. GBF preserves such weak textures better.

The GBF for the *Y* channel is calculated as:

$$Y_f(r,c) = \frac{\sum_{i=-S/2}^{S/2}\sum_{j=-S/2}^{S/2} Y(r+i,c+j) \times V(i,j) \times W(D(P(r+i,c+j),P(r,c)))}{\sum_{i=-S/2}^{S/2}\sum_{j=-S/2}^{S/2} V(i,j) \times W(D(P(r+i,c+j),P(r,c)))},$$

where *S* is the spatial kernel size; *V*(*i*, *j*) is the spatial kernel; *W*(*x*) is the edge-stop function; *D*(*P*(*r* + *i*, *c* + *j*), *P*(*r*, *c*)) is the similarity (or distance) function between blocks; *P*(*r*, *c*) is a squared block of pixels.

There is an additional generalization of an idea to apply filtration depending on the similarity between blocks of various sizes. We named the following two filters *Multi-Block Bilateral Filters* (MBBFs):

$$Y_{f+}(r,c) = \frac{\sum_{i=-S/2}^{S/2}\sum_{j=-S/2}^{S/2} Y(r+i,c+j) \times V(i,j) \times \sum_k W(a_k D_k(P_k(r+i,c+j),P_k(r,c)))}{\sum_{i=-S/2}^{S/2}\sum_{j=-S/2}^{S/2} V(i,j) \times \sum_k W(a_k D_k(P_k(r+i,c+j),P_k(r,c)))},$$

$$Y_{f\times}(r,c) = \frac{\sum_{i=-S/2}^{S/2}\sum_{j=-S/2}^{S/2} Y(r+i,c+j) \times V(i,j) \times \prod_k W(a_k D_k(P_k(r+i,c+j),P_k(r,c)))}{\sum_{i=-S/2}^{S/2}\sum_{j=-S/2}^{S/2} V(i,j) \times \prod_k W(a_k D_k(P_k(r+i,c+j),P_k(r,c)))},$$

where *S* is the spatial kernel size; *V*(*i*, *j*) is the spatial kernel, *W*(*x*) is a weight function; $D_k(P_k(r + i, c + j), P_k(r, c))$ is a similarity function between blocks of size

Table 1.2 Comparison of filters for suppression of AWGN ($\sigma = 5$)

Filter	PSNR (dB)
BF (*Gaussian*)	35.8398
BF (*E6*)	35.8129
BF (*E8*)	35.8274
GBF (*Gaussian*, 3 × 3 block)	35.9707
GBF (*E6*, 3 × 3 block)	36.5896
GBF (*E8*, 3 × 3 block)	**36.6454**
MBBF (*Gaussian*, 1 × 1 * 2 × 2 * 3 × 3)	**36.6593**
MBBF (*E8*, 1 × 1 * 2 × 2 * 3 × 3)	36.6406
MBBF (*Gaussian*, 1 × 1 + 2 × 2 + 3 × 3)	36.1830
MBBF (*E8*, 1 × 1 + 2 × 2 + 3 × 3)	36.1925

Bold is used in order to emphasize the highest values.

k by k; $P_k(r, c)$ is a squared block of pixels of size k by k; a_k is a weight for a block of size k by k.

Table 1.2 contains a comparison of different filters for the suppression of Additive White Gaussian Noise (AWGN) in the image from Fig. 1.5. The Peak Signal-to-Noise Ratio (PSNR) is a figure-of-merit. One can see that GBF and MBBF provide better outcomes.

The computational complexity of bilateral filtering with a domain kernel size S for an N by N image is estimated as $O(N^2S^2)$ and requires significant computational resources. Dispersion of the Gaussian σ_D^2 in the spatial kernel of the BF is large enough. It is preferable to use a flat kernel u instead of a Gaussian. Also, it is reasonable to apply a BF to process the image first row-wise and then column-wise (Pham and van Vliet 2005; Safonov et al. 2006):

$$I_r(r,c) = \frac{\sum_{j=-S/2}^{S/2} I(r,c+j) \times u(j) \times w(I(r,c+j) - I(r,c))}{\sum_{j=-S/2}^{S/2} u(j) \times w(I(r,c+j) - I(r,c))},$$

$$I_f(r,c) = \frac{\sum_{i=-S/2}^{S/2} I_r(r+i,c) \times u(i) \times w(I_r(r+i,c) - I_r(r,c))}{\sum_{i=-S/2}^{S/2} u(i) \times w(I_r(r+i,c) - I_r(r,c))},$$

$$u(x) = \begin{cases} \frac{1}{S}, & |x| \le \sigma_D \\ 0, & |x| > \sigma_D \end{cases}.$$

By analogy to convolution, such a filter can be named "separable", although separability of the BF in a mathematical sense has not been proved rigorously. The computational complexity of the separable BF is $O(2N^2S)$, and for $S < 100$, a separable filter is faster than a piecewise-linear BF (Durand and Dorsey 2002).

1.4.6 Results of Shadow Correction

We used two approaches for evaluation of the quality of our method. First, we compared the proposed method with existing software and hardware solutions for correction of shadows. Second, we collected subjective opinions about the visual quality of correction.

We suggest that the best method produces the smallest number of undesirable artefacts. The following unwanted effects were revealed:

- a halo along strong edges;
- colour distortions, that is, alterations of hue and/or saturation;
- growth of the noise level in dark areas or for an entire image;
- over-amplification of dark tones, which leads to black colour becoming an unpleasant dirty-grey tint;
- over-amplification of middle and/or light tones;
- a cartoon effect.

Figure 1.23 shows a test pattern for assessment of the enumerated unwanted effects. Visual analysis is used for detection of various artefacts and the cartoon effect. In particular, a halo was estimated on the steep edges of region #4, and the horizontal gradient in region #3 was used to reveal a cartoon effect. The right half of the test pattern is intended for measuring the colour values and noise levels in several different regions. In the right half of the test pattern, Gaussian noise is added. The correction parameters for all tested solutions were selected so that the outcomes of correction were approximately identical in region #1. The average standard deviations for the R, G, and B channels in region #1 indicate the noise levels. For assessment of the colour invariance and absence of colour distortions, we measured the hue and saturation in region #2. The brightness in region #5 is very low; it is almost black in colour. It is presumed that region #5 should not

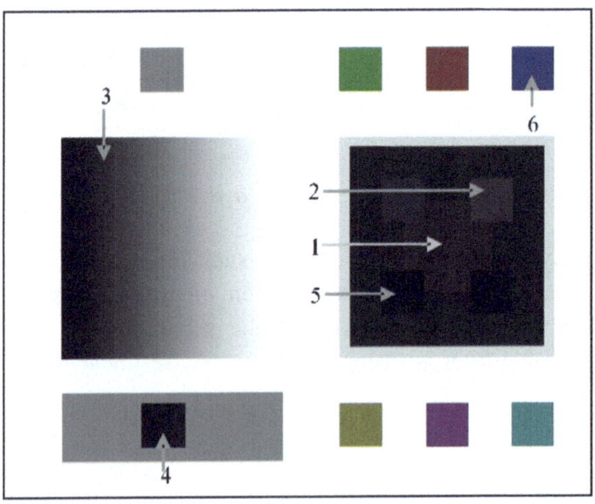

Fig. 1.23 Test pattern for assessment of approaches for enhancement of shadows

Table 1.3 Software tools for correction of shadows

#	Software tool	Comments
1	Adobe Photoshop	Shadow/Highlight menu item
2	Kodak Digital SHO Professional	Plug-in for Adobe Photoshop by Austin Development Center of Kodak
3	Chroma Shadow Control	Plug-in for Adobe Photoshop
4	FixerLabs Shadow Fixer	Plug-in for Adobe Photoshop
5	IntrigueTek Shadow Illuminator	Plug-in for Adobe Photoshop (Brajovic 2004)
6	Nikon Capture	Nikon D-Lighting (Chesnokov 2007)
7	Akvis Enhancer	Plug-in for Adobe Photoshop
8	TruView PhotoFlair	Retinex (Rahman et al. 2004)

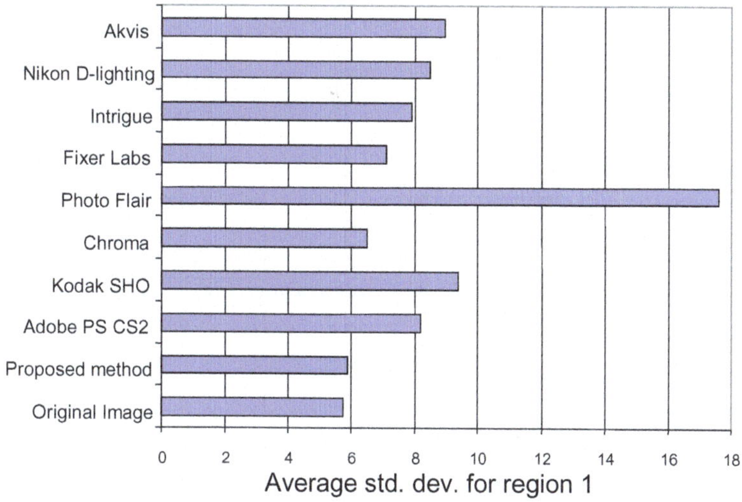

Fig. 1.24 Comparison of the noise level

become black due to correction. Region #6 is used for measuring the amplification in the midtones. Presumably, region #6 should not change its brightness much.

Table 1.3 enumerates the software tools aimed at the correction of shadows that were used for comparison. Figures 1.24, 1.25, 1.26 and 1.27 show diagrams of the measurement results. Our approach demonstrates a negligible increase of noise

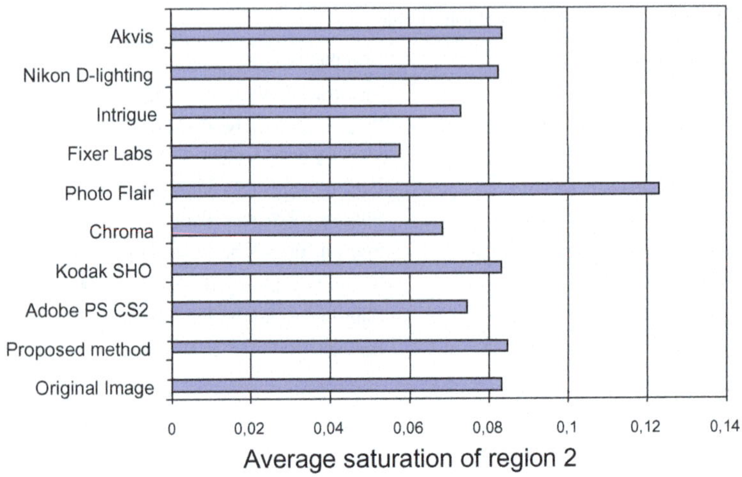

Fig. 1.25 Comparison of saturation

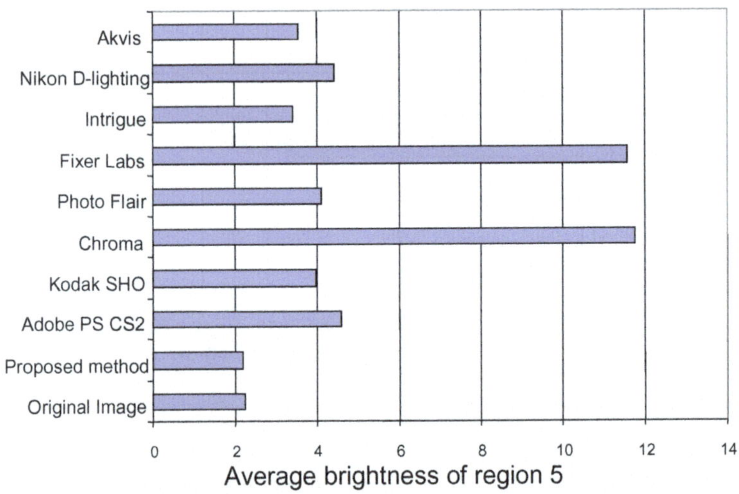

Fig. 1.26 Comparison of amplification of extremely dark tones

level in comparison with the original image. Other methods amplify the noise more significantly. The considered techniques modify the hue insignificantly, but some of them change the saturation greatly. Adobe, Chroma, and IntrigueTek give modest desaturation. FixerLabs provides noticeable desaturation. PhotoFlair yields large

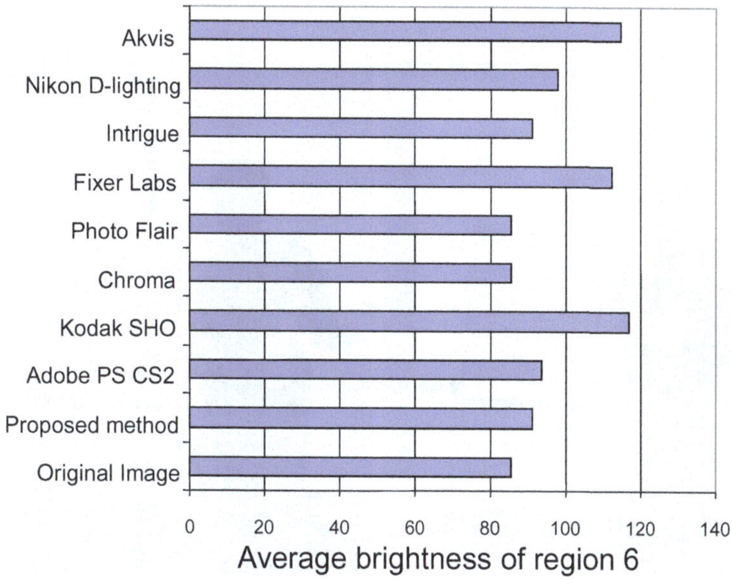

Fig. 1.27 Comparison of the amplification of middle tones

oversaturation. Our method does not influence extremely dark tones. Other methods perform brightening of these tones. Chroma and FixerLabs demonstrate unacceptable results. Our method modifies the middle tones insignificantly. Kodak, FixerLabs, and Akvis amplify middle tones by 15–20%. Every considered method except for IntrigueTek leads to a cartoon effect. Our algorithm does not lead to a noticeable halo. A halo appears along the edges after the application of Adobe, Nikon D-lighting, and Photo Flair. Figure 1.28 demonstrates the advantages of our method over Adobe Photoshop's shadow/highlight tool. Adobe's approach produces a strong halo effect, whereas a halo is absent in the image processed by our method.

HDR compression techniques can be applied for the correction of photos damaged by backlighting as well. However, the methods usually lead to deterioration of highlights. Figure 1.29 shows a comparison of our approach and HDR gradient domain compression (Fattal et al. 2002). The approach for HDR compression visualizes dark areas very well but damages bright areas. Sky in the photo becomes darker and desaturated. Details of the roof of Notre Dame become less distinguishable. Our method lightens dark tones slightly more weakly but does not modify highlights.

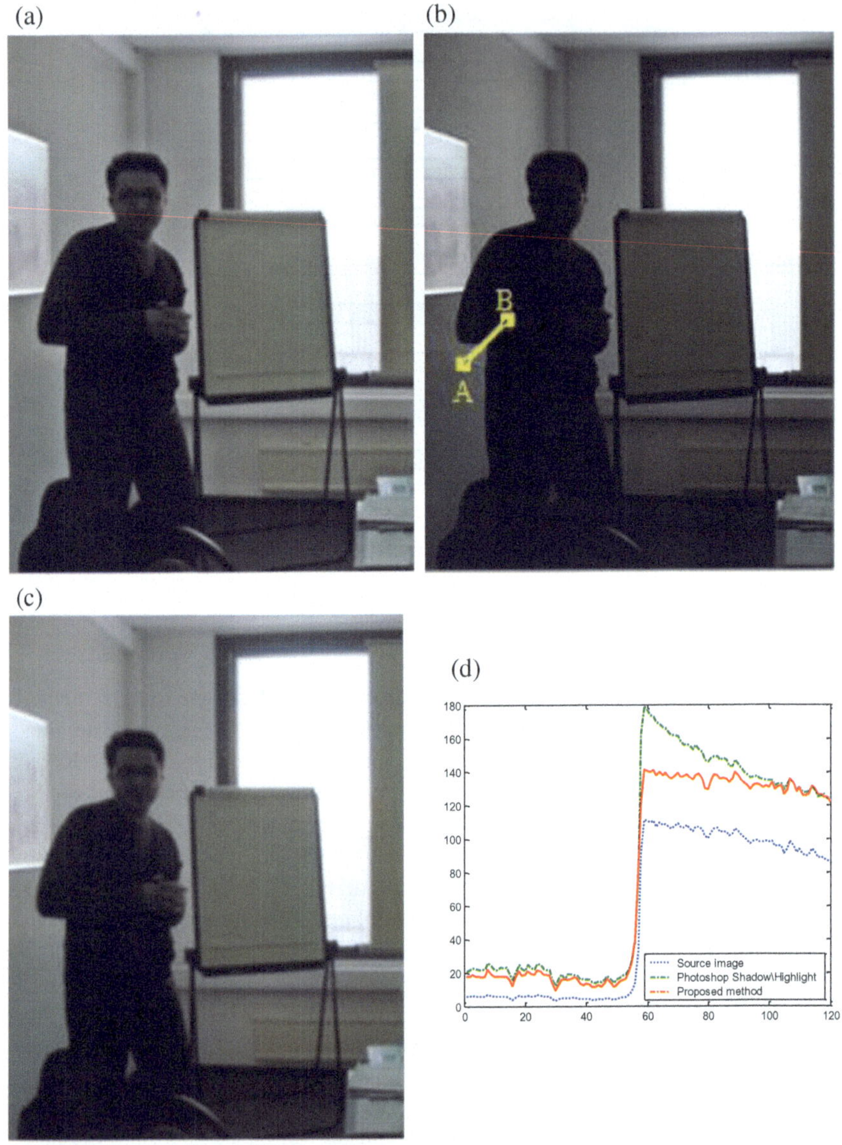

Fig. 1.28 **a** Photo enhanced by Photoshop Shadow/Highlight; **b** initial image; **c** photo enhanced by proposed method; **d** plots of brightness profiles along line segment AB

For estimation of the quality of processing from the point of view of the observer, we corrected five photos damaged by backlighting and also printed original and enhanced photos. A survey of opinions among 28 observers was carried out. Each interviewee selected the original photo, the enhanced photo, or

Table 1.4 Results of opinion survey

Photo	Enhanced (%)	Original (%)	No difference (%)
"Nature"	82	11	7
"Sculpture"	64	29	7
"Night shot"	71	25	4
"Party"	82	18	0
"Museum"	89	11	0

both as the most pleasing. Ninety per cent of observers selected all of the enhanced photos as more pleasing (Table 1.4). Figure 1.30 shows the image "Nature" that was used in the survey and its correction outcome.

1.5 Visibility Improvement

1.5.1 A Tone Mapping Curve Controlled by a Local Histogram

Sometimes photos have strong differences in contrast between the background and foreground details in various areas of the image. Parts of the objects in the foreground are emphasized while others have poor visibility. To improve the contrast in each local area, a locally adaptable tone-mapping function should be applied. Our concept is the following (Kurilin et al. 2015):

- an S-shaped transformation curve that is able to change shape adaptively depending on the local brightness distributions will be defined;
- smooth alteration of a curve shape between adjacent pixels and regions will be provided;
- the family of transformation functions will be applied to R, G, and B channels.

We utilize a cubic Hermite spline as an S-shaped curve. The spline in a parametric form is:

$$f(t) = (1 - t^2 + 2t^3)P_0 + (3t^2 - 2t^3)P_1 + (t - 2t^2 + t^3)Q_0 + (t^3 - t^2)Q_1,$$

where the parameter t changes in the range of [0, 1]. The curve begins at the point P_0 and finishes at the point P_1. Vectors Q_0 and Q_1 are tangent to the curves at points P_0 and P_1 respectively. The vectors change the slope of the curve.

To calculate the x coordinates of the starting point P_{0x} and ending point P_{1x} we analyse the global brightness histogram H as described in Sect. 1.2. Figure 1.31 illustrates the dependence of starting and ending points of the spline on the brightness histogram of an image. The starting point $P_{0x} = l$ corresponds to the

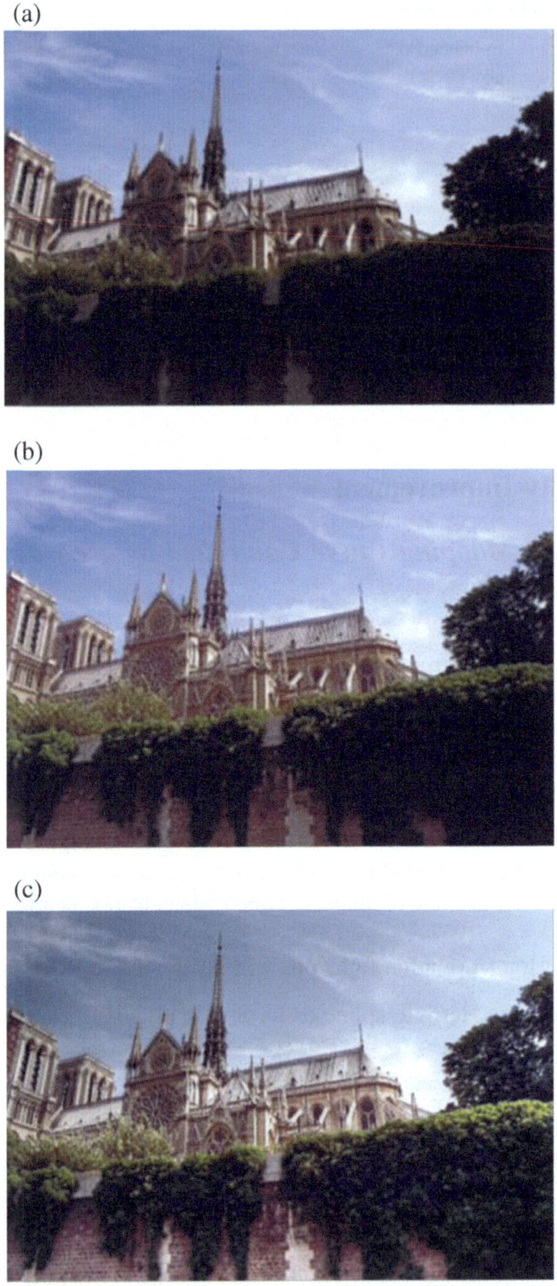

Fig. 1.29 Comparison with HDR compression: **a** Initial image; **b** our outcome; **c** outcome of gradient domain HDR compression

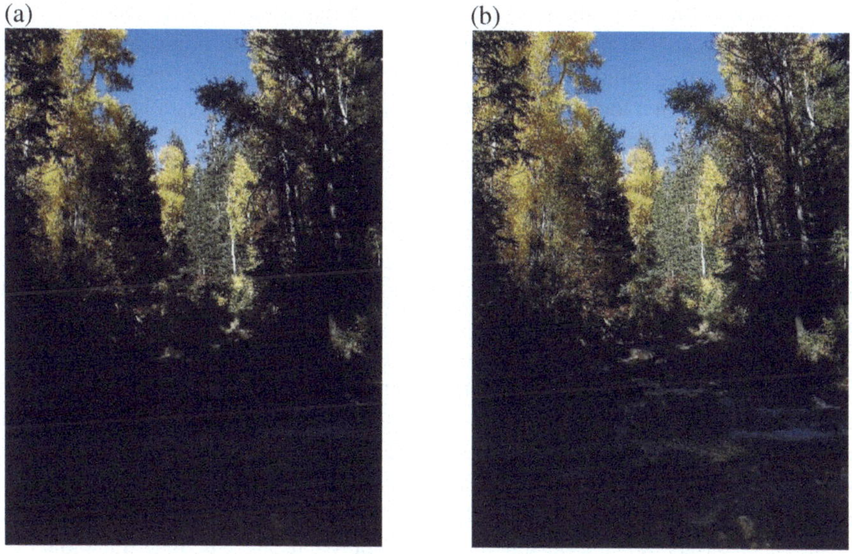

Fig. 1.30 Example of an image from survey: **a** Initial "Nature" photo; **b** enhanced image

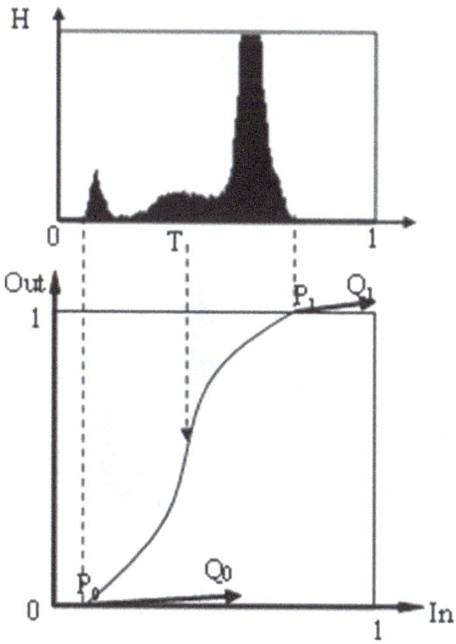

Fig. 1.31 Starting and ending points of spline depending on the global brightness histogram

darkest tones (the so-called black point) on an image. The ending point $P_{1x} = u$ corresponds to the lightest tones (the so-called white point) on an image.

The shape of a spline depends on the local distribution of background and foreground pixels in a local region. We propose to use Otsu's threshold (Otsu 1979) for the brightness channel in order to set up Q_0 and Q_1 properly:

$$\begin{array}{ll} \textit{if}\ K > 0.5: & Q_{0x} = 1 + A(K - 0.5);\ Q_{0y} = 0;\ Q_{1x} = 1;\ Q_{1y} = 0; \\ \textit{else}: & Q_{0x} = 1;\ Q_{0y} = 0;\ Q_{1x} = 1 + A(0.5 - K);\ Q_{1y} = 0; \end{array}$$

where A is a coefficient in the range from 1 to 6, and K is Otsu's threshold normalized to the range [0, 1].

If the tonal width of a local histogram is narrow or unimodal, then K is calculated for a global brightness histogram; otherwise K is calculated for a local histogram. Figure 1.32 demonstrates the adjustment of spline shape for two local areas. We obtain a convex curve for a dark area and a concave curve for a light one. To prevent the formation of visible artefacts, it is necessary to provide smooth alteration of the spline shape for adjacent local areas and neighbouring pixels. Figure 1.33 shows the general scheme of our approach. We calculate thresholds K_i for overlapped blocks and store K_i in the matrix M. Further, we apply a low-pass filter for smoothing the matrix of thresholds. Each RGB pixel of the source image is transformed by a unique curve that has tangents depending on the parameter K that is extracted from the matrix M by means of bilinear interpolation. Zavalishin et al. (2017) describe additional advances of the approach by using segmentation for definition of a local area instead of squared blocks.

There are many approaches for effective implementation of the algorithm. Calculation of the cubic Hermite spline for each pixel has a high computational

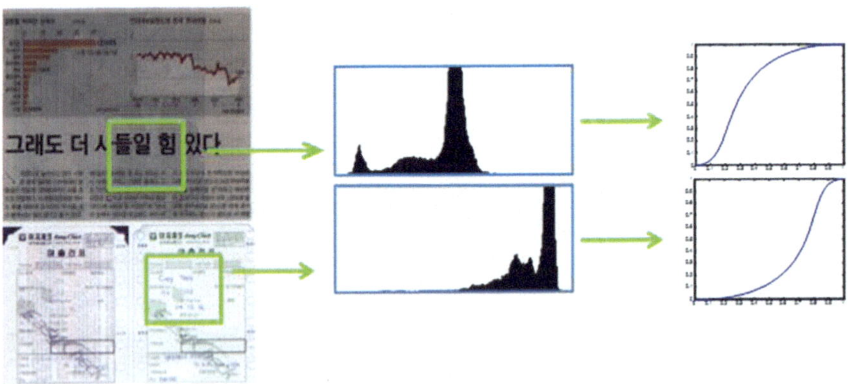

Fig. 1.32 Alteration of the spline shape depending on a local histogram

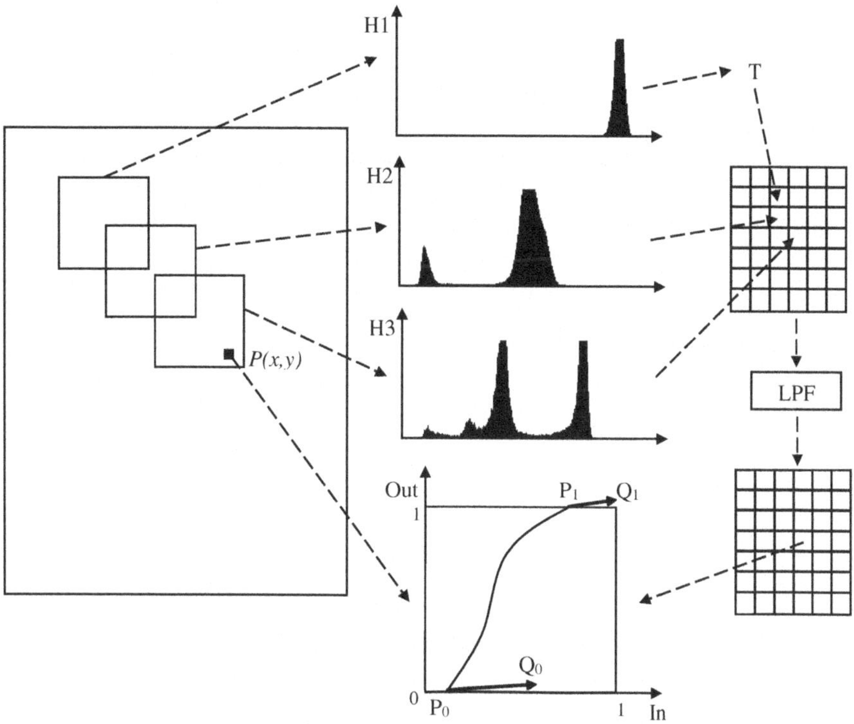

Fig. 1.33 An illustration of the local contrast improvement algorithm

cost. We pre-calculate the 2D lookup table with a size of 256 × 256 pixels for the starting point $P_0 = 0$, ending point $P_1 = 1$, and A = 3. The size of the 2D LUT is 64 kB. This may be too big for an embedded implementation. We decimate the LUT four times, store the 2D LUT with a size of 64 × 64, and take values from the table by bilinear interpolation. To take into account the global contrast enhancement, we carry out contrasting of the 2D LUT in the first processing stage. To decrease the computational complexity, we carry out an estimation of the global and local thresholds for a downsampled greyscale copy of the initial RGB image. We apply an optimized version of the Otsu algorithm for threshold estimation (Lin 2005). To smooth matrix M, we utilize a box-filter with a kernel size of 5 × 5. Hensley et al. (2005) describe an effective algorithm for a box-filter based on a summed area table (also well-known as an integral image). All calculations are performed using a fixed-point arithmetic. In addition, the algorithm can be parallelized effectively for a modern multi-core CPU by means of OpenMP.

1.5.2 Results of Visibility Enhancement

Figure 1.34 shows a test pattern for quality estimation of our approach in comparison with existing techniques. It contains dark grey and light grey background regions. There are eight grey and colour text areas in the image. Neither global nor local contrast is too low; all text areas are distinguishable. However, it is preferable to improve the contrast from the point of view of human perception.

For original and processed images, we calculated the Michelson contrast C_m for eight local areas on the test pattern. For each image, Table 1.5 lists the number of improved areas, the mean C_m, the minimal C_m, which is the worst case, and the existence of visual artefacts. One can see that the proposed method outperforms all analysed solutions: it improves all local areas, significantly increases the average and minimal Michelson contrast, and does not lead to any visual artefacts. The proposed technique operates similarly to an optical polarization filter. Due to the

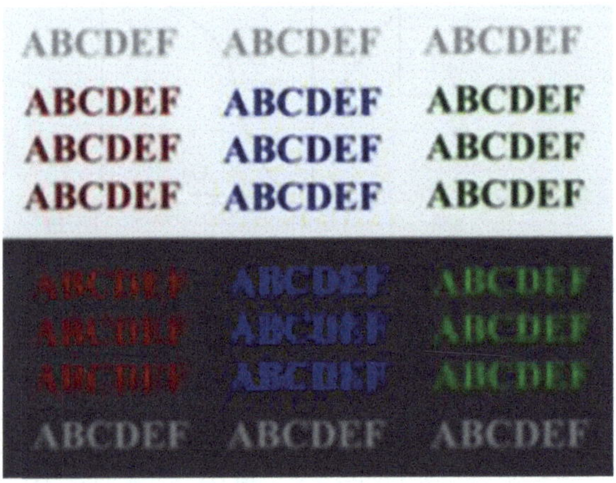

Fig. 1.34 Test pattern for the estimation of visibility enhancement

Table 1.5 Comparison of algorithms for visibility enhancement

Method	Number of improved areas	Mean C_m	Min. C_m	Visual artefacts
Original image	N/A	0.40	0.05	N/A
PS auto contrast	0	0.40	0.05	No
PS auto levels	2	0.40	0.06	No
MSR (Rahman et al. 2004)	4	0.42	0.17	Yes, halo
Moroney (2000)	6	0.43	0.05	Yes, halo
Safonov et al. (2006)	6	0.43	0.01	No
CLAHE (Zuiderveld 1994)	4	0.40	0.09	Yes, stains
Our approach	8	0.54	0.26	No

(a) (b)

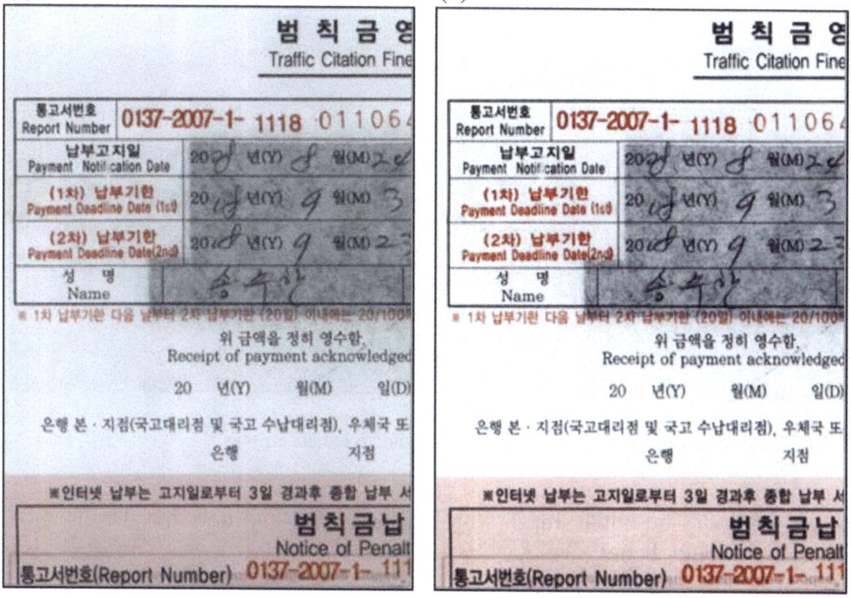

Fig. 1.35 An example of scanned image enhancement: **a** Original image; **b** enhanced one

(a) (b)

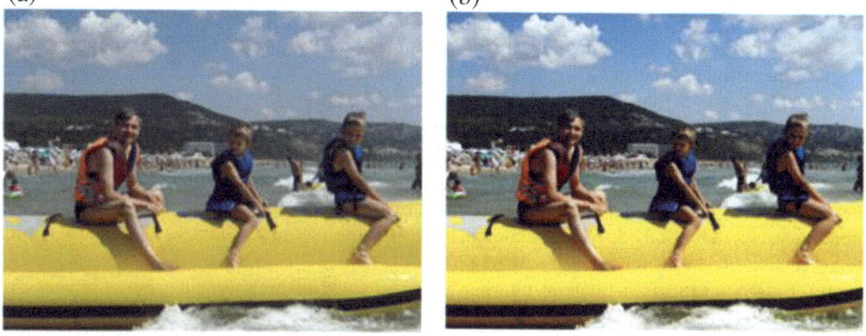

Fig. 1.36 An example of photo enhancement: **a** Original photo; **b** enhanced one

increase in local contrast, the saturation increases too and the processed photos look nice and vivid. Figure 1.35 shows an example of visibility enhancement of a scanned image. Figure 1.36 does the same for a photo.

Nowadays, one of the important problems is the correction of photos and video affected by fog/haze and smoke (Tarel et al. 2012; Yao et al.2009) as well as underwater photos (Schettini and Silvia 2010). A visibility enhancement algorithm can be applied for the correction of such photos. Figure 1.37 demonstrates dehazing

(a) (b)

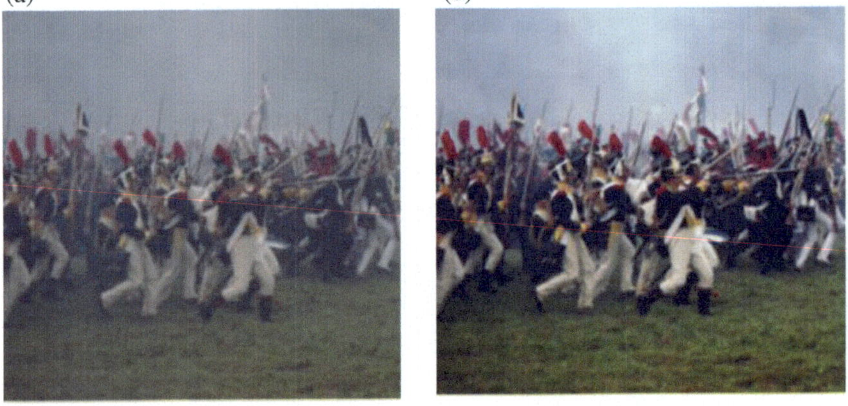

Fig. 1.37 An example of dehazing: **a** Original photo; **b** enhanced one

by our method. The initial image with fog looks dull and ugly but the enhanced image looks much better.

References

Brajovic, V.: Brightness perception, dynamic range and noise: a unified model for adaptive image sensors. In: Proceedings of the 2004 IEEE Computer Society Conference on Computer Vision and Pattern Recognition (2004)

Breiman, L.: Random forests. Mach. Learn. **45**(1), 5–32 (2001)

Buades, A., Coll, B., Morel, J.M.: A non local algorithm for image denoising. In: Proceedings of IEEE International Conference on Computer Vision and Pattern Recognition, vol. 2, pp. 60–65 (2005)

Chesnokov, V.: Image enhancement methods and apparatus therefor. US patent 7,302,110 (2007)

Choudhury, P., Tumblin, J.: The trilateral filter for high contrast images and meshes. In: ACM SIGGRAPH Courses, p. 5. ACM (2005)

Durand, F., Dorsey, J.: Fast bilateral filtering for the display of high-dynamic range images. In: SIGGRAPH 2002 (2002)

El-Fallah, A.I., Ford, G.E.: Structure preserving inhomogeneous diffusion in image filtering. In: IEEE 27th Conference on Signals, Systems and Computers (1993)

Fattal, R., Lischinski, D., Wierman, M.: Gradient domain high dynamic range compression. ACM Trans. Graph. **21** (2002)

Fechner, G.T.: Elemente der Psychophysik. band 2. Breitkopf und Härtel, Leipzig (1860)

Foley, J.D., van Dam, A., Feiner, S.K., Hughes, J.F.: Computer Graphics: Principles and Practice. Addison Wesley (1990)

Hensley, J., Scheuermann, T., Coombe, G., Singh, M., Lastra, A.: Fast summed-area table generation and its applications. Comput. Graph. Forum **24**(3), 547–555 (2005)

Horn, B.K.P.: Robot Vision. MIT Press (1986)

Jobson, D.J.: A multiscale Retinex for bridging the gap between colour images and the human observation of scenes. IEEE Trans. Image Process. **6**(7) (1997)

Kurilin, I.V., Safonov, I.V., Rychagov, M.N., Zavalishin, S.S., Han, D.H., Kim, S.H.: Fast algorithm for visibility enhancement of the images with low local contrast. In: Proceedings of IS&T/SPIE Electronic Imaging, p. 9395 (2015)

Lin, K.C.: On improvement of the computation speed of Otsu's image thresholding. J. Electron. Imaging **14**(2) (2005)

Michelson, A.A.: Studies in Optics. University of Chicago (1927)

Moroney, N.: Local color correction using non-linear masking. In: 8th Color Imaging Conference (2000)

Orsini, G., Ramponi, G., Carrai, P., Di Federico, R.: A modified Retinex for image contrast enhancement and dynamics control. In: Proceedings 8th Euromicro Conference on Digital System Design (DSD'05) (2005)

Otsu, N.: A threshold selection method from grey level histogram. IEEE Trans. Syst. Man Cybern. **9**(1), 62–66 (1979)

Pham, T.Q., van Vliet, L.J.: Separable bilateral filtering for fast video preprocessing. In: Proceedings of IEEE International Conference on Multimedia (ICME) (2005)

Poynton, C.A.: Rehabilitation of gamma. In: Proceedings of IS&T/SPIE Electronic Imaging, pp. 232–249 (1998)

Rahman, Z., Jobson, D.J., Woodell, G.A.: Retinex processing for automatic image enhancement. J. Electron. Imaging **13** (2004)

Safonov, I.V.: Automatic correction of amateur photos damaged by backlighting. In: Proceedings of 16th International Conference on Computer Graphics. GraphiCon, pp. 80–89 (2006)

Safonov, I.V., Rychagov, M.N., Kang, K.M., Kim, S.H.: Automatic correction of exposure problems in photo printer. In: Proceedings of IEEE Tenth International Symposium on Consumer Electronics (ISCE), pp. 13–18 (2006)

Schettini, R., Silvia, C.: Underwater image processing: state of the art of restoration and image enhancement methods. EURASIP J. Adv. Signal Process. (2010)

Sobol, R.: Improving the Retinex algorithm for rendering wide dynamic range photographs. J. Electron. Imaging **13**(1) (2004)

Tao, L., Asari, V.: Modified luminance based MSR for fast and efficient image enhancement. In: Proceedings 32nd Applied Imagery Pattern Recognition Workshop (AIPR'03) (2003)

Tarel, J.P., Hautiere, N., Caraffa, L., Cord, A., Halmaoui, H., Gruyer, D.: Vision enhancement in homogeneous and heterogeneous fog. IEEE Intell. Transp. Syst. Mag. **4**(2), 6–20 (2012)

Tomasi, C., Manduchi, R.: Bilateral filtering for gray and color images. In: Proceedings of IEEE Conference on Computer Vision (1998)

Tumblin, J., Hodgins, J., Guenter B.: Two methods for display of high contrast images. ACM Trans. Graph. **18** (1999)

Watanabe, T., Kuwahara, Y., Kojima, A., Kurosawa T.: An adaptive multi-scale Retinex algorithm realizing high color quality and high-speed processing. J. Imaging Sci. Technol. **49** (2005)

Yao, B., Huang, L., Liu, C.: Adaptive defogging of a single image. In: Proceedings of IEEE Second International Symposium on Computational Intelligence and Design, vol. 1, pp. 56–59 (2009)

Zavalishin, S.S., Safonov, I.V., Bekhtin, Y.S., Gurov, V.S.: Parameters adaptation framework for local contrast enhancement. In: Proceedings of 6th Mediterranean Conference on Embedded Computing (MECO) (2017)

Zuiderveld, K.: Contrast limited adaptive histogram equalization. In: Graphic Gems IV. Academic Press Professional, pp. 474–485 (1994)

Chapter 2
Fusion of Photos Captured with Exposure Bracketing

2.1 Introduction

2.1.1 What Is High Dynamic Range Imaging?

Usually the terms "High Dynamic Range" (HDR) and "High Dynamic Range Imaging" (HDRI) are used whenever the lightness intensity values span four or more orders of magnitude. A related term, "low dynamic range (LDR)", refers to digital systems supporting only eight bits per colour channel or intensity ratios between 0 and 255.

HDRI is an emerging field. It is guided by recent electronic industry trends that lead to further miniaturization of compact Digital Still Cameras (DSCs) and increase the image pixel resolution, making the sensor smaller and causing it to suffer from a lower dynamic range due to its reduced capacity. On the other hand, consumer demand for increased image quality and attractiveness is forcing manufacturers to develop more sophisticated image-processing algorithms.

The visual quality of high dynamic range images is much higher than that of conventional low-dynamic-range images. With the development of this technology, it will become the norm rather than the exception.

The main challenges of HDR technology are capturing, storing, and displaying/printing images. When taking photographs of an outdoor scene, the brightness ratio, or dynamic range, between an area in full sunlight and another area in deep shade can exceed 10,000:1. Cameras with charge-coupled device (CCD) imaging sensor are able to deal with dynamic ranges of up to 2500:1, and when a reflective print of this scene is created, the print is typically limited to a dynamic range of about 100:1. The human eye can accommodate luminance in a single view over a range of about 10,000:1, is capable of distinguishing about 10,000 colours at a given brightness, deals with such high dynamic range scenes by adapting locally to each part of the scene, and thus is able to retrieve details from low-luminance as well as high-luminance areas.

I.V. Safonov et al., *Adaptive Image Processing Algorithms for Printing*,
Signals and Communication Technology,
https://doi.org/10.1007/978-981-10-6931-4_2

In still photography, HDR can be used to create images from scenes that have a broad range of tonal values from shadows to highlights. This situation is typical of landscapes and scenes with illuminants, since such scenes may have a dynamic range of 100,000:1 or more. Besides traditional landscapes, other outdoor images that might benefit from HDR include those with significant highlights such as strongly lit reflective surfaces. HDR can also be applied to good effect with indoor or outdoor architectural photography, where natural and artificial light combined with shadows can produce a wide dynamic range over various materials and surfaces.

However, there is no guarantee that the HDR technique by itself can produce a "better" interpretation of a given scene. For decades, photographers made decisions on what is really important, by composing and exposing, and leaving the rest behind.

As can be seen, HDR is not only the bit size of images to be stored. In the process of acquiring digital images, sensors have to possess higher capacities so that more gradations of light can be represented and true gradations of scene lightness are not spoilt by noise spikes.

Now, the problem consists of compressing the dynamic range so that highlights are not overexposed and shadows are not too dark. This problem becomes more essential in photo printing technology. Because of the large mismatch between the dynamic range of the print paper and the original scene, there needs to be some mapping or transformation of the dynamic range of the scene onto the available dynamic range of the print paper to create a pleasing reproduction. Simple scaling will not produce the desired result, because a large volume of lightness gradations will be lost.

2.1.2 Hardware Solutions

Fujifilm Super CCD SR. In January 2003, Fujifilm announced the new SuperCCD SR sensor type. The SuperCCD SR (Super Dynamic Range) offers almost two stops more dynamic range than conventional CCD. Beneath each microlens on the sensor surface (a photosite) are two photodiodes: the primary one captures dark and normal light levels (more sensitive) and the secondary one captures brighter details (less sensitive). The signals from the two photodiodes are intelligently combined by the camera to deliver an image with an extended dynamic range. This combination of primary and secondary photodiodes produces an image that is more richly detailed than conventional CCDs, resolving more detail in the highlight and dark areas of the image.

BrightSide Technology HDR Display. BrightSide Technology introduced the DR37-P that uses an array of individually modulated LED backlights to provide 10 times the brightness and 100 times the contrast of existing televisions and computer monitors. BrightSide's Extreme Dynamic Range display delivers more vibrant images and makes it possible to see the data in vivid detail. In 2007, BrightSide

Tech was acquired by Dolby Laboratories. Currently, the modern technology Dolby Vision allows the creation of 12-bit depth content that is ready to be viewed on compatible displays and TVs.

Modern trends in the TV industry include the creation of the HDR10 standard supporting 10-bit depth contents. In 2017, in Consumer Electronics Show (CES) all the major TV manufactures demonstrated TVs that allow 10-bit video content to be viewed; some of them, such as LG and Sony, are able to deal with Dolby Vision content.

The Canon PROGRAF iPF5000 colour inkjet printer appeared on the market in May 2006. It was the first high-end inkjet printer to offer a 16-bit printer driver and is a non-standard plug-in for Photoshop because Windows printer drivers are restricted to an eight-bit path. The plug-in processes the image in 16-bit mode and then sends it to the printer as 12-bit data. The goal of this is to expand the dynamic range of the reproduced image and surprisingly also the gamut. There is some debate about whether the Canon 16-bit mode truly offers a benefit or not. Even as display technology pushes along towards HDR, ink on paper will remain an LDR medium. To make great prints from truly "high" dynamic range images will take a lot of effort in tone mapping.

Epson Surecolor P series printers were announced in 2015 and incorporate a reformulated Epson UltraChrome HDX 10-colour ink set for commercial and flexographic printers, as well as graphic designers, photographers, and fine art reproduction houses.

2.1.3 Main Challenges

When the dynamic range is less than the contrast in an actual scene, the image sensor may become saturated. When that happens, the image sensor cannot capture details from bright and dark areas of such an image at the same time. Adjusting or increasing the sensitivity of the dark areas may result in over-saturation of the bright areas, and conversely, adjusting or increasing the sensitivity of the bright areas may result in over-saturation of the dark areas. One of the possible approaches is to use exposure bracketing to take several shots: underexposed, normal, and overexposed.

During photo shooting, the camera may occasionally be translated or rotated; therefore, prior to processing, the images must be co-aligned, or registered, taking into account optical distortions of the camera's lenses.

In the case of linear scaling of the image to fit eight bits, a serious loss of contrast may occur. To avoid this, a sophisticated tone-mapping technique should be applied.

Another problem is adequately processing colour channels to avoid colour distortions.

The preservation of the scene's dynamic range is a critical problem in panorama creation. Very often outdoor panoramas have very large variations of lightness, so mapping to LDR image helps to preserve details in overall scene.

2.2 Proposed Method of Dynamic Range Compression

In this work, we considered the problem of lost details in the images captured with low dynamic range sensors. The main idea is to use several images captured with exposure bracketing. This means that one of the images has normal exposure, and several other images are over- or underexposed. The method is aimed at photos shot by regular consumer cameras. Recent trends in the electronic industry have led to the miniaturization of compact DSCs, and thus the sensors have become smaller and suffer from a low dynamic range due to their reduced capacity. A side effect of this process is the amplification of noise, especially in shadow areas, so that details of the image in shadows are completely lost. Oversaturation of image sensors leads to flat white patches. One way to avoid a loss of details in shadow and highlight areas is to capture several images with exposure bracketing to preserve highlight details in underexposed images and shadow details in overexposed images. Note here that colour representation also suffers in shadowed and highlighted areas. The final image should contain as much detail as possible from the entire set of bracketed images.

Usually, exposure bracketing mode is present in modern consumer cameras, so it is possible to shoot a set of differently exposed images in a relatively small period of time. However, usually the camera will shake or move during capture, leading to some translations between captured images. Additionally, some distortion may be present due to the imperfect optical system of a DSC. Hence the first necessary step is to align all images of the set, taking into account optical distortion.

The main problem in preserving details of the image is matching the dynamic range of the scene being captured to the display or printing device. The application of simple linear stretching will inevitably lead to a loss of local contrast, and the final image will look flat. The main challenge is therefore to compress the dynamic range while preserving local contrast, that is, to apply a special tone mapping technique. A very common problem of various local mapping operators is the formation of a halo around very steep edges in cases when the subject to be captured has some light emitting sources behind. This problem is addressed by Safonov et al. (2006).

In summary, the following problems exist.

- Translations between images, which also lead to different distortions due to the optical system.
- Insufficient data for camera profile reconstruction and HDR image formation.
- Loss of details in low dynamic range images.
- Loss of contrast as a result of dynamic range compression.
- Halo formation around very steep edges.
- Preservation of colours from all images of the set.

2.2.1 *Registration*

The first step in image formation is registration of all images in the set with one chosen reference image. Possible image transformations in the case of hand shooting are translation and rotation as well as correction of optical distortion.

A detailed description of the registration method is outside the scope of this work.

2.2.2 *Fusion*

The next step is fusion of the registered images. Images with different exposures can be fused together using a sophisticated method of camera calibration and recovery of the true HDR image. One of the first papers dedicated to the reconstruction of a nonlinear camera response profile was published by Mann and Picard (1995) and Mann and Mann (2001). The authors addressed the problem of a nonlinear response curve but provided a rather limited method of recovering it. They used several fully aligned images captured with exposure bracketing and reconstructed the curve using polynomial representation (Fig. 2.1).

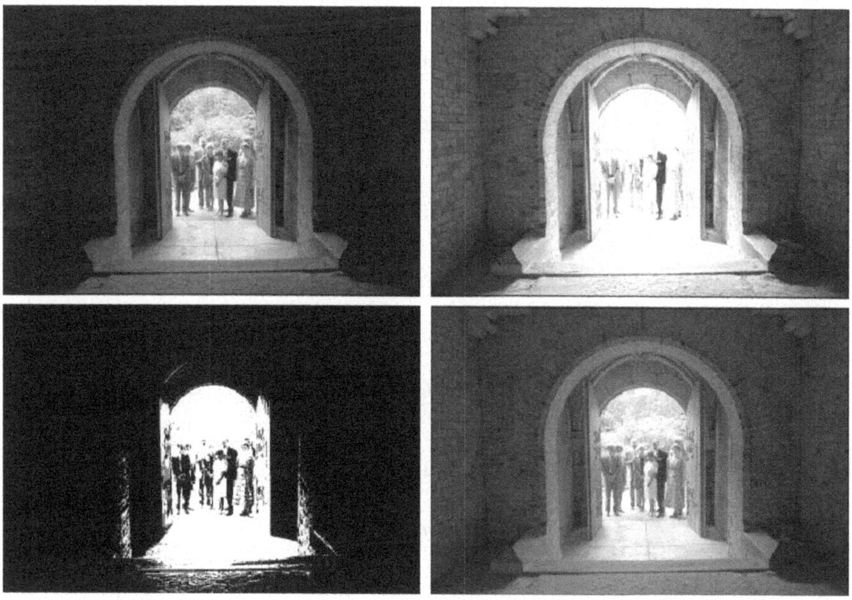

Fig. 2.1 Mann and Picard (1995) used several fully aligned images captured with exposure bracketing and reconstructed the curve using polynomial representation

In the two reference methods described above, an HDR image itself is not formed but a calibration step is necessary. Work on recovering the camera response profile was continued by Debevec and Malik (1998). Their method has become widespread among researchers as well as among users who work with HDR still images, having sets with images captured with exposure bracketing. Instead of some abstract units for lightness, here the authors propose a physically based model for recovering the camera response profile. It is assumed that known pixel values Z are the result of exposure of the sensor to the irradiance E during some known exposure time t using the mapping function f.

$$Z_{ij} = f(E_i \Delta t_j),$$

$$g(Z_{ij}) = \ln E_i + \ln \Delta t_j, \quad \text{where} \quad g \equiv f^{-1},$$

and the problem of response curve reconstruction is reduced to the least squares problem of recovering unknown g and E_i.

To recover the camera response profile, it is necessary to have multiples of images with known priors, such as the exposure time, in order to have enough data for smooth curve recovery. If the number of images is insufficient to calibrate the camera, the problem of high dynamic compression consists of preserving details that are present in all images.

Very interesting works have been done on the problems of combining differently exposed images or even combining images of a different nature. Perez et al. (2003) extended an idea proposed by Fattal et al. (2002) concerning the recovery of an image using its gradient. Basically, the problem consists of approximating an image using its gradient, that is, by minimizing the misfit between a known gradient G and the gradient of the target image I in the least squares sense:

$$\iint \|\nabla I - G\|^2 dxdy \to \min.$$

Here I is an unknown image and G is its known gradient. If the norm of the difference is represented as

$$\|\nabla I - G\|^2 = \left(\frac{\partial I}{\partial x} - G_x\right)^2 + \left(\frac{\partial I}{\partial y} - G_y\right)^2,$$

this equation leads to the well-known Poisson equation:

$$\Delta I = divG.$$

Usually, a Laplace operator is approximated with a finite difference operator, and this in turn leads to a system of linear equations, which can be solved using various methods.

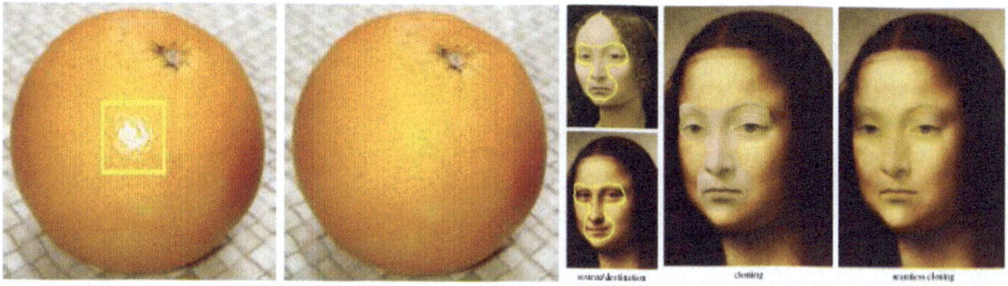

Fig. 2.2 Seamless cloning by Fattal et al. (2002)

Fig. 2.3 Night-time and daytime images are fused together using gradients (Raskar et al. 2004)

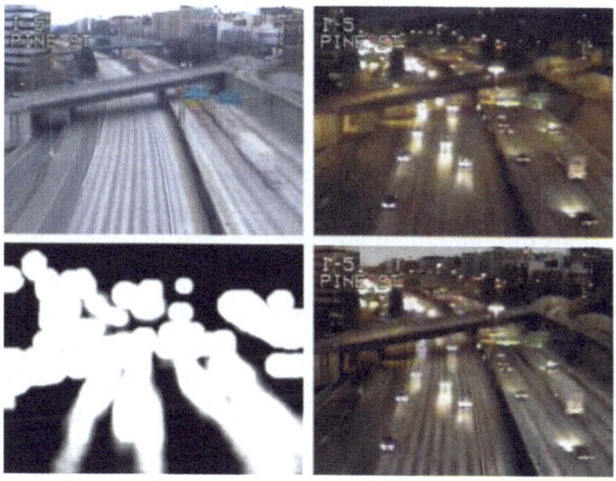

Based on this method, Fattal et al. (2002) developed various techniques of image processing such as seamless cloning and image editing (see Fig. 2.2). Note here that more operations are possible using gradients such as rotation, affine transforms, noise suppression, combining images with different depths of field, and so on.

The idea of employing image gradients was elaborated by Raskar et al. (2004) and Ilie et al. (2005). The authors describe a method of enhancing images, obtained from, for example, surveillance video cameras, where objects are captured under different lighting conditions (daylight and night-time images); see Figs. 2.3 and 2.4.

They first construct the image using the weighted sum of gradients of input images based on the "pixel importance" computed from the local variance. After that, they apply the technique described by Fattal et al. (2002) in the gradient domain or methods from Durand and Dorsey (2002) and Reinhard et al. (2002) in the image domain to compress the dynamic range of the resulting image.

In the paper by Socolinsky (2000), the problem of recovering an image using a known gradient is addressed in more detail. The author points out the very important problem of local contrast reduction that occurs as a result of dynamic range compression. The author proposes to perform optimization under constraints for luminance values.

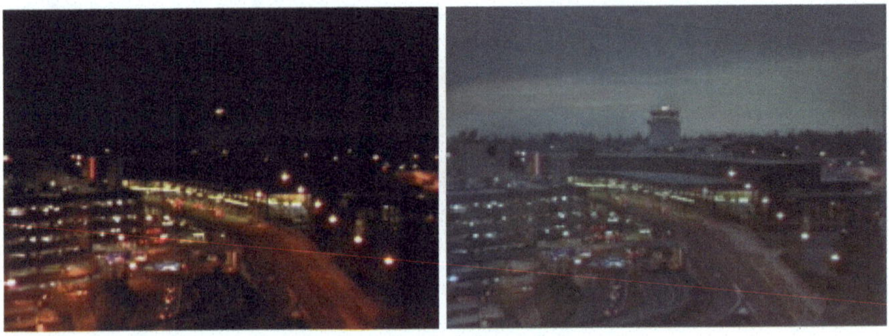

Fig. 2.4 Night-time and daytime images are fused together using gradients (Raskar et al. 2004)

The method presented by Fattal et al. (2002) was originally applied to the HDR image. The authors first computed gradients of the input HDR image, then performed the reduction of gradients by multiplying them by the gradient attenuation function, and finally reconstructed the image using processed gradients.

In a set of differently exposed photos, some images may look dark but have details of highlighted areas of the captured scene, while some images may look too bright but have details in shadowed areas of the scene. Since the human visual system is mostly aimed at details, that is, local changes of lightness, as mentioned by DiCarlo and Wandell (2000), the gradient of the input images must be properly preserved. The main idea is to take highlight details from dark images and shadow details from bright images. For this purpose, an image gradient consisting of maximal magnitudes of gradients of input images is created, and after that (following some tone mapping), lightness values are reconstructed using the Poisson equation (Tolstaya et al., 2009).

Suppose we have a set of images $\{I\}_{k=1}^{K}$ of the same size, M by N pixels. Let $\{Y\}_{k=1}^{K}$ be the corresponding lightness channels of the input images. A fused gradient $\boldsymbol{G}^{\mathrm{HDR}}(i, j)$ is constructed using the gradients of each individual image $\boldsymbol{G}^{k}(i, j)$ in the following way:

$$\begin{cases} G_x^{HDR}(i,j) = \left\{ G_x^p(i,j) \middle| \left|G_x^p(i,j)\right| = \max\limits_{k=1..K} \left|G_x^k(i,j)\right| \right\} \\ G_y^{HDR}(i,j) = \left\{ G_y^p(i,j) \middle| \left|G_y^p(i,j)\right| = \max\limits_{k=1..K} \left|G_y^k(i,j)\right| \right\} \end{cases},$$

where

$$\mathbf{G}^k(i,j) = \left\{ G_x^k, G_y^k \right\}_{ij}.$$

If there is no shift or rotation between images, gradients are computed using the following formulas:

$$\begin{cases} G_x^k(i,j) = Y^k(i+1,j) - Y^k(i,j) \\ G_y^k(i,j) = Y^k(i,j+1) - Y^k(i,j) \end{cases}.$$

If images are not registered, it is necessary to register them.

2.2.3 Mapping

Tone reproduction (also known as tone mapping) provides a method of scaling (or mapping) luminance values in the real world to a displayable range. Tone reproduction is necessary to ensure that a wide range of light in a real-world scene is conveyed on a display with limited capabilities.

Tone mapping was developed for use in television and photography, but its origins lie in the field of art, where artists make use of a limited palette to depict high contrast scenes.

Usually, the image is calibrated in advance to determine the illumination in physically based units in order to assure accurate reproduction of tone. This is especially true when speaking about night-time shots, which can be mapped as daylight scenes, if uncalibrated. However, it is possible to use heuristics to infer lighting conditions for scenes depicted by uncalibrated images. In particular, a histogram can reveal whether an image is dark overall irrespective of the actual values in the image.

The problem of tone reproduction is still a matter of intensive research interest. Many solutions have been proposed but still none of them can be universally applied. The problem of tone reproduction remains a vital one, especially in the printing area, because paper is not capable of emitting light and this will never become possible.

The majority of tone reproduction operators can be classified into the following two types: global, also known as single-scale or spatially uniform; and local, also known as multi-scale or spatially varying (see Ferwerda 1998; DiCarlo and Wandell 2000; Devlin 2002). Global operators apply the same transformation to every pixel. A global operator may depend upon the contents of the image as a whole, as long as the same transformation is applied to every pixel. Conversely, local operators apply a different scale to different parts of an image.

As noted by Fattal et al. (2002), high intensity drops lead to large gradients, so to compress the dynamic range, it is necessary to attenuate its magnitude. Here the authors propose to use different image scales to identify large gradients. They construct a Gaussian pyramid of an input HDR image and identify a large gradient at each scale. As a result, they construct a gradient attenuation function.

We propose to use bilateral filtering of an image Y^s constructed as the sum of lightnesses of all input images divided by their number:

$$Y^s = \frac{1}{K}\sum_{k=1}^{K} Y^k.$$

A bilateral filter proposed by Tomasi and Manduchi (1998) will smooth small variations of lightness and preserve strong edges.

$$Y^f = \frac{\sum_{k,l} Y^s_{i+k,j+l} w\left(\left|Y^s_{i+k,j+l} - Y^s_{i,j}\right|\right) h(i,j)}{\sum_{k,l} w\left(\left|Y^s_{i+k,j+l} - Y^s_{i,j}\right|\right) h(i,j)},$$

where $w(x) = \left(1 - \left|\frac{x}{\sigma_R}\right|^2\right)^2$ is the kernel for range processing, $h(i,j) = \exp\left(-\frac{i^2+j^2}{2\sigma_D^2}\right)$ is a window function, and σ_D and σ_R are the standard deviations for the domain and range, respectively. Here it is proposed to set σ_D as large as 100 to achieve smoothing of small variations of lightness. As for σ_R, here it is controlled by the size of the input images and varies between 10 and 50.

After that, a gradient $\mathbf{G}^f = \left\{G^f_x, G^f_y\right\}$ of the filtered image is computed. This gradient will have a large magnitude at strong edges and a small magnitude in smooth or textured areas. The magnitude of the gradients of the filtered image is processed so that it spans an interval between 0.1 and 1. The gradient attenuation function is formed using computed values, as follows:

$$\Phi = \left(1 - \left(\frac{1}{1 + |G^f_x| + |G^f_y|}\right)^{0.5}\right)^{\gamma},$$

where gamma is usually set to 1, but in some extreme cases can reach values greater than 1 and smaller than 3. The greater gamma is, the more intense the gradient attenuation that occurs: strong edges will be attenuated more than weak edges. Constant multipliers of the new gradient field will not affect the results. This means that the scale of the gradient attenuation function does not matter.

The new gradient field ("low dynamic") is the product of this function and the initial fused gradient:

$$\mathbf{G}^{LD}(i,j) = \Phi(i,j)\mathbf{G}^{HDR}(i,j).$$

This transformation allows attenuation of strong edges, thus compressing the dynamic range of an image.

To recover the image intensities, a Poisson equation has to be solved. Here an image is approximated using its gradient; that is, the misfit between the known gradient $\mathbf{G}$ and the gradient of a target image ∇I is minimized in a least squares sense, which leads to the Poisson equation (as mentioned above):

$$\Delta Y = div\mathbf{G}^{LD}.$$

The Laplace operator is approximated with a finite difference operator, and this in turn leads to a system of linear equations that can be solved using various methods.

The Laplace operator is approximated as follows:

$$\Delta Y(i,j) \approx Y(i+1,j) + Y(i-1,j) + Y(i,j+1) + Y(i,j-1) - 4Y(i,j).$$

The gradient (as explained above) is computed using the forward difference:

$$\nabla Y(i,j) \approx \{Y(i+1,j) - Y(i,j); Y(i,j+1) - Y(i,j)\},$$

and for *div*, the backward difference is applied:

$$div\mathbf{G} \approx G_x(i,j) - G_x(i-1,j) + G_x(i,j) - G_x(i,j-1).$$

Note here that the integration of gradients involves scale and translation ambiguity. That is why the resulting image *I* is first scaled between minimum and maximum over all images in the set. After that, every pixel of the resulting image is forced between the corresponding pixels in the entire set:

$$Y^{scaled}(i,j) = \begin{cases} M(i,j), & \text{if } \mathrm{Y(i,j)} > \mathrm{M(i,j)} \\ m(i,j), & \text{if } \mathrm{Y(i,j)} < m(\mathrm{i,j}), \\ Y(i,j), & \text{otherwise} \end{cases}$$

where $M(i,j) = \max\limits_{k=1...K}\{Y^k(i,j)\}$ and $m(i,j) = \min\limits_{k=1...K}\{Y^k(i,j)\}$.

Figure 2.5 illustrates the process in a one-dimensional (1D) case: first, gradients having maximal magnitudes are combined to get $\mathrm{G}^{\mathrm{HDR}}$. After that, $\mathrm{G}^{\mathrm{HDR}}$ is tone mapped to find $\mathrm{G}^{\mathrm{LDR}}$. When it has been recovered and scaled, the new gradient preserves lightness variations.

2.2.4 Colour Processing

Most tone reproduction operators attempt to reduce an image in dynamic range, while keeping the response of human visual system to the reduced set of intensities constant. This leads to various approaches that aim at preserving brightness, contrast, and so on.

However, it is a common practice for tone reproduction operators to exclude comprehensive treatment of colours. Most operators derive a luminance based channel from input RGB values and then compress the luminance channel. However, this may change in the near future, since the fields of colour-based

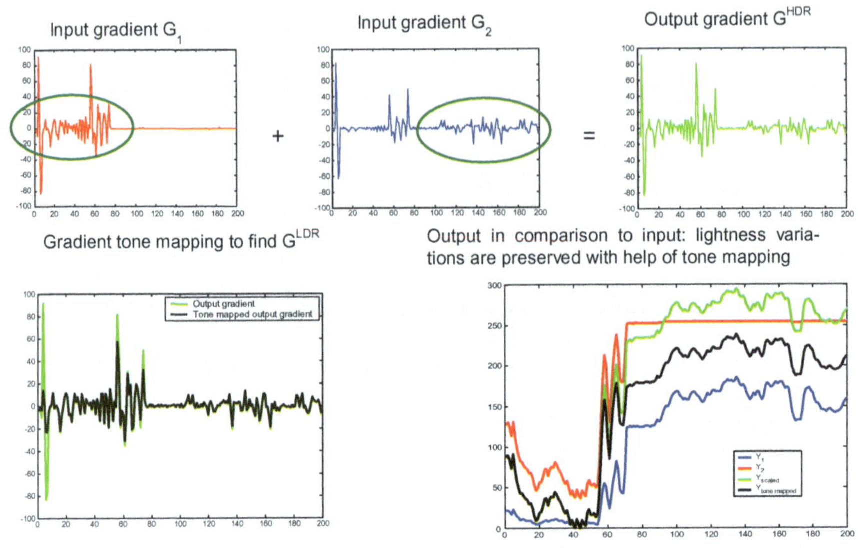

Fig. 2.5 The process in the 1D case: first, gradients having maximal magnitudes are combined to get G^{HDR}. After that G^{HDR} is tone mapped to find G^{LDR}

modelling and tone reproduction are growing closer together. This can be seen in the multiscale observer model by Pattanaik et al. (1998) and in the more recent iCAM (image Colour Appearance Model) in Fairchild and Johnson (2002, 2003), and Akyuz and Reinhard (2006).

Usually, colour processing can be attributed to two main types:

(I) Colour processing of such a type is usually applied in methods involving camera calibration and construction of an HDR image with an enlarged bit rate.

Colour processing is performed in the following way: the lightness component is compressed and colour channels are computed by preservation of the ratio between them (as in Fattal et al. 2002; Mann and Pickard 1995; Mann and Mann 2001; and others).

$$\begin{bmatrix} R_{out} \\ G_{out} \\ B_{out} \end{bmatrix} = \begin{bmatrix} \frac{R_{in}}{Y_{in}} Y_{out} \\ \frac{G_{in}}{Y_{in}} Y_{out} \\ \frac{B_{in}}{Y_{in}} Y_{out} \end{bmatrix},$$

where R_{in}, G_{in}, B_{in}, and Y_{in} are the input colour and lightness channels respectively, and R_{out}, G_{out}, B_{out}, and Y_{out} are the processed colour and lightness channels respectively. This technique makes it possible to minimize colour translations before and after compression. Fattal et al. (2002) also included some parameter s to control the saturation:

$$C_{out} = \left(\frac{C_{in}}{Y_{in}}\right)^s Y_{out},$$

where C takes values of R, G, or B. When s decreases, the saturation decreases too.

An alternative method is to convert an image to YCrCb colour space, compress the luminance, and convert the image back to RGB, if necessary. In Chap. 1 we described a method of adjusting the chromaticity channels in YCrCb, which makes it possible to avoid colour desaturation as a result of lightness channel modification.

(II) This type of colour processing is more typical for methods of fusion of colour images. Since the actual HDR image is not computed, three colour channels are processed independently and similarly to each other, as in Perez et al. (2003), Raskar et al. (2004), and so on.

A solution proposed in US patent 6,078,357 for an "image mixing circuit" by Yamamoto et al. (Matsushita Electric Industrial Co., Ltd.) uses a similar method for mixing colour-difference channels as for luminance mixing, that is, weighted summation of overexposed and underexposed images:

$$Y_{mix} = Y'_{short} \times Y_{cont} + Y_{long} \times (1 - Y_{cont}),$$

where Y_{mix} is the output luminance component, Y_{long} and Y_{short} are long and short exposure luminance components, respectively, Y_{cont} is the control signal, and Y'_{short} is the short exposure luminance component, boosted up to the level of the Y_{long} component. Y_{cont} is obtained by application of LPF to S_{long} (long exposure image signal), and then a constant and clipping are added. Note again that colour components are treated similarly.

The disadvantages of these methods are that in the first case an HDR image is required to compute ratios between colour channels or in the second case colour translations are possible due to inaccuracy of computation.

Therefore, a different method is proposed that performs linear (or quadratic or cubic) interpolation between input chromaticity channels for every pixel and then computes the new chromaticity values using the new lightness values (Fig. 2.6):

$$C_r = \frac{C_r^p - C_r^q}{Y^p - Y^q}\left(Y^{scaled} - Y^p\right) + C_r^p,$$

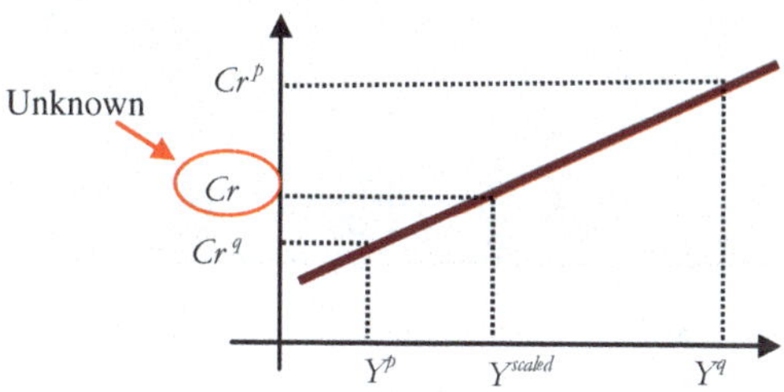

Fig. 2.6 Computation of chromaticity Cr using input lightness values Y^p, Y^q, and Y^{scaled} and known Cr^p and Cr^q

$$C_b = \frac{C_b^p - C_b^q}{Y^p - Y^q}\left(Y^{scaled} - Y^p\right) + C_b^p,$$

In the case when $Y^p = Y^q$, chromaticity is computed as the mean between C^p and C^q. This approach makes it possible to smoothly interpolate chromaticity for the output image, since Y^{scaled} always lies between the minimum and maximum of the input images.

This method is quite effective, does not require significant computational effort, and produces pleasing results.

2.2.5 Results

This section presents the results of the described method as well as some discussions and future directions.

Figure 2.7 shows an example of initial panorama images captured with exposure bracketing (Vogl):

Figures 2.8 and 2.9 present the final coloured image of the panorama and the gradient attenuation function for this panorama.

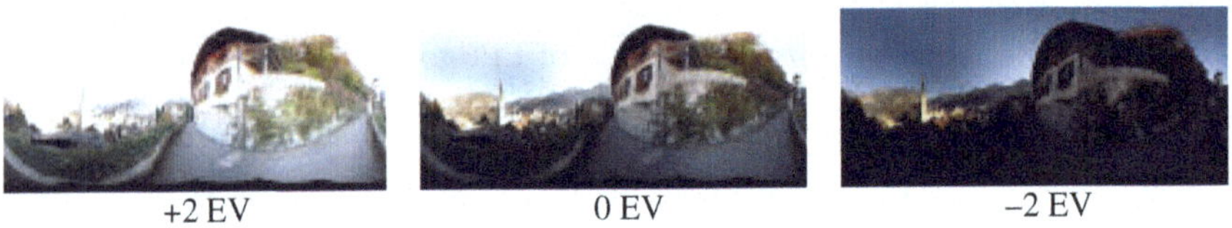

Fig. 2.7 Panorama images captured with exposure bracketing (Vogl)

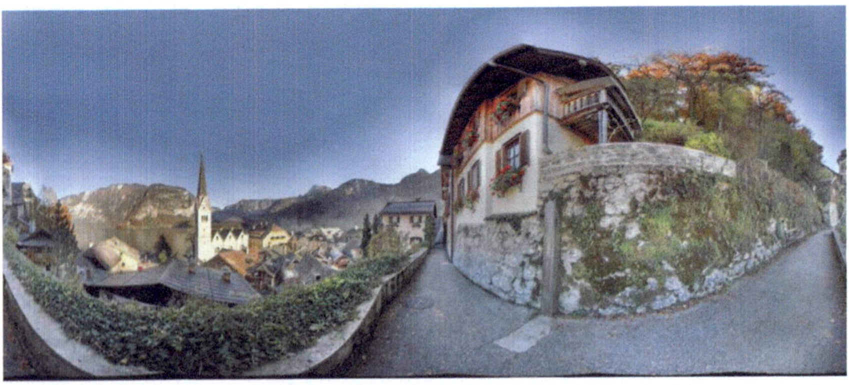

Fig. 2.8 Final coloured image of panorama obtained by the described method

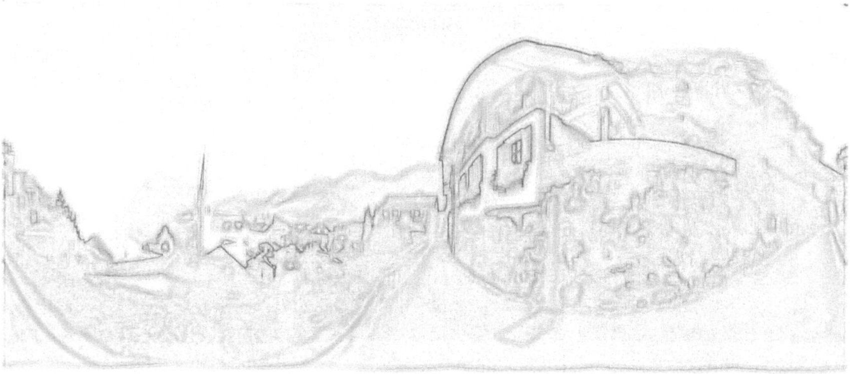

Fig. 2.9 Gradient attenuation function for the panorama image

As can be seen, strong edges will be attenuated more than weak edges. Constant multipliers of the new gradient field will not affect the results. This means that the scale of the gradient attenuation function does not matter.

Capturing waterfalls during bright days while preserving the texture and surroundings is a very challenging problem. Figure 2.10 shows the results of combining waterfall images.

Figures 2.11 and 2.12 show the results of mapping of a thermal area, including bright vapour trails and a waterfall.

Figure 2.13 is an example provided by the Photomatix website.

In general, as we can see, a method involving calibration can result in light areas in the final image that are brighter than the same areas in all input images. This drawback can be seen in the window curtain. The same is true for bright areas: if an area is bright throughout the whole of the image set, in the final image it can appear darker, which can be seen in the area where light from the window falls on the floor.

In the proposed method, the final image brightness is controlled by the input images: in the case of an underexposed set of input images, the result will look dark as well. (However, in this case histogram stretching is usually applied, although this step can be omitted.) Uneven distribution of exposures in input images will result in unnatural-looking images with over-brightened and over-darkened parts, since the method is aimed at images sets produced by consumer cameras in "exposure bracketing" mode, which is usually characterized by constant exposure steps.

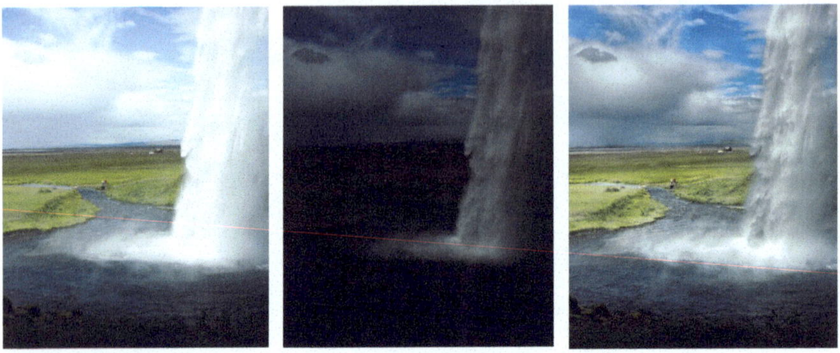

Overexposed image

Underexposed image

Proposed method

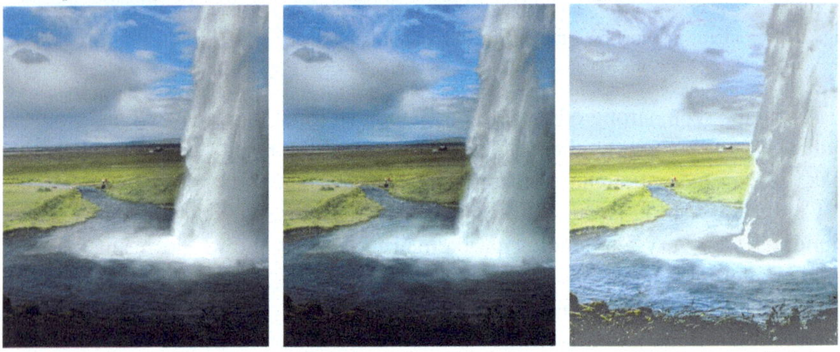

Fused image without tone mapping – looks flat

Fused image with colour processing with maximal saturation

True HDR image failed because of lack of data for HDR image generation.

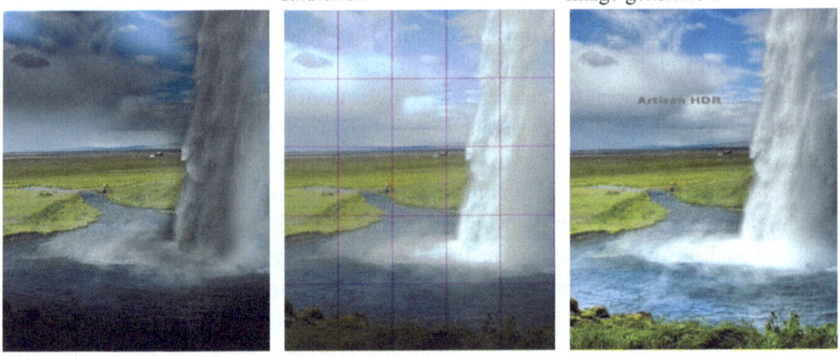

Exposure blending with PhotoMatix. Results are unsatisfactory.

Tone mapping with FDRTools. Some areas have completely lost contrast

Tone mapping in ArtizenHDR, using the Fattal operator (default parameters).

Fig. 2.10 Results of combining waterfall images

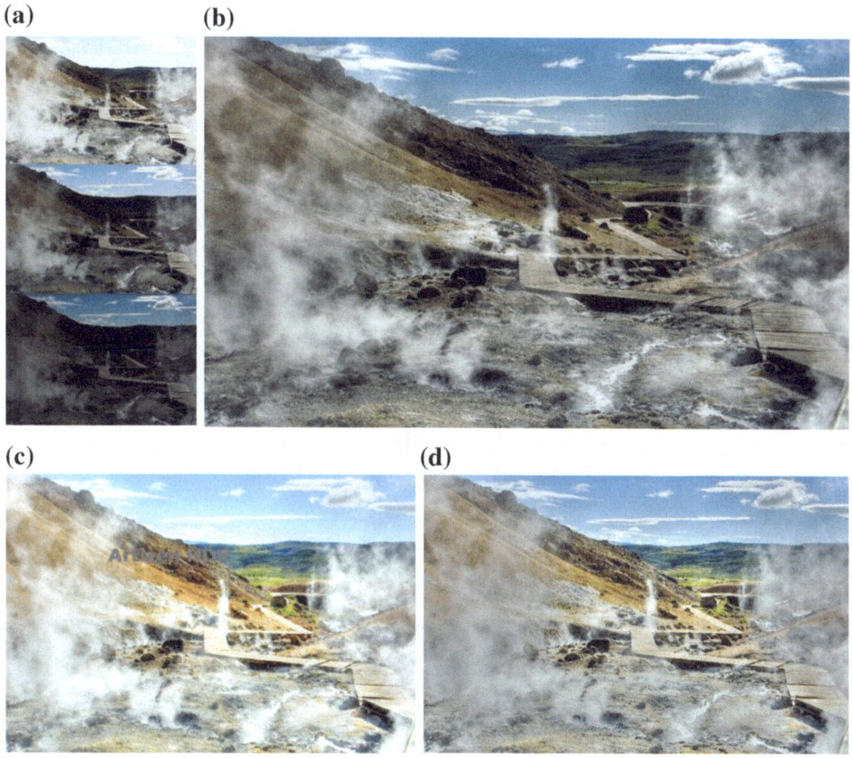

Fig. 2.11 Results of mapping of a thermal area, including bright vapor trails: **a** input images; **b** resulting fused image; **c** result produced by Artizen HDR (default settings); **d** result produced by Photomatix (default settings)

The developed method has several advantages and disadvantages. The following are among the advantages.

- There is no need for calibration: the number of input images is arbitrary, from 2 to 10 (as opposed to the majority of cited methods).
- An HDR image is not constructed.
- Image interpolation for aligning is performed using image gradients.
- The minimization problem is solved just once.

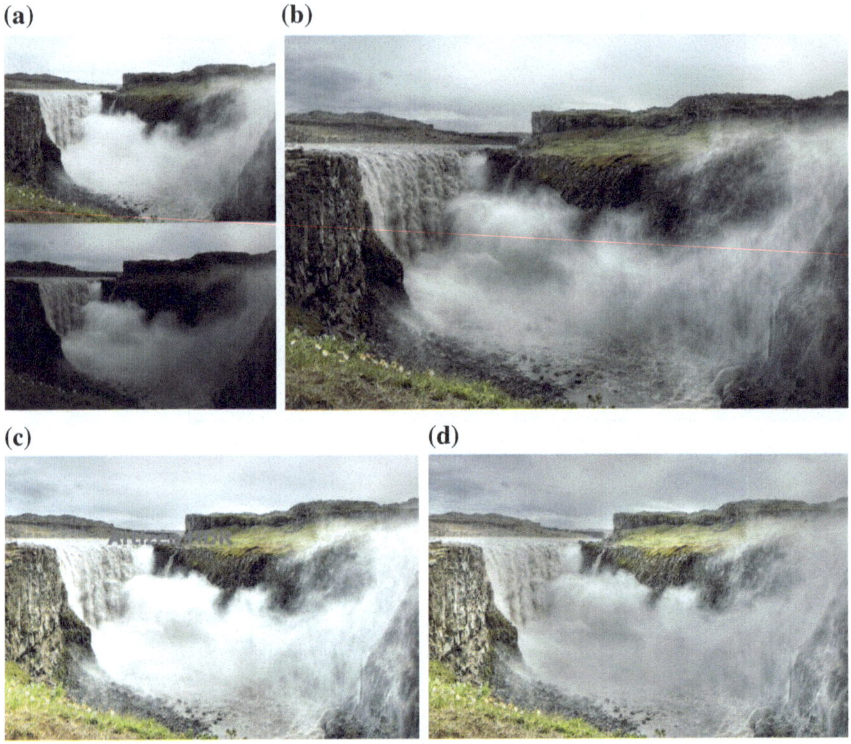

Fig. 2.12 Results of mapping of waterfall: **a** input images; **b** resulting fused image; **c** result produced by Artizen HDR (default settings); **d** result produced by Photomatix (default settings)

- Processing is performed in YCrCb colour space therefore no extra conversions in case input images are JPEG images.
- New simple scheme of colour mapping.
- Fully automatic.
- The final image is kept as close as possible to the original.

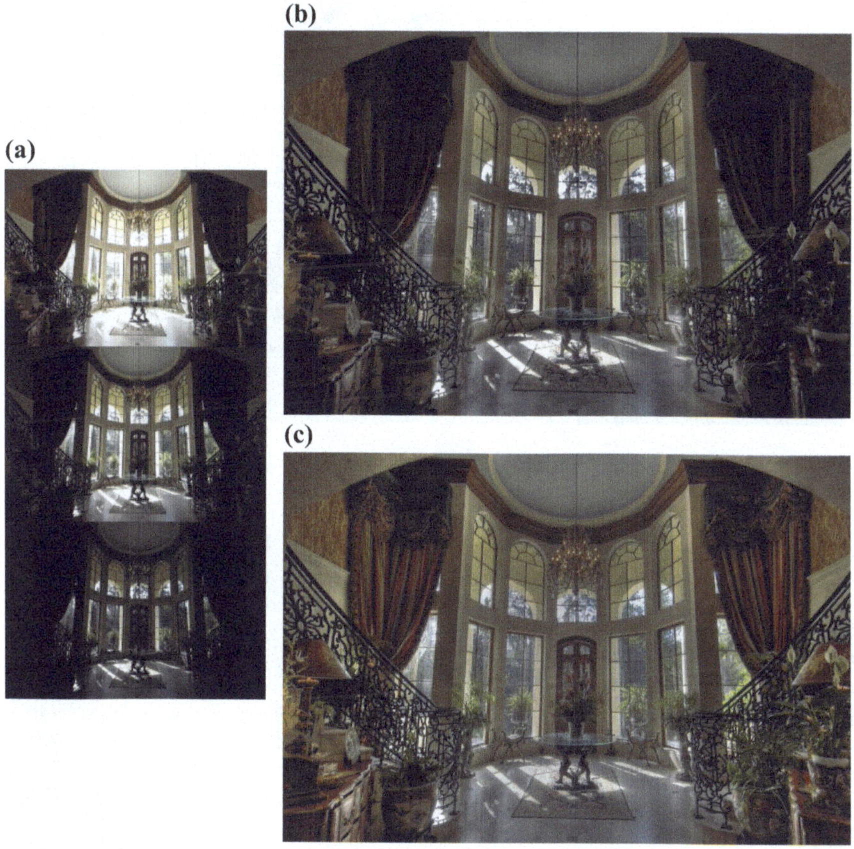

Fig. 2.13 Example from Photomatix website: **a** input images; **b** results of proposed method; **c** results of Photomatix

The method has the following disadvantages.

- There is a slight halo around strong edges when parameter γ is not set precisely.
- The minimization problem has to be solved.
- An existing halo (for example, around light bulbs) will not be suppressed.

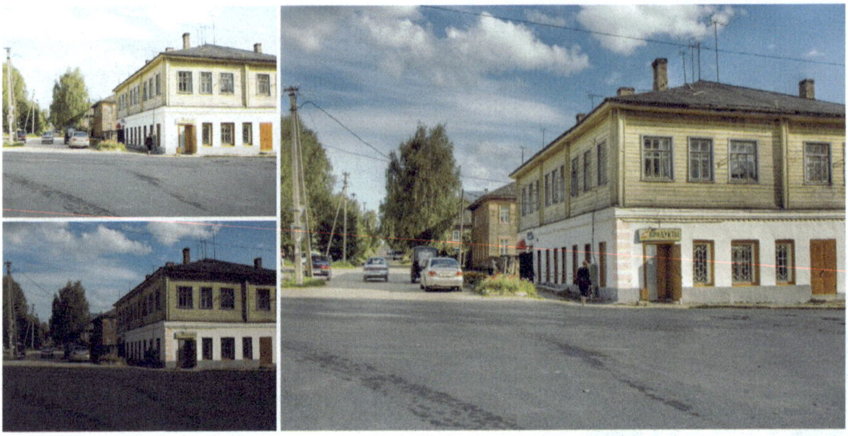

Fig. 2.14 Example 1 of image processing by proposed method

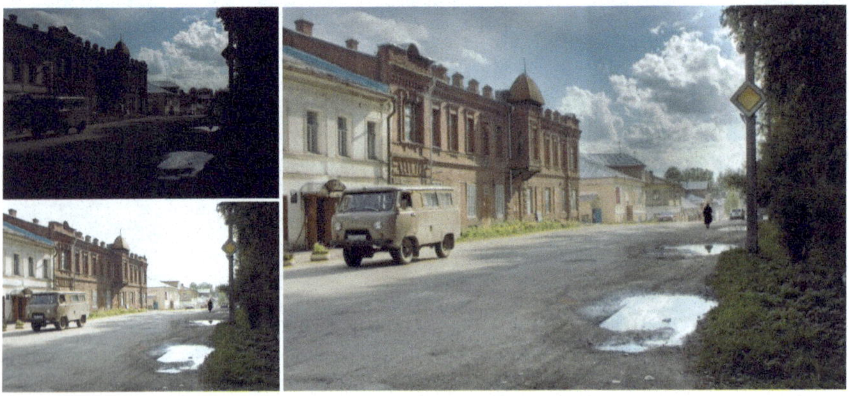

Fig. 2.15 Example 2 of image processing by proposed method

Figures 2.14, 2.15, 2.16 and 2.17 present some more results of image fusion. The quality characteristics of the output image depend, of course, on the number of input images. The greater the quantity of input information, the more details are

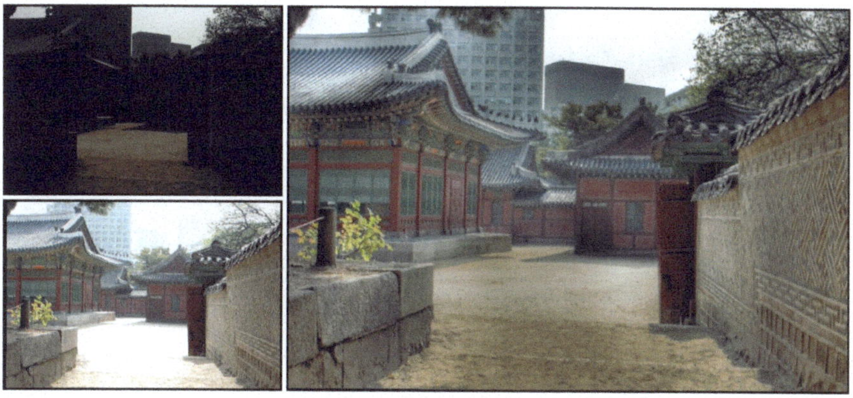

Fig. 2.16 Example 3 of image processing by proposed method

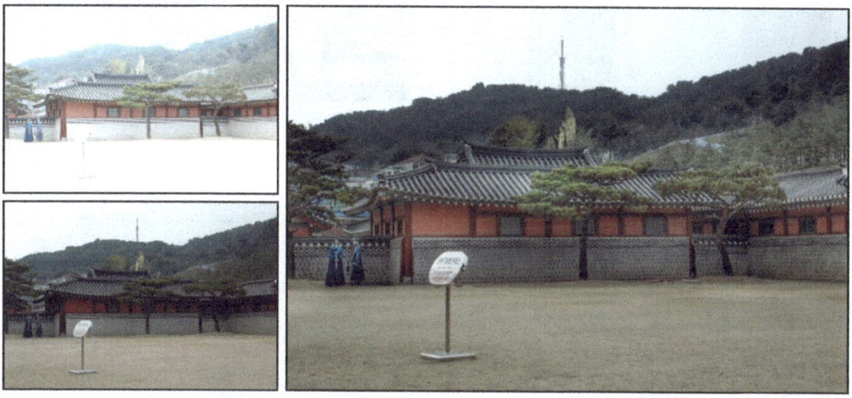

Fig. 2.17 Example 4 of image processing by proposed method

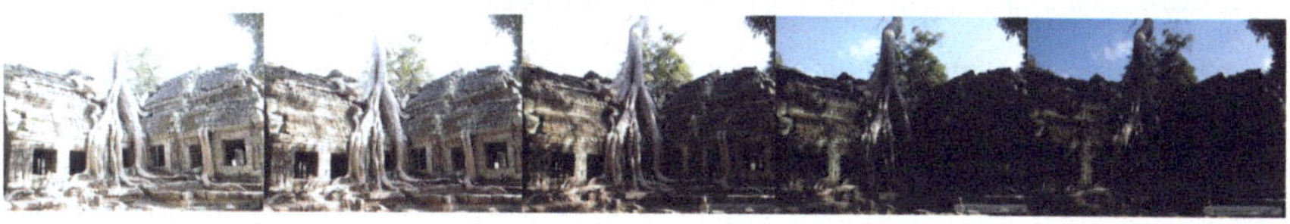

Fig. 2.18 Input images

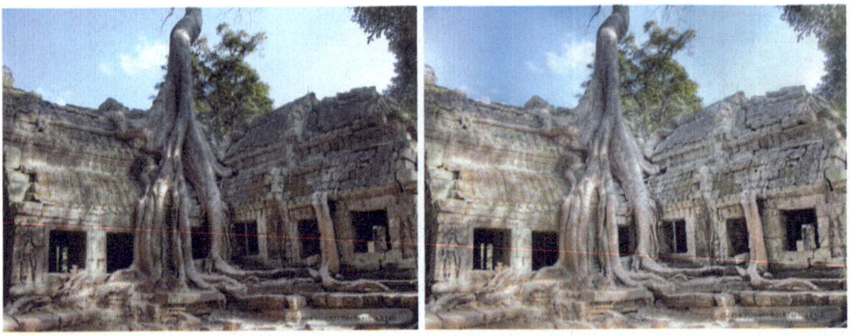

Fig. 2.19 Left: result with two input images (second and fourth); right: result with five input images (demonstrates significant loss of contrast)

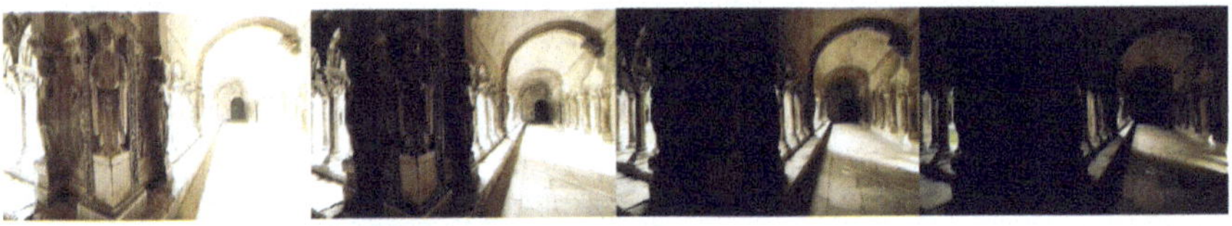

Fig. 2.20 Input images

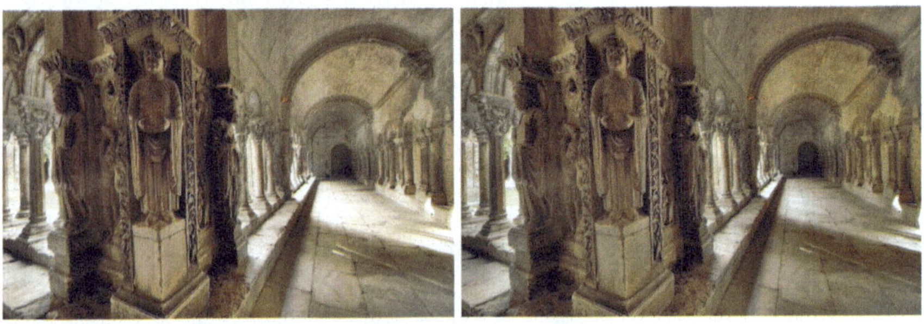

Fig. 2.21 Left: result with two input images (first and third); right: result with four input images

preserved. However, it results in some loss of contrast. Figures 2.18 and 2.19 demonstrate results with two and five input images. In Figs. 2.20 and 2.21, the difference is not so noticeable.

References

Akyuz, A.O., Reinhard, E.: Color appearance in high dynamic range imaging. SPIE J. Electron. Imaging **15**(3) (2006)

Debevec, P., Malik, J.: Recovering high dynamic range radiance maps from photographs. In: ACM SIGGRAPH 2008 classes, p. 31. ACM (1998)

Devlin, K.: A review of tone reproduction techniques. University of Bristol, Tech. Rep. CSTR-02-005 (2002)

DiCarlo, J., Wandell, B.: Rendering high dynamic range images. In: Proceedings of the SPIE: Image Sensors, vol. 3965, pp. 392–401 (2000)

Durand, F., Dorsey, J.: Fast bilateral filtering for the display of high-dynamic-range images. ACM Trans Graph (TOG) **21**(3), 257–266 (2002)

Fairchild, M.D., Johnson, G.M.: Meet iCAM: an image color appearance model. In: IS&T/SID 10th Color Imaging Conference (2002)

Fairchild, M.D., Johnson, G.M.: Image appearance modelling. In: Proceedings of IS&T/SPIE Electronic Imaging, vol. 5007, pp. 149–160 (2003)

Fattal, R., Lischinski, D., Werman, M.: Gradient domain high dynamic range compression. ACM Trans. Graph. **21**(3), 249–256 (2002)

Ferwerda, J.: Fundamentals of spatial vision. In: Applications of visual perception in computer graphics, SIGGRAPH'98 Course Notes (1998)

Ilie, A., Raskar, R., Yu, J.: Gradient domain context enhancement for fixed cameras. Int. J. Pattern Recognit. Artif. Intell. **19**(4), 533–549 (2005)

Mann, S., Mann, R.: Quantigraphic imaging: estimating the camera response and exposures from differently exposed images. In: Proceedings of IEEE Conference on Computer Vision and Pattern Recognition, pp. 842–849 (1998)

Mann, S., Pickard, R.W.: Being undigital with digital cameras: extending dynamic range by combining differently exposed pictures. In: Proceedings of IS&T 48th Annual Conference, pp. 422–428 (1995)

Pattanaik, S.N., Ferwerda, J.A., Fairchild, M.D., Greenberg, R.W.: A multiscale model of adaptation and spatial vision for realistic image display. In: ACM SIGGRAPH Computer Graphics Proceedings, Annual Conference Series, pp. 287–298 (1998)

Perez, P., Gangnet, M., Blake, A.: Poisson image editing. ACM Trans. Graph. **22**(3), 313–318 (2003)

Raskar, R., Ilie, A., Yu, A.: Image fusion for context enhancement. NPAR 2004: Third International Symposium on Non Photorealistic Rendering (2004)

Reinhard, E., Stark, M., Shirley, P., Ferwerda, J.: Photographic tone reproduction for images. In: Proceedings of SIGGRAPH 2002, ACM SIGGRAPH, pp. 267–276 (2002)

Safonov, I.V., Rychagov, M.N., Kang, K.M., Kim, S.H.: Automatic correction of exposure problems in photo printer. In: Proceedings of IEEE Tenth International Symposium on Consumer Electronics (ISCE), pp. 13–18 (2006)

Socolinsky, D.: Dynamic range constraints in image fusion and visualization. In: Proceedings on Signal and Image Processing (2000)

Tolstaya, E.V., Rychagov, M.N., Kang, K.M., Kim, S.H.: Fusion of high dynamic range scene photos. In: Proceedings of IS&T/SPIE Electronic Imaging (2009)

Tomasi, C., Manduchi, R.: Bilateral filtering for gray and color images. In: Proceedings of IEEE Conference on Computer Vision (1998)

Vogl, B.: Overview of tone mapping operators and algorithms. http://dativ.at/logmap/#overview

Chapter 3
Image Enhancement Pipeline Based on EXIF Metadata

3.1 Introduction

Modern digital still cameras (DSCs) store the accompanying data about a captured photograph in a standard format called the *Exchangeable Image File Format* (EXIF). This standard was introduced in 1998 by the Japan Electronics and Information Technology Industries Association as the standard for storing information about a digital image into an image file during photo acquisition (EXIF Version 2.2 2002).

The metadata tags defined in the EXIF standard cover a broad spectrum: (a) date and time information; (b) camera settings, including both static information such as the camera manufacturer, model etc. and information that varies from image to image such as orientation (rotation), aperture, shutter speed, focal length, metering mode, ISO speed information; (c) a thumbnail for previewing the picture on the camera's screen, in file managers or in photo editing software; and (d) copyright information.

In 1991, the International Press Telecommunications Council (IPTC) and Newspaper Association of America (NAA) proposed the proprietary *Information Interchange Model* (IIM) as a standard for storing descriptive metadata information within digital images. The IIM information about the digital image consists of datasets known as IPTC headers for digital image files.

Adobe Systems invented their own platform for inserting metadata structures into Photoshop, JPEG, and TIFF files (the *Extensible Metadata Platform* (XMP), recognised as an ISO standard); at the same time, they adopted the IIM data structure and several of its metadata elements such as author, caption, keywords, categories, comments and some free-format fields.

In 2010, the Metadata Working Group, formed of a consortium of companies such as Adobe Systems Inc., Apple Inc., Canon Inc., Microsoft Corp., Nokia Corp. and Sony Corp., released recommendations concerning the use and best

I.V. Safonov et al., *Adaptive Image Processing Algorithms for Printing*,
Signals and Communication Technology,
https://doi.org/10.1007/978-981-10-6931-4_3

practice for EXIF, IPTC and XMP metadata in images (Guidelines for Handling Image Metadata 2010).

If EXIF is a DSC image file format standard, the IPTC header standard is a standard for storing and accessing metadata information in digital images, since many commercial applications have adopted the Adobe mechanism for inserting and reading the IPTC metadata headers (Tešić 2005).

EXIF information is widely used in several applications: (a) the identification of forged images (Kee et al. 2011; Fan et al. 2013); (b) the development of effective photo management systems (Jang et al. 2007; Ryu et al. 2010; Gu et al. 2013); (c) scene and image classification (Zhao et al. 2012; Li et al. 2016; Boutell and Luo 2004); and (d) high-performance reading and modification of metadata itself (Amornpashara et al. 2015; Toevs 2015).

It is also reasonable to analyse the EXIF tags for an estimation of the probability of the presence of defects in photos and for the selection of correction parameters. Yen et al. (2007) proposed the use of several EXIF tags for backlit scene detection in HP cameras. There are numerous regulations for the values of EXIF tags for each specific camera model; for each camera model, the collection of tags and their values vary significantly, especially for cameras produced by different manufactures.

We investigate the applicability of and identify the more common EXIF-tags for different camera models, finding a correlation between the tags and the typical defects of digital photos. The primary goal was to construct a general EXIF-based image enhancement pipeline for automatic printing.

3.2 Analysis of Typical Artefacts and EXIF Tags

3.2.1 Typical Defects of Photos

As previously noted in Chap. 1, the following typical defects are identified in digital photos.

1. Exposure problems (*E*). This category includes under-exposed and over-exposed photos as well as photos with low global contrast and backlit scenes.
2. Noise (*N*). For DSCs, the high level of visual noise is caused by the nature of the sensor (CCD or CMOS) at long exposure times.
3. Colour cast (*C*). Colour imbalance is caused by external lighting conditions and incorrect white balance settings.
4. Blur (*B*). Incorrect focusing or shaking of the camera are typical causes of blurred photos.
5. JPEG artefacts (*J*). These artefacts appear for images stored with a high compression ratio.

6. Red eye artefacts (*R*). This is a widespread problem caused by the reflection of flash light from the eye fundus, which appears red due to the blood vessels.

For the classification of above mentioned defects, an EXIF-related training dataset of images was created and labelled manually. In the particular case considered in this chapter, the training dataset includes 1054 images captured by 10 camera models produced by different manufacturers.

3.2.2 Presence of Tags in the Files

Table 3.1 contains the percentages of tags existing in the files in the created picture database. The results correlate completely with those of substantial investigations by Dietmar and Reiner (2008) carried out on 4.8 million photos intended for photofinishing. In addition, for each tag in Table 3.1, defects are indicated whose

Table 3.1 EXIF-Tags presence in photos captured by different camera models

Tag	Presence of tag (%)	Related to defects
Flash	100	*E, N, C, R*
FNumber	99.6	*E, N, B*
ExposureTime	99.6	*E, N, B*
ExposureMode	92.3	*E, N*
ExposureProgram	60.5	*E*
ApertureValue	55.7	*E*
SceneCaptureType	91.5	*C*
SceneType	56.1	*E, B*
BrightnessValue	8.80	*E, N*
ISOSpeedRatings	51.0	*E, N*
GainControl	29.1	*E, N*
MeteringMode	99.1	*E*
ExposureBiasValue	99.7	*E*
ShutterSpeedValue	55.7	*N, B*
LightSource	60.8	*C*
WhiteBalance	91.5	*C*
ColourSpace	99.8	*C*
DigitalZoomRatio	62.3	*B*
SubjectDistanceRange	26.4	*B*
FocalLengthIn35mm	30.6	*B*
CompressedBitsPerPixel	87.3	*J*
FocalPlaneXResolution	45.2	*J*
FocalPlaneYResolution	45.2	*J*
Thumbnail	99.5	*E*

presence and weights could relate either to nonzero value of a specific tag or to the value of corresponding tag parameter.

It is easy to see that the importance of the tags for subsequent defect detection varies in magnitude.

For example, although the fields *ExposureTime* and *FNumber* represent the actual values of the exposition time and focal length used in the DSC, the *BrightnessValue* tag frequently contains values indicating that the true value of the brightness was inaccessible for measurement. Accordingly, the field *BrightnessValue* has no informative value for 8.8% of images.

It should be particularly noted that some part of the useful information about shooting conditions is often stored in the proprietary *MakerNotes* tag. This generally means that the information cannot be applied; instead, for a specific camera model, the application of proprietary tags may be valuable in the analysis of the picture quality. Accordingly, for effective classification, it is necessary to select and analyse the most informative tag/feature combination for each defect class.

3.3 Detection of Defects and Artefacts

3.3.1 Classification of Exposure Problems

For classification of the exposure problem, the following tags were selected after empirical data analysis: (a) *Flash*; (b) *FNumber*; and (c) *ExposureTime*. The tags *FNumber* and *ExposureTime* correspond to intensity of the luminous flux falling onto the CCD matrix of the DSC. The tags used most frequently, i.e. *MeteringMode* and *ExposureMode*, are non-numerical in nature. They correspond to the mode of DSC operation. The tag *ExposureBiasValue* is in practice equal to zero in most cases, and hence is less informative.

In the learning database, 33.6% of the total number of images were attributed to the general class {Exposure problem}. Of these, sub-classes were identified as follows (Table 3.2).

Figures 3.1 and 3.2 illustrate the data distribution depending on the values of *FNumber* and corresponding $ShutterSpeed = -\log_2(ExposureTime)$ when the flash is on or off.

Table 3.2 Statistics of photos with exposure problems

Sub-classes	Within class {Exposure problem} (%)	Total database (%)
Under-exposure	45.9	17.5
Over-exposure	8.5	3.2
Low contrast	17.8	6.8
Back light	27.8	10.6

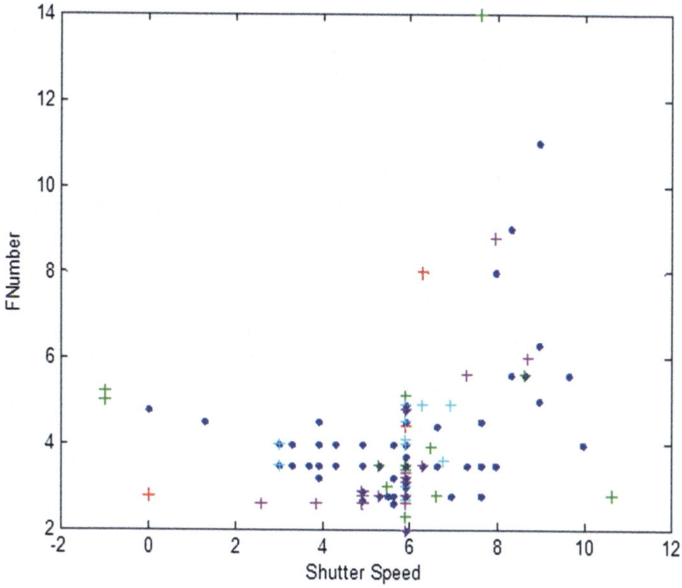

Fig. 3.1 Data distribution within the {Exposure problem} class for the two features of *FNumber* and *ShutterSpeed* (flash on)

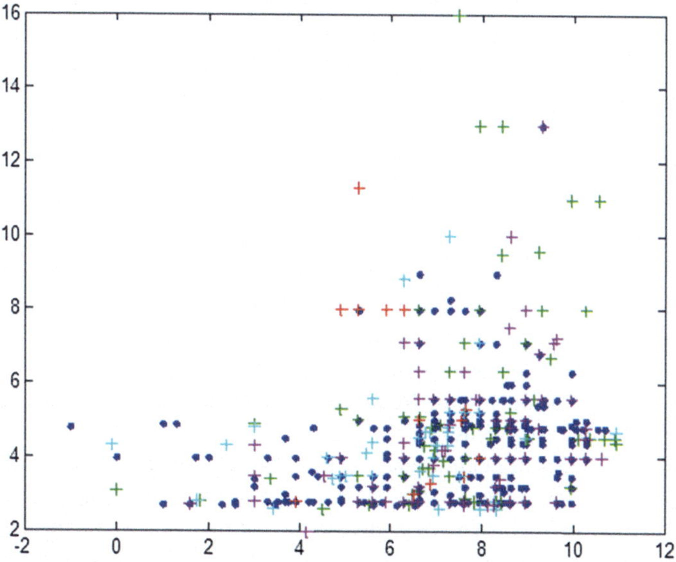

Fig. 3.2 Data distribution in the {Exposure problem} class for the two features of *FNumber* and *ShutterSpeed* (flash off)

Blue dots represent the images classified as {Undistorted}; green crosses represent the {Under-exposure} class; red represents {Over-exposure}; cyan represents {Low contrast}; and purple represents {Back light}.

It can therefore be seen that the separation of the classes in terms of the two chosen features is difficult.

The construction of a *decision tree classifier* for the three classes, i.e. for the combined class {Under-exposure}, {Over-exposure}, {Low contrast}, and the {Back light} and {Normal} classes, provides cross-validation errors of about 34% ($FlashFired = 1$) and 29% ($FlashFired = 0$). Overall, poor cluster reparability and a high error rate is noted for photos captured by various cameras.

The solution suggested by Yen et al. (2007) was also used for the image classification of {Back light} on the basis of the information about EXIF tags. The rule relies upon the very rarely occurring EXIF tag *BrightnessValue*, and is given in the following form:

$$\begin{aligned} &\frac{ExposureTime}{FNumber^2} < 0.005 \text{ and } BrightnessValue < 4.9823, \quad \text{if } FlashFired = 0; \\ &\frac{ExposureTime}{FNumber^2} < 0.005 \text{ and } BrightnessValue > -3.0, \quad \text{if } FlashFired = 1. \end{aligned}$$

Testing of the abovementioned rule on the EXIF-related dataset gives the following results:

- $FlashFired = 0$: altogether, 64 images were processed. Of these, six images were marked as corresponding to the {Back light} class. The rule attributed 50 images to the class {Back light} and 14 images were classified as {Normal}. The errors in this case were as follows: errors of the type I (FP): 90%; errors of type II (FN): 7%;
- $FlashFired = 1$: altogether, six images were processed; of these, one image was distorted. All images were classified as {Normal}.

In view of these experiments, the conclusion can be drawn that the construction of a classifier which operates solely on EXIF meta information leads to serious classification errors. This is because the image distortions caused by the incorrect adoption of exposure characteristics are determined primarily by the complexity of external conditions in the process of image capture.

We emphasise that only most frequently used EXIF tags were selected for the analysis of exposure problems.

Most of the images containing an EXIF header also include a *thumbnail*, i.e. a reduced image used for indexing the basic images or for recognising and organising them, playing the same role for these images as a normal text index does for words. In our experimental database of 1054 images, 98.5% contained thumbnails. The histogram calculation for the thumbnail is much faster than for the entire image. For detection of *under-exposed* and *over-exposed* images and *images with low contrast*, the approach described in Efimov et al. (2007) may be applied. A technique for the detection of backlit scenes presented in Chap. 1 can also be used.

3.3.2 Detection of Noisy Photos

Theoretically, image noise arises from long exposure times, low illumination on the capturing matrix, high ISO values and high signal gain. In view of this, the following EXIF tags were selected: *ExposureTime*, *FNumber*, *Flash*, *ISOSpeedRating* and *GainControl*.

Overall, 13% of the photos were identified as containing visual noise.

Figure 3.3 shows the relationship between $\log_2$(*ExposureTime*) and *FNumber* for the {Noisy} (red crosses) and {Normal} (blue dots) photos. It is apparent that there is a certain area where small values of *Exposure Time* (large values of $\log_2$(*Exposure Time*)) and for small values of *FNumber* (where the diaphragm is open) for the noisy images is unlikely. For a particular DSC model (and even for a particular type of DSC) this area may be relatively large; however, if all models are analysed together, this area can be determined as:

$$-\log_2(\mathit{Exposure\,Time}) > 6 \text{ and } \mathit{FNumber} < 4;$$
$$\mathit{Exposure\ Time} < 0.015 \text{ and } \mathit{FNumber} < 4.$$

The given classification conditions allow the detection of about 15% of the normal photos; these may be not subjected to noise reduction in the general case in order to reduce the total processing time.

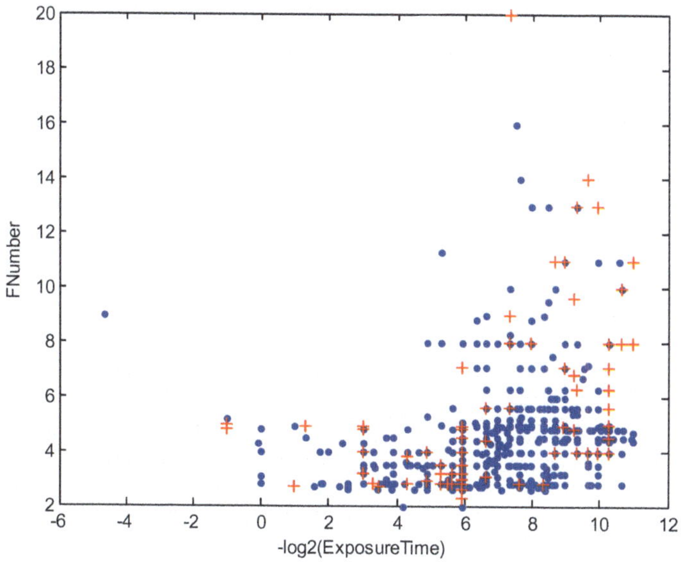

Fig. 3.3 Data distribution in the {Noisy} (red crosses) and {Normal} (blue dots) classes for the two features of *Exposure Time* and *FNumber*

The detection of noisy photos with an error of less than 30–40% seems to be impossible. Figures 3.4 and 3.5 demonstrate the same dependency when the flash is both on and off. It is clear that information on the flash does not give rise to new data in the detection of noisy photos.

Theoretically, high values of ISO correspond to a high noise level. Nevertheless, for low-cost compact DSCs, a value of 100 ISO can be considered high, while for professional DSCs a normal image is provided for 400 ISO and higher.

Figure 3.6 shows that the tag *ISOSpeedRatings* cannot be used for the classification of noisy images in the general case where information about camera type is unavailable.

The tag *ISOSpeedRatings* appears in 29% photos from the training dataset. The value of *GainControl* = 2 (high gain) corresponds to a higher noise level. According to the abovementioned dependencies, a rule for image detection can be suggested for which there is no need to carry out noise detection and correction:

$$\begin{aligned}&\textit{Exposure Time} < 0.015 \text{ and } \textit{FNumber} < 4\\&\textit{ISOSpeedRatings} \leq 800 \text{ (if } \textit{ISOSpeedRatings} \text{ is present) and}\\&\textit{GainControl} \mathrel{!=} 2 \text{ (if } \textit{GainControl} \text{ is present).}\end{aligned}$$

Application of this rule allows the elimination of the estimation of noise level and noise suppression stages for about 20% of the photos, thus reducing total processing time.

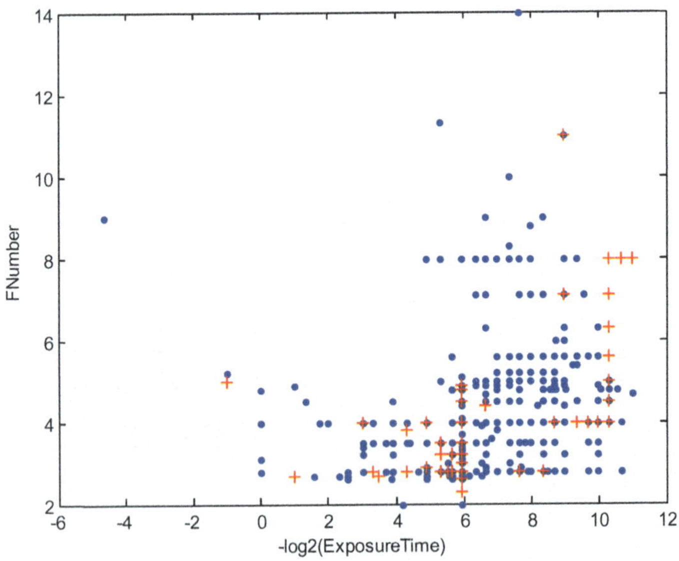

Fig. 3.4 Data distribution in the {Noisy} (red crosses) and {Normal} (blue dots) classes for the two features of *Exposure Time* and *FNumber* (flash fired)

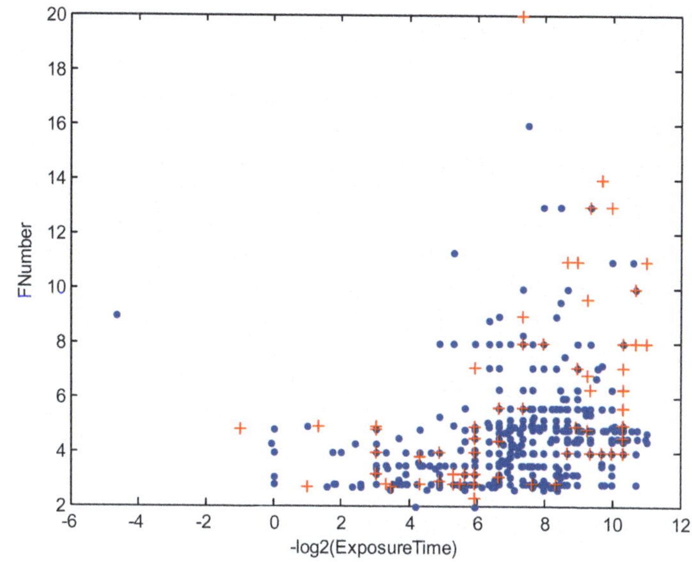

Fig. 3.5 Data distribution in the {Noisy} (red crosses) and {Normal} (blue dots) classes for the two features of *Exposure Time* and *FNumber* (flash not fired)

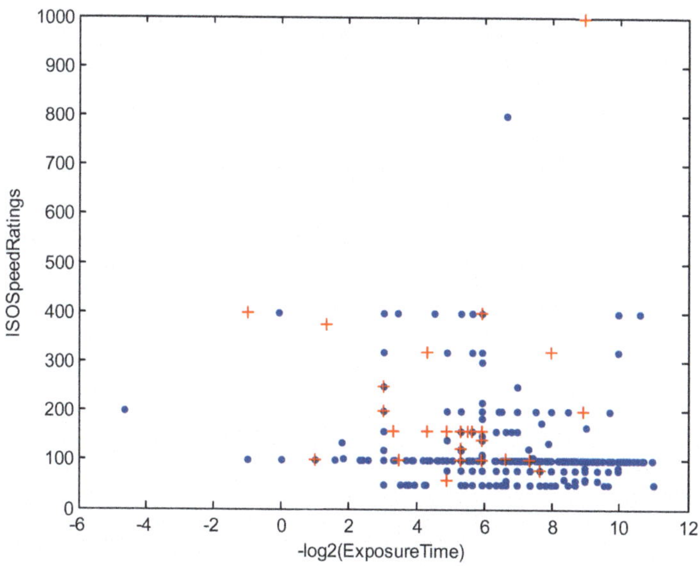

Fig. 3.6 Data distribution in the {Noisy} (red crosses) and {Normal} (blue dots) classes for the two features of *Exposure Time* and *ISOSpeedRatings*

3.3.3 Colour Cast Detection

To explore the possibility of colour imbalance detection using only information from EXIF, about 10% of the images from the EXIF-related image database with a colour cast were analysed. The following EXIF tags were selected, which directly affect the appearance of the colour cast: *ColourSpace*, *WhiteBalance*, *SceneCaptureType* and *LightSource*. However, in most cases these turned out to be non-informative. For example, the *LightSource* tag which is intended to store important information about scene illumination was set as "unknown" in 81.7% of cases. The *WhiteBalance* tag corresponded to the automatic mode in 96.3% of cases and the *SceneCaptureType* tag was set to the value "Standard" in 91.4% of cases. It therefore seems to be impossible to construct a generally reliable classifier for colour cast detection based solely on EXIF information.

3.3.4 Blurred Photo Classification

Damaged photos made up 16.8% of the training set {Blurred photo}, including two subclasses: {Out-of-focus} and {Motion blur}. The images were classified into the following sub-classes:

- photos affected by motion blur or defocused as a class {Blurred};
- photos not affected by blur, i.e. class {Normal}.

Statistics for the photos affected by blur with respect to their presence in the EXIF-related database are given in Table 3.3.

The most important key tags are *FNumber* and *ExposureTime*. The *ExposureTime* tag is very informative for the {Blurred photo} class. When exposure time is long, the probability of the camera shaking or undergoing other undesirable motion is relatively high. The *ShutterSpeedValue* and *ExposureTime* tags are mutually dependent, as shown in the expression $ShutterSpeed = -\log_2(ExposureTime)$. The *DigitalZoomRatio* tag is equal to one in most cases, indicating that there is no digital zoom. An analysis of the

Table 3.3 Statistics of photos with blur

Class	Statistics (%)
Flash	100
ExposureTime	99.6
FNumber	99.6
ShutterSpeedValue	53.2
DigitalZoomRatio	40
FocalLengthIn35mmFilm	30.6
SubjectDistanceRange	12.1
SceneCaptureType	7.9

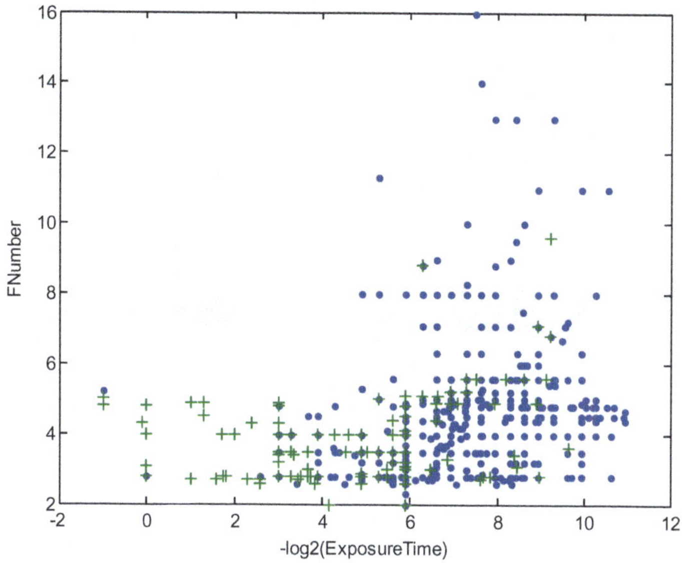

Fig. 3.7 Data distribution in the {Normal} (blue dots) and in {Blurred} (green crosses) classes for the two features of *Exposure Time* and *FNumber*

FocalLengthIn35mmFilm tag showed that this tag is insignificant, and it was therefore excluded from consideration. The *SubjectDistanceRange* tag does not contain sufficient information to estimate whether or not the focus distance was set correctly. As example, Fig. 3.7 shows distribution of {Normal} (not defect) photos (dots) and photos, labelled as {Blurred photo} (cross). The x and y axes are *FNumber* and $\mathit{ShutterSpeed} = -\log_2(\mathit{ExposureTime})$ respectively.

It can be seen from the plot above that photos corrupted by blur are located in the region where the value of *ShutterSpeed* corresponds to higher values of exposure. To support subjective differentiation, histograms of $-\log_2(\mathit{ExposureTime})$ and *FNumber* are depicted in Fig. 3.8. A lighter colour of the histogram beans corresponds to the class {Normal}, and a darker colour to {Blurred}. The histogram values are not normalized here.

The full classification tree contains six levels and is too complex for implementation. After the excess branches were pruned with the minimisation of cross-validation error, the following tree was constructed, which realises one very straightforward rule:

$$\mathit{ExposureTime} > 0.0678\,\mathrm{s}.$$

On testing, this rule gives a classification error of 12.9%.

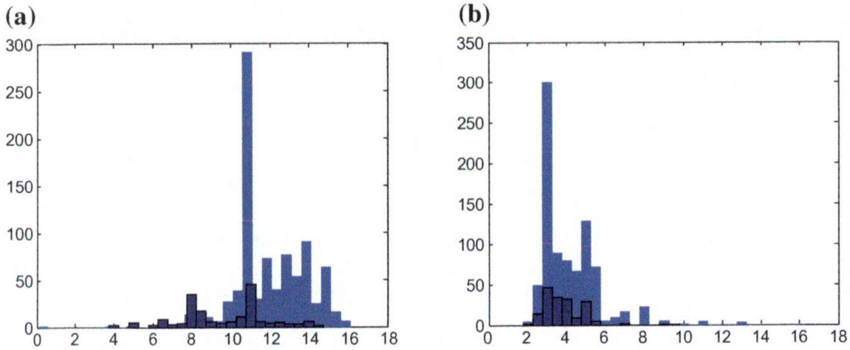

Fig. 3.8 Histograms of **a** *ShutterSpeed* and **b** *FNumber*

3.3.5 Classification of Images with JPEG Artefacts

For classification of the images with compression, i.e. JPEG-originated artefacts, the EXIF tag *CompressedBitsPerPixel* was analysed. This tag appears in 87% of the photos in the image database. Figure 3.9 illustrates the correspondence of tag values with the actual bit per pixel (BPP) rate, which is calculated as follows:

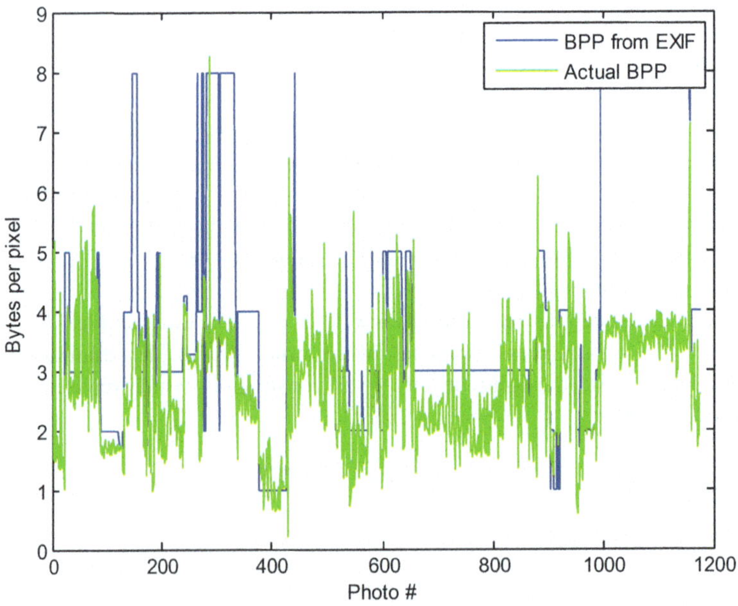

Fig. 3.9 Correspondence of data from EXIF tags with actual BPP

$$Actual\ BPP = \frac{File\ size\ in\ bits}{Width \times Height}.$$

Sometimes, a rather low correlation of the data extracted from the EXIF tag with the actual BPP is observed. This indicates that the actual BPP depends on image content. It is also the case that a smoother image content corresponds to a higher JPEG compression rate. For instance, photos affected by motion blur will have a higher compression rate.

As can be seen from Fig. 3.10, it is possible to separate the good and bad photos (in terms of redundant compression) using two attributes: *BPP from EXIF* and *Number of pixels in image*. It is thus possible to minimise errors of type II, i.e. *False Negatives* (FN).

It seems to be reasonable to consider the problem of minimisation of the errors of type II: this classifier can exclude situations when the defect is present, but classifier reports this photo as a good one. In the EXIF enhancement pipeline, the classifier will report when artefacts are possible and can prevent the JPEG enhancement of good images.

The JPEG compression method is based on the division of an image into non-overlapping 8 × 8 pixel squares and the subsequent application of a discrete cosine transform. Thus, with increasing image size (and increasing printing

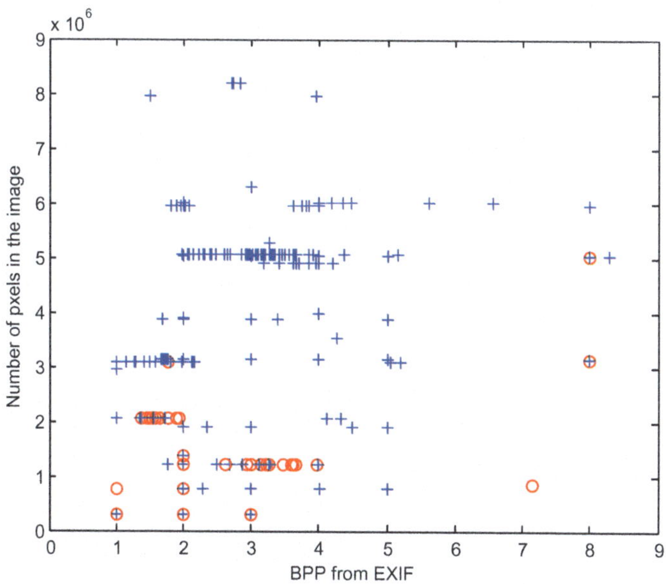

Fig. 3.10 Data distribution in the {Normal} (blue crosses) and {Redundant Compression} (red circles) classes for the two features of *BPP from EXIF* and *Number of pixels in image*

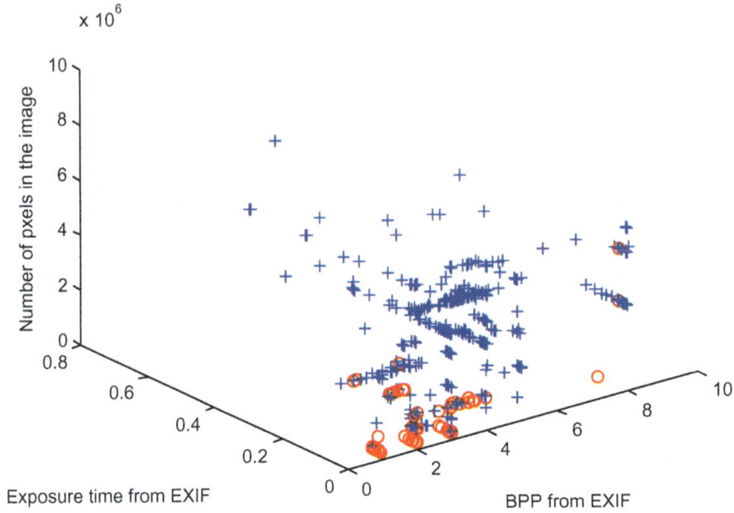

Fig. 3.11 Visualisation of JPEG compression artefacts in consumer photos for printing

resolution) compression artefacts become less and less visible, especially when the image is printed. This implies that the image size in pixels can serve as an additional informative attribute for image classification (Fig. 3.11).

Thus, for the robust classification of images affected by JPEG compression artefacts, the use of the following EXIF tags is proposed:

1. *CompressedBitsPerPixel (CBPP);*
2. *PixelYDimension (PYD);*
3. *PixelXDimension (PXD).*

These tags are widely used by DSC manufacturers and consequently appear in all the images in the image database. Bad photos with visible JPEG artefacts make up 11% of the database. As result of linearly separating the classifier construction, an appropriate decision rule can be obtained:

$$\mathbf{a} = \left[\begin{pmatrix} 476.85 & -24.18 \\ 0 & -832.74 \times 10^{-6} \end{pmatrix} \times \begin{pmatrix} CBPP - 1.95 \\ PXD \times PYD - 601557.75 \end{pmatrix}\right] \times 10^{-3},$$

$$C_1 = -\frac{1}{2}\left(\mathbf{a}^{\mathrm{T}} \times \mathbf{a}\right) + 29.48.$$

$$\mathbf{b} = \left[\begin{pmatrix} 476.85 & -24.18 \\ 0 & -832.74 \times 10^{-6} \end{pmatrix} \times \begin{pmatrix} CBPP - 4.45 \\ PXD \times PYD - 3736583.04 \end{pmatrix}\right] \times 10^{-3},$$

$$C_2 = -\frac{1}{2}\left(\mathbf{b}^{\mathrm{T}} \times \mathbf{b}\right) + 29.48.$$

If $C_1 > C_2$, then the image is affected by JPEG artefact compression; if $C_1 < C_2$, then post-processing is unnecessary.

3.3.6 *Red Eye Artefacts*

Appearance of red eye artefacts is possible only when the flash is fired. It is only worthwhile to check the *Flash* tag and to perform the red eye detection procedure when the flash is fired. This approach significantly decreases the total processing time (Safonov et al. 2008).

3.4 The EXIF-Based Image Enhancement Pipeline

3.4.1 *Pipeline Flowchart*

Figure 3.12 shows the proposed automatic image enhancement pipeline, based on the analysis of EXIF tags. Typical defects of photos are estimated and corrected (if necessary) in the following order: (a) JPEG artefacts; (b) noise; (c) exposure problems; (d) sharpness and (e) red eye artefacts.

Due to the existence of photos with an intrinsic colour cast which should not be corrected, it was not possible to find a reliable method for the detection of colour cast based both on metadata and image data.

3.4.2 *Characteristic Results*

Examples of the classification of ready-to-print images using the EXIF-based image enhancement pipeline are given in Figs. 3.13, 3.14, 3.15, and 3.16.

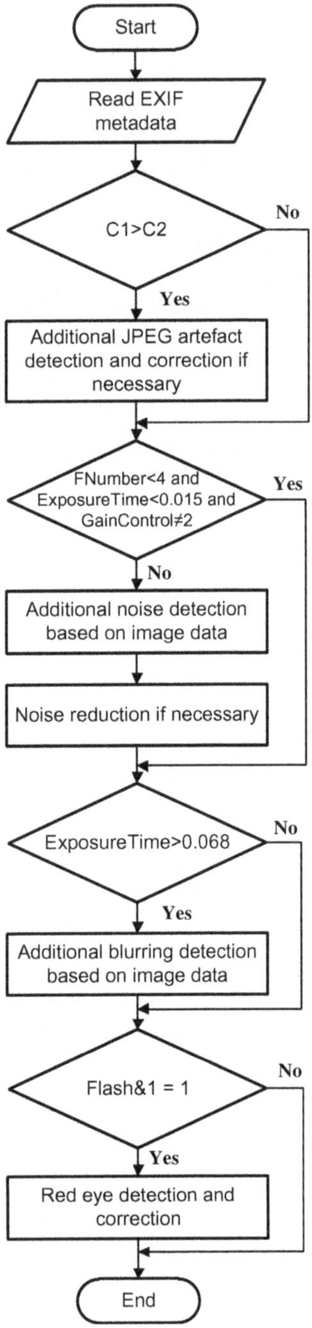

Fig. 3.12 Flowchart of the EXIF-based image enhancement pipeline

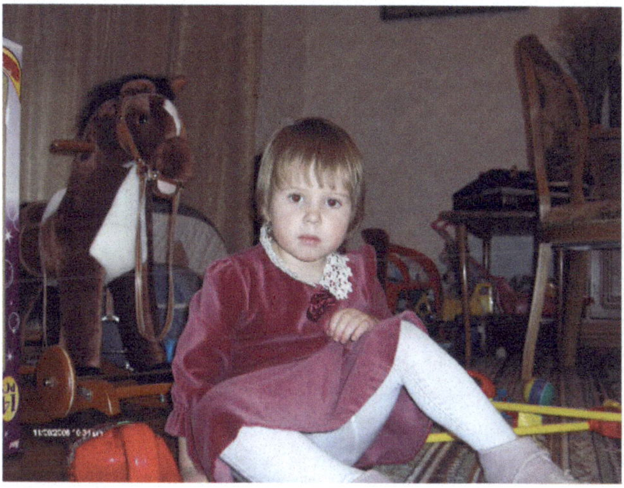

J correction	*N* correction	*B* correction	*R* correction
+	+	-	+

Fig. 3.13 EXIF-processing of photo with *J*, *N* and *R* defects ("+" indicates correction is needed; "–" indicates the opposite)

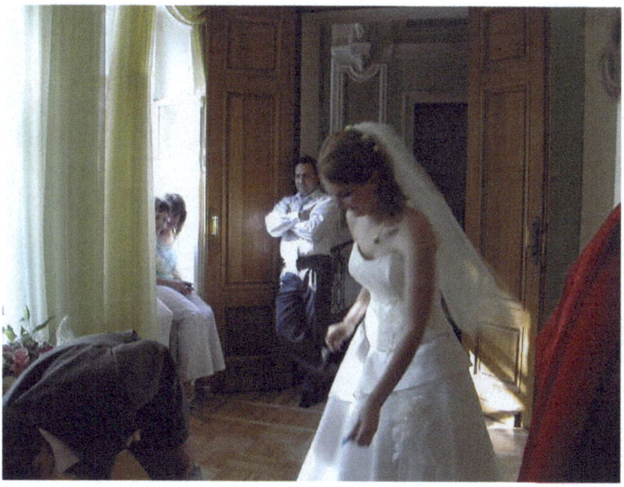

J correction	*N* correction	*B* correction	*R* correction
+	+	-	-

Fig. 3.14 EXIF-processing of photo with *J* and *N* defects ("+" indicates correction is needed; "–" indicates the opposite)

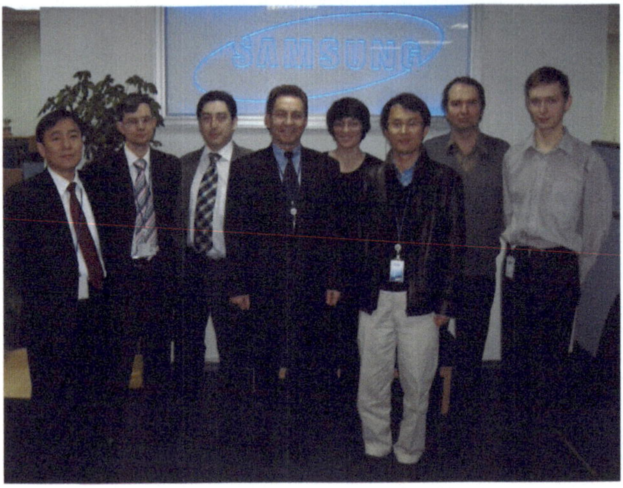

J correction	*N* correction	*B* correction	*R* correction
+	+	-	+

Fig. 3.15 EXIF-processing of photo with *J*, *N* and *R* defects ("+" indicates correction is needed; "–" indicates the opposite)

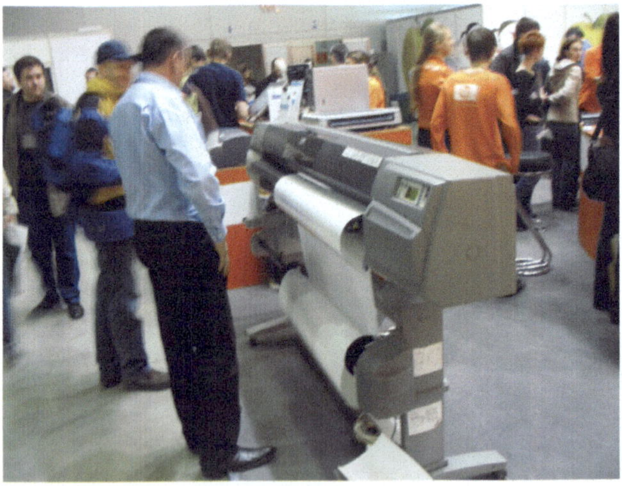

J correction	*N* correction	*B* correction	*R* correction
-	+	+	-

Fig. 3.16 EXIF-processing of photo with *N* and *B* defects ("+" indicates correction is needed; "–" indicates the opposite)

References

Amornpashara, N., Arakawa, Y., Tamai, M., Yasumoto, K.: Landscape photo classification mechanism for context-aware photography support system. In: IEEE International Conference on Consumer Electronics (ICCE), pp. 663–666 (2015)

Boutell, M., Luo, J.: Photo classification by integrating image content and camera metadata. In: Proceedings of the 17th International Conference on Pattern Recognition, ICPR 2004, vol. 4, pp. 901–904 (2004)

Dietmar, W., Reiner, F.: Statistic analysis of millions of digital photos. Proc. IS&T/SPIE Electron. Imaging **6817**, 68170L (2008)

Efimov, S.V., Nefyodov, A.V., Rychagov, M.N.: Block-based image exposure assessment and indoor/outdoor classification. In: Proceedings of 17th International Conference on Computer Graphics, GraphiCon, pp. 84–87 (2007)

Exif Version 2.2: Japan Electronics and Information Technology Industries Association (2002)

Fan, J., Cao, H., Kot, A.C.: Estimating EXIF parameters based on noise features for image manipulation detection. IEEE Trans. Inf. Forensics Secur. **8**(4), 608–618 (2013)

Gu, J.M., Wu, Y.L., Hung, W.C., Tang, C.Y.: Personal photo organization using event annotation. In: Proceedings of 9th International Conference on Information, Communications & Signal Processing, pp. 1–4 (2013)

Guidelines for Handling Image Metadata: Metadata working group. 2010-11-01. www.metadataworkinggroup.org (2010). Retrieved 2015-05-11

Jang, C.J., Lee, J.Y., Lee, J.W., Cho, H.G.: Smart management system for digital photographs using temporal and spatial features with EXIF metadata. In: Proceedings of 2nd International Conference on Digital Information Management, vol. 1, pp. 110–115 (2007)

Kee, E., Johnson, M., Farid, H.: Digital image authentication from JPEG headers. IEEE Trans. Inf. Forensics Secur. **6**(3), 1066–1075 (2011)

Li, H., Kot, A.C., Li, L.: Color space identification from single images In: IEEE International Symposium on Circuits and Systems (ISCAS), pp. 1774–1777 (2016)

Ryu, D.S., Kim, S.Y., Cho, H.G.: A photo visualization system considering region of interest. In: 10th IEEE International Conference on Computer and Information Technology, pp. 560–566 (2010)

Safonov, I.V., Rychagov, M.N., Kang, K.M., Kim, S.H.: Automatic red eye correction and its quality metrics. Proc. IS&T/SPIE Electron. Imaging **6807** (2008)

Tešić, J.: Metadata practices for consumer photos. IEEE MultiMedia **12**(3), 86–92 (2005)

Toevs, B.: Processing of metadata on multimedia using Exif tool. A programming approach in Python. In: Proceedings of the Annual Global Online Conference of Information and Computer Technology, pp. 26–30 (2015)

Yen J., Wu P., Tretter, D.: Knowledge discovery for better photographs. In: Proceedings of IS&T/SPIE Electronic Imaging, Multimedia Content Access: Algorithms and Systems, 65060B. (2007). doi:10.1117/12.702891

Zhao, K., Cao, C., Zhang, H., Song, Y.: A dataset for scene classification based on camera metadata. In: Proceedings of IEEE International Conference on Network Infrastructure and Digital Content, pp. 443–447 (2012)

Chapter 4
Adaptive Sharpening

4.1 Introduction

Sharpness is one of the main characteristics of image quality assessment (Keelan 2002). Until now, there is no evident, unified definition of sharpness or blurriness. Each individual has his or her own notions and preferences about the sharpness of an image. Moreover, some photo regions can be blurred intentionally. There are several kinds of scenes in which photographers blur the background and/or foreground knowingly, for example in portraits. An observer appraises such photos positively. That type of blurring is not a defect. So, the development of an adaptive sharpening procedure that correlates with a subjective assessment is a challenge.

The perception of image sharpness depends on the viewing conditions. There is no need to estimate sharpness and to perform sharpness enhancement under abstract viewing conditions. Figure 4.1 demonstrates two scanned hardcopies of the same photo. Is the initial image sharp? There is no proper answer. It depends on the size of the hardcopy, printing resolution, and technology: ink-jet, laser, sublimation, and so on. Is image (a) sharp? No, it is definitely not. Is image (b) sharp? Yes, probably.

When printing from a personal computer or mobile device, it is difficult to estimate the sharpness of a photo adequately due to differences in the printer and display resolutions. An image on a display may seem to be sharp enough but the printed image is blurred. The fact that the printer changes the size of an image is not taken into account during editing using software tools. Frequently, obtaining a good sharpening result for a hardcopy is a complex task for the user. Some capture devices and photo printers have a small display for viewing photos; it is hard to estimate the sharpness of a photo prior to printing. Thus, sometimes a poor quality photo affected by a strong blur is printed. This is an unpleasant situation and the problem remains unsolved so far.

In this chapter, we describe an adaptive sharpening algorithm intended for a dye-sublimation photo printer that is capable of printing photos with a resolution of 300 pixels per inch (ppi). The size of a hardcopy is 6×4 inches (15×10 cm).

I.V. Safonov et al., *Adaptive Image Processing Algorithms for Printing*,
Signals and Communication Technology,
https://doi.org/10.1007/978-981-10-6931-4_4

(a)

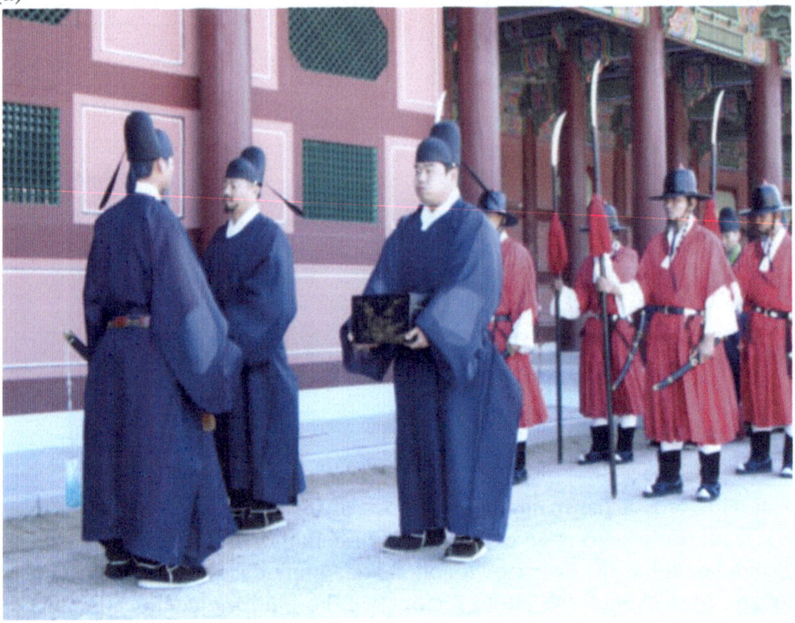

(b)

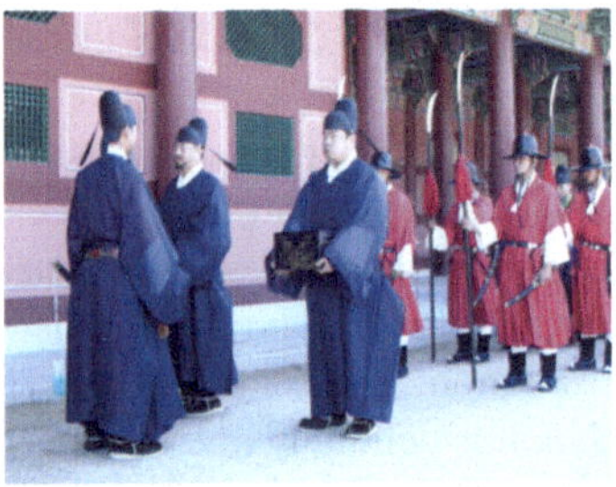

Fig. 4.1 Two hardcopies of the same photo: **a** the photo looks blurry; **b** the photo looks sharper

Some adaptation of the algorithm is required for other sizes and printing resolutions.

4.2 Sharpening Methods

All techniques dealing with the improvement of blurred images fall into two main categories: image restoration and image enhancement. Image restoration methods estimate the Point Spread Function (PSF) or Optical Transfer Function (OTF) and then perform deconvolution assisted by regularization.

Several factors have a joint influence on a photo. The more significant ones are the following: a fault of the focusing system; shaking of the camera during long exposure times; the movement of objects in the scene during shooting; upscaling of the photo; and suppression of noise and artefacts of compression. So, a mathematical model of blurring for real-world photos is rather complex, which prevents the wide usage of image restoration techniques for the correction of blurred photos.

Image enhancement methods for image sharpening are more practical. The techniques are divided into two groups.

- Those that increase the local contrast at the ends of the edge transition slope.
- Those that decrease the length of the edge transition slope.

Figures 4.2 and 4.3 illustrate both methods and emphasize the difference between them.

Usually methods from the first group consist of well-known and popular techniques that work through the addition of high-pass frequencies to an image. The most widespread method is the use of an unsharp mask:

$$Ie = I + k \times (I - Ib),$$

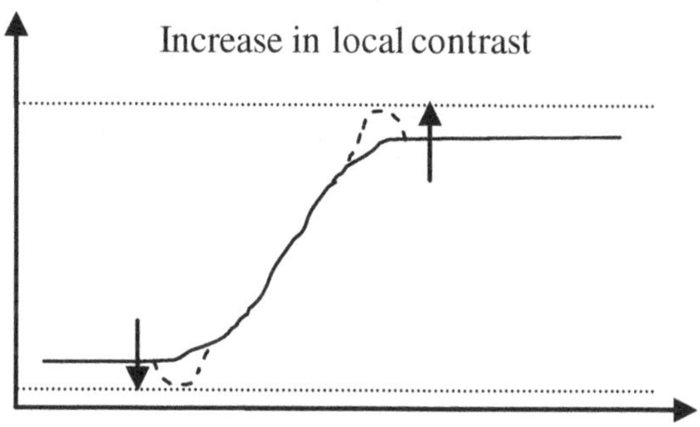

Fig. 4.2 Illustration of the increase in local contrast at the ends of the edge transition slope for the profile of the intensity through the edge

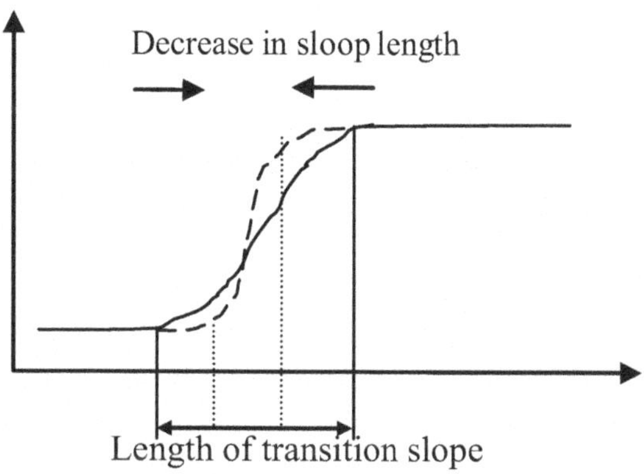

Fig. 4.3 Illustration of decrease in transition slope length for the profile of the intensity through the edge

where I is the initial image, Ib is I filtered by Gaussian blur, and k is an amplification factor. The unsharp mask improves the visual perception of an image by increasing the local contrast near edges. The following are the drawbacks of the traditional unsharp mask filter: noise amplification, the formation of halo artefacts, and the fact that increasing the local contrast depends on the edge magnitude. If the length of an edge transition slope is notably large, then it is impossible to obtain a sharp image by applying an unsharp mask. A large number of publications are devoted to various modifications of unsharp masks (Kotera and Wang 2005; Polesel et al. 2000; Kim and Allebach 2005).

Techniques intended for decreasing edge transition slopes are applied rarely in comparison with unsharp mask filters. Unlike algorithms, which increase the local contrast, the second group of methods makes it possible to enhance strongly blurred photos with a relatively large slope length. A morphological toggle filter transforms a blurred edge into an ideal edge that is a step function (Schavemaker et al. 2000). Half of the slope is substituted by a minimum value found in the vicinity of the current pixel and the other half of the slope is substituted by the maximum value found in the vicinity. Edges produced by the filter are absolutely sharp, but large regions near edges have the same colour values, and the photo looks unnatural, like a painted cartoon.

A rational elaboration of morphological sharpening is local tone mapping, where pixels of the edge slope are substituted with values from minimum to maximum from the vicinity according to an S-shaped function that is typical for contrast stretching. Wang et al. (2005) and Luong and Philips (2005) describe local tone mapping for sharpening after image upscaling.

4.3 General Pipeline of Adaptive Sharpening

We base our method on the following psychophysical features of the human visual system.

- **Resolving ability**: it is well-known that the human visual system is able to discern objects up to a size of 1/4 to 1/3 mm when observed from 30–40 cm distance; for 300 ppi this is approximately equal to 3–4 pixels, which is a sharp edge having a length of transition slope of less than 3 pixels.
- **Human vision is anisotropic**: the diagonal resolving ability is notably lower than the vertical and horizontal ones.

In addition, we assume that an observer evaluates even well-focused photos positively with a slightly increased local contrast.

Figure 4.4 demonstrates the general pipeline of the adaptive sharpening algorithm. Since the proposed method is aimed at a printed photo with a size of 6×4 inches and a resolution of 300 ppi, the input image is resampled to the target size. The next step is a non-reference sharpness assessment. The measure is based on the

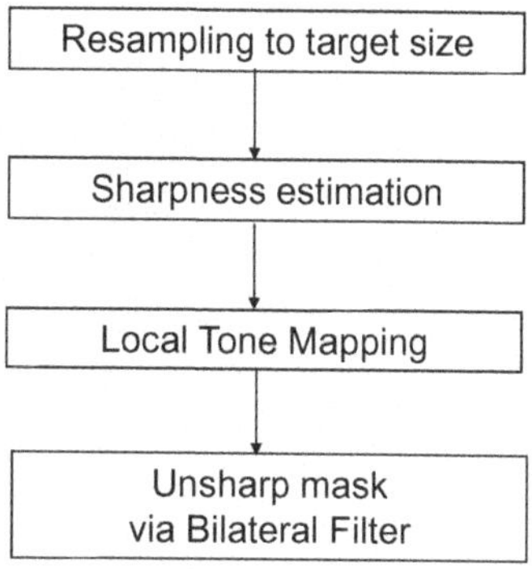

Fig. 4.4 A general pipeline of adaptive sharpening

analysis of variations of histograms of outcomes of high-pass filters with various kernel sizes; the array of sums of logarithmic histograms characterizes the sharpness of photo. The result of assessment is a sharpness level on a discrete four-grade scale. Further, *local tone mapping* reduces the length of the edge transition slope. The final stage is the use of an *unsharp mask* via *a bilateral filter* to increase the local contrast. The parameters of both stages of sharpening are adjusted according to the sharpness level.

In contrast to the majority of existing approaches, which either increase local contrast on the ends of the edge slope or decrease the edge transition slope, we employ those two techniques one by one. This makes it possible to achieve better sharpness without forming noticeable artefacts.

The approach is intended for both greyscale and colour photos. For colour photos, we process a brightness channel in any colour space.

4.4 Blind Sharpness Metrics

During many experiments with artificially blurred images, we discovered a form of histogram of image filtered by edge-detection, for example, the Finite Impulse Response (FIR) High-Pass Filter (HPF) depends on the size of the sliding window of the filter. If the size of a convolution kernel of HPF is smaller than the size of the blurring filter, then the form of histogram will change strongly as the size of the HPF kernel increases. If the size of the convolution kernel of the edge detection filter is greater than the size of the blurring filter, then the form of the histogram will change insignificantly with growth in the size of the kernel (Safonov et al. 2008).

Let us consider a simple test pattern with Gaussian blurring with variance equal to 1; that is, the size of the convolution kernel is 3 (see Fig. 4.5). Figure 4.6 shows the set of histograms of absolute values of high-pass filtration with convolution kernels of increasing sizes: [1 −1], [1 0 −1], [1 0 0 −1], and [1 0 0 0 −1]. One can see that the form of histogram changes significantly when the size of kernel of the HPF is smaller than or equal to the size of kernel of the Gaussian blur.

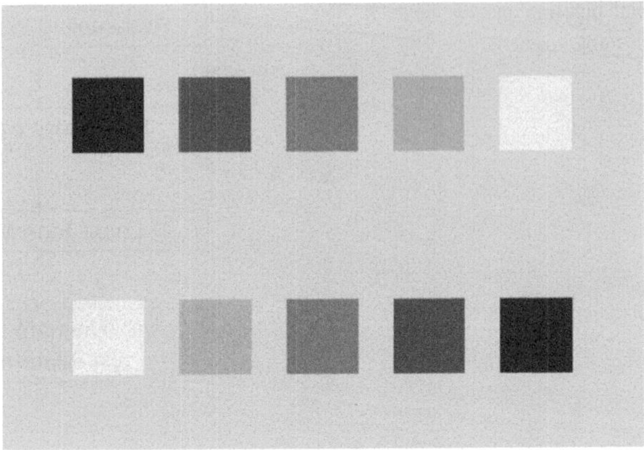

Fig. 4.5 Test pattern filtered by Gaussian blur with $\sigma^2 = 1$

The entropy *En* characterizes the ‘flatness’ or ‘peakedness’ of a histogram:

$$En = -\sum H_i \log H_i,$$

where H_i are bins of a histogram.

Indeed, the entropy of a histogram of an image processed by HPF can serve as a sharpness measure for identical images. However, such a metrics depends on the edges in the image (namely the response of the HPF); that is, it depends on the photo’s content. We normalize *En* by dividing it by the minus sum of H_i:

$$A = \sum \frac{-H_i \log(H_i + 1)}{-H_i} = \sum \log(H_i + 1),$$

where 1 is added to avoid a calculation error.

The dependence of *A* on image content is significantly less than *En*. In addition, *A* has a simple geometric interpretation: *A* is approximately equal to the area under the envelope of histogram in semi-logarithmic scale, which is equal to the sum of the logarithm of the histogram of the HPF outcome. Figure 4.7 shows plots of *A* depending on the size of kernel *S* of a high-pass filter for a blurred test pattern without noise and with Additive White Gaussian Noise (AWGN). Function *A*(*S*) is approximately constant when the size of the kernel of the edge detection filter is greater than 3, where 3 is the size of the blurring kernel. This dependence is not accidental. For other sizes of blurring windows as well as for convolution kernel coefficients this dependence is similar. Moreover, *A*(*S*) changes insignificantly for a test pattern affected by white noise. An analysis of *A*(*S*) makes it possible to estimate the size of the blurring kernel.

Obviously, the shape of the *A*(*S*) for real-world photos differs from the shape for the simple test pattern. Various low-pass filters affect real photos. For the majority of photos, it is impossible to measure the blurring radius precisely. However, there

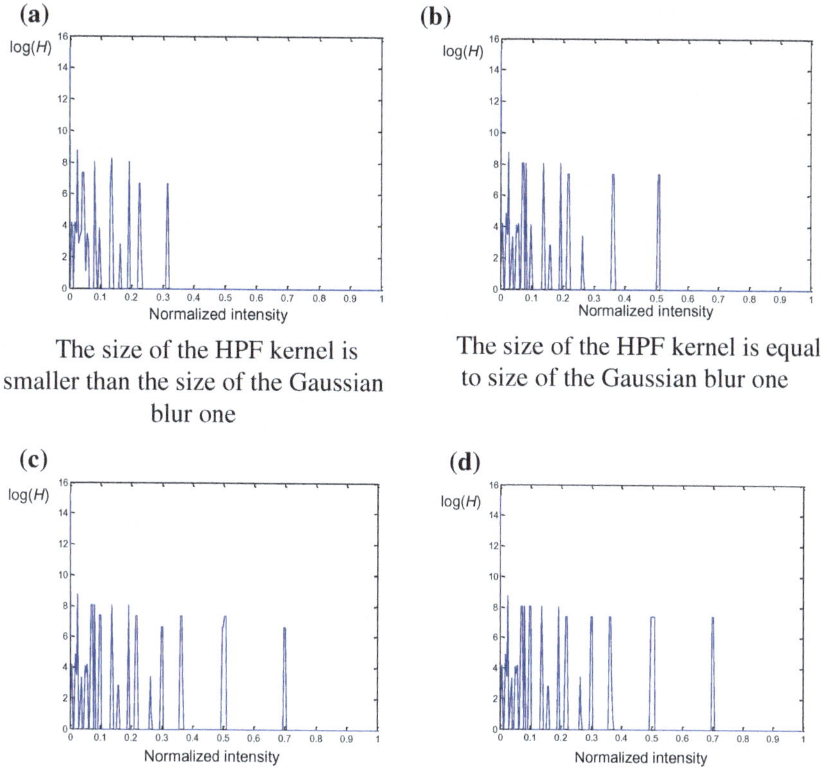

(a) The size of the HPF kernel is smaller than the size of the Gaussian blur one

(b) The size of the HPF kernel is equal to size of the Gaussian blur one

(c) The size of the HPF kernel is greater than size of the Gaussian blur one

(d) The size of the HPF kernel is greater than size of the Gaussian blur one

Fig. 4.6 The histograms in semi-logarithmic scale of the test pattern convolved with the following kernels: **a** [1 −1], **b** [1 0 −1], **c** [1 0 0 −1], **d** [1 0 0 0 −1]

are some trends and these issues can be applied for rough estimation of blurring radius: on a sharp photo, $A(S)$ grows rapidly for $S \leq 2$ and sometimes for $S \leq 3$; for $S > 3$, the first derivative of $A(S)$ is almost constant (e.g., see Fig. 4.9). In the case of a blurred photo, $A(S)$ grows slowly (e.g., see Fig. 4.8).

Let us try to find features $\{Fi\}$ to characterize the shape of the $A(S)$ function. We selected the following three features:

$$F1 = An(3) - An(2),$$

$$F2 = \sum_{i=2}^{11} An(i) - 5,$$

$$F3 = A(2),$$

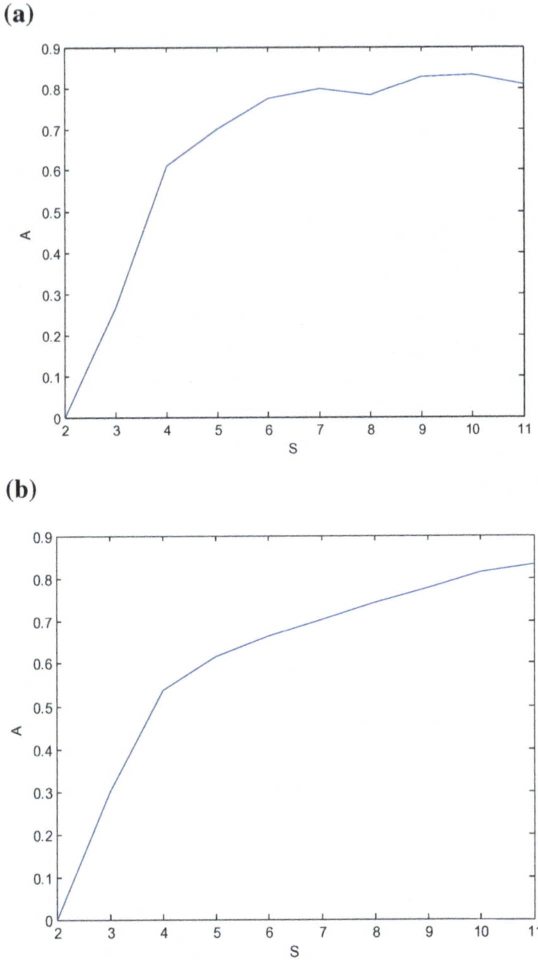

Fig. 4.7 Plots of sums of logarithms of histograms (*A*) depending on the HPF kernel size (*S*) for the test pattern: **a** without noise, **b** with AWGN

where *An*(*i*) is *A*(*i*) normalized to the range of [0, 1] by dividing by the maximum of *A*(*i*).

We found that thresholding of features $\{F_i\}$ makes it possible to detect blurred photos with an error rate of 20–40%; thus, it is worth applying the features as a weak classifier in terms of machine learning. A weak classifier is any classifier that can achieve a slightly better error rate than a random guess. The conjunction of weak classifiers into an ensemble makes it possible to obtain a much higher classification outcome in comparison with the usage of the weak classifier separately. Adaptive boosting (AdaBoost) is a family of learning algorithms that produce a classifier that is a weighted sum of weak classifiers (Schapire and Singer 1999). The outputted sum can be treated as a confidence level. The sum is compared with a

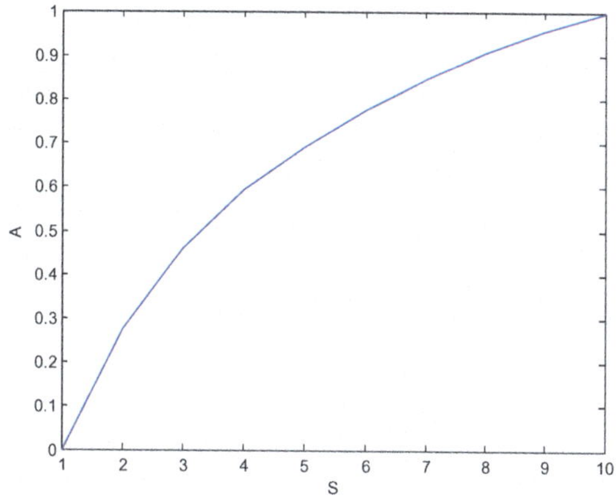

Fig. 4.8 Plot of sums of logarithms of histograms (A) depending on the HPF kernel size (S) for the blurred photo

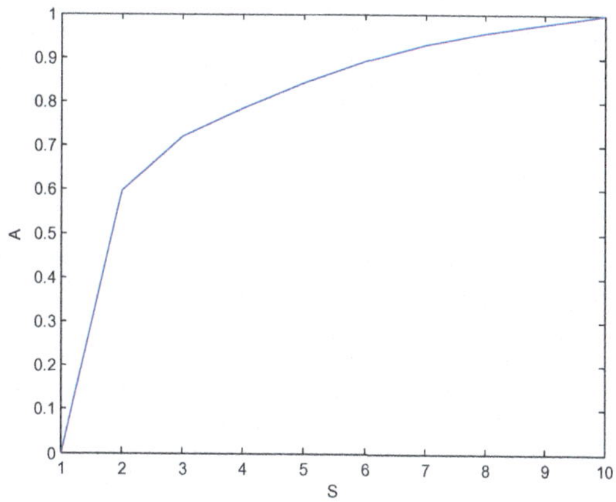

Fig. 4.9 Plot of sums of logarithms of histograms (A) depending on HPF kernel size (S) for a sharp photo

threshold T. Alteration of T adjusts the ratio between False Positive (FP) and False Negative (FN) errors.

We decided to use a supervised machine learning technique to transform features $\{F_i\}$ to a sharpness measure. Initially, we collected a training set of photos of various qualities and labelled it manually. A labelling procedure assigns a sharpness level on a ten-class scale. After that, we attempted several approaches for training 10-class classifiers. In addition, we tried to solve the task by regression. All those attempts were unsuccessful: the quality of the sharpness level estimation was low for the cross-validation and test sets. There are two main reasons for this.

- Three features are not enough to build classifiers for 10 classes.
- The labelling procedure is very subjective and depends on the viewing conditions; an expert who sets labels are sure about labels in the ends of the scale for sharp and blurry photos but has doubts about labels in the middle of the sharpness scale containing 10 grades, so the labels are inaccurate.

Finally, we trained a binary classifier for detection of strongly blurred images, because labelling of such photos can be done without errors. The quality of the classifier was fairly high. We use the confidence level of the binary classifier for classification on a discrete four-grade sharpness scale: *sharp, slightly blurred, blurred,* and *strongly blurred.*

4.5 Local Tone Mapping

Local tone mapping (LTM) by means of an S-shaped curve is an efficient tool for decreasing the edge transition slope. However, a drawback limits its application for sharpening of photos. The disadvantage is caused by the normalization of the current pixel from a range from minimum to maximum in some vicinity of [0, 1], whereas the minimum and maximum values correspond to noise or outliers. Application of LTM to a noisy image leads to amplification of the noise level along the edges. To overcome this drawback, we propose to perform normalization from the range [L, H] and suppress pixels with values outside of the range. For calculation of L and H, we apply techniques similar to order-statistic filters: L is the mean among the first 25 percentiles of pixels in a sliding window; H is the mean among the last 25 percentiles.

The square sliding window moves across the image. The image is filtered non-recursively. For processing of pixel P(r, c), the following algorithm is used.

1. Calculate L and H.
2. Suppress pixels with values outside the range [L, H] for noise filtering as follows: if P(r, c) < L then P(r, c) = L and if P(r, c) > H then P(r, c) = H.
3. If the difference between L and H is greater than the threshold *Tltm* then:
   a. Normalize P(r, c) from the range [L, H] to [0, 1]:

$$x = (\mathrm{P}(\mathrm{r},\ \mathrm{c}) - \mathrm{L})/(\mathrm{H} - \mathrm{L}).$$

   b. Carry out local contrast mapping via an S-shaped function, for example:

$$y = \frac{x^2}{x^2 + (1 - x)^2}.$$

   c. Scale the target pixel back:

$$\mathrm{Pe}(\mathrm{r}, \mathrm{c}) = \mathrm{L} + y(\mathrm{H} - \mathrm{L}).$$

The comparison with *Tltm* in Step 3 is intended to prevent correction in flat regions. Figure 4.10 demonstrates the advantages of our approach over scaling to range [minimum, maximum]. In min/max LTM, arbitrary noise outliers near the

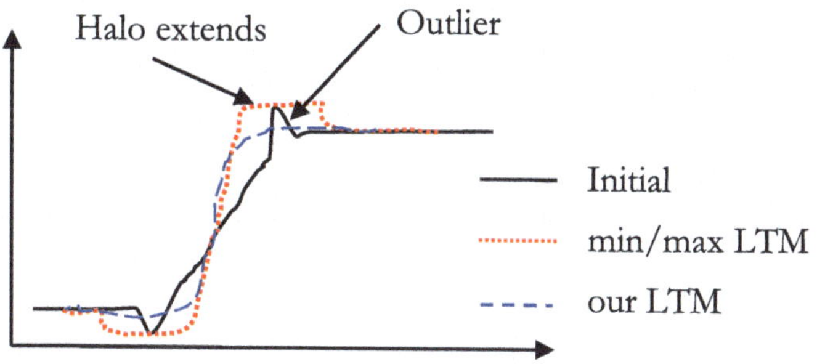

Fig. 4.10 Profile of intensity processed by LTM with mapping between local minimum and maximum and with mapping between means among the first and last 25 percentiles

edges are enlarged, creating a noticeable flat halo, and the noise level increases. Our method is more robust to any noises and artefacts. It approximates the initial signal better.

4.6 Unsharp Mask via Bilateral Filter

The use of a conventional unsharp mask filter via Gaussian blur has the following drawbacks:

- noise amplification;
- formation of halo artefacts; sometimes the halo is wide enough;
- the amplitude of the increase in local contrast depends on the edge magnitude.

To overcome the disadvantages, we propose to apply a bilateral filter in an unsharp mask instead of a Gaussian:

$$Ie = \begin{cases} I + k \times (I - If) & : |I - If| > T_{\text{usm}} \\ I + k \times (I - If)/2 : & |I - If| \leq T_{\text{usm}} \end{cases},$$

where I is the initial image, If is the outcome of the bilateral filter for image I, k is the amplification factor, and T_{usm} is the threshold for preventing noise level amplification.

The bilateral filter by Tomasi and Manduchi (1998) uses a Gaussian function as an edge-stop or photometric distance function. In Chap. 1, we proposed to apply the function $E\mu$ instead of a Gaussian. On one hand, the function is similar to a Gaussian function and on the other hand it does not tend to zero so rapidly. Variation of μ in the range from 1 to 5 allows the sharpening to be controlled. Our algorithm produces a sharper image with a narrow and almost undistinguishable halo; the value of local contrast gain is insignificantly dependent on the edge magnitude in comparison with an unsharp mask via Gaussian blur. One can

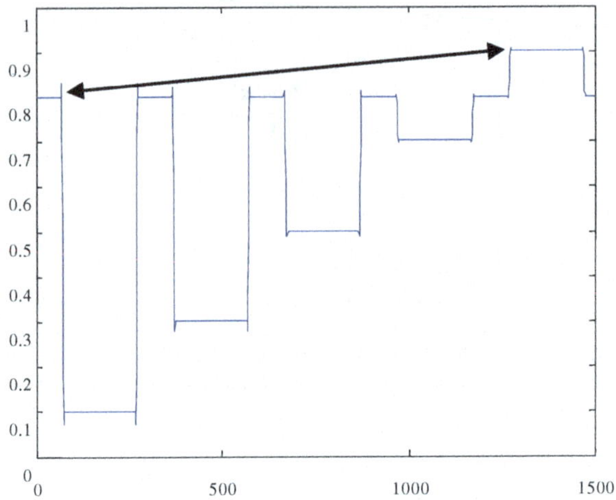

Fig. 4.11 Profile of intensity for test pattern after application of unsharp mask with Gaussian blur

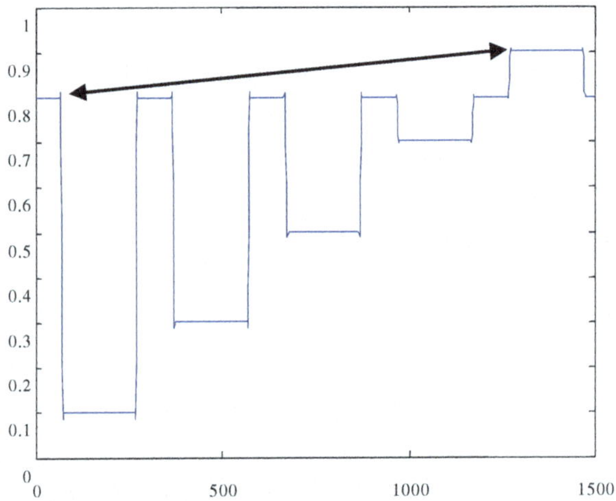

Fig. 4.12 Profile of intensity for test pattern after application of unsharp mask via bilateral filter

compare the uneven peaks near the edges in Fig. 4.11 with almost identical peaks in Fig. 4.12. It is worthwhile to implement a bilateral filter in a row-wise and column-wise manner, especially taking into account the anisotropy of human vision.

4.7 Results

4.7.1 Design of Survey

The sharpness metrics and sharpening algorithm were evaluated using a subjective test procedure. The survey was held among 31 volunteers. We collected 10 photographs at random from a dataset containing about 2000 images, both sharp and blurry. We processed the images using our sharpening algorithm in automatic mode. Original and processed photos were printed on paper with a size of 6 × 4 inches, applying dye-sublimation photo printer with a resolution of 300 dpi, and were shown to observers pairwise. The observers were asked two questions.

- Which photograph do you consider more pleasant visually?
- Which photograph do you consider sharper?

The answer "I don't see any difference" was acceptable. The third question required that the sharpness of the original image be evaluated according to the following four-grade scale: sharp, slightly blurred, blurred, and strongly blurred. As these notions are not strictly formalized and each observer can have an understanding of his or her own, the following explanations were provided as well (see Table 4.1).

4.7.2 Sharpness Measure

For the third question, we calculated the Mean Opinion Score (MOS) and an error range for a confidence probability of 0.95. Figure 4.13 shows the dependency of the proposed sharpness measure on the MOS. Some points are spread near their discrete values for better visualization. For comparison, we calculated the Digital Sharpness Scale (DSS) (Zhang et al. 2005) and Crete's (Crete et al. 2007) blurriness metrics (see Figs. 4.14 and 4.15 correspondingly) for these ten images and computed a Pearson correlation coefficient between the subjective MOS and numerical metrics (see Table 4.2). Preliminarily, all photos were resampled to the target printing size and resolution.

Table 4.1 Four-grade sharpness metrics

Quality	Score	Explanation
Sharp	0	Sharp and clear, definitely high quality
Slightly blurred	1	Blurriness is perceptible but not annoying
Blurred	2	Definitely blurred, which irritates; sharpening is necessary
Strongly blurred	3	Definitely blurred, which irritates; we do not want to print a photo of such poor quality

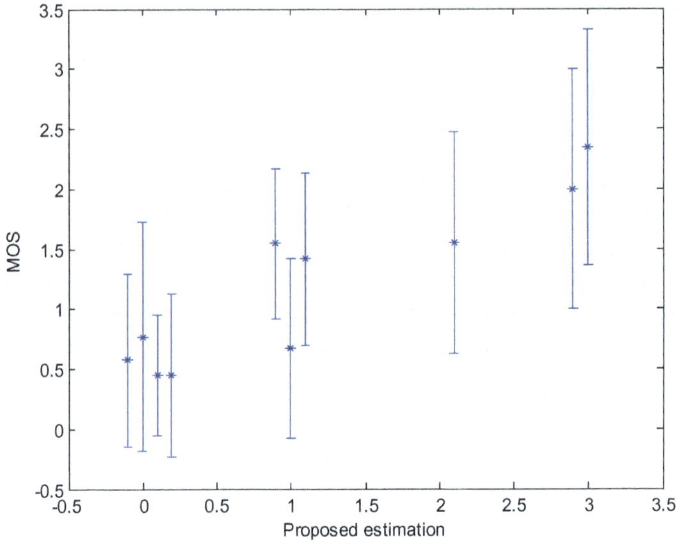

Fig. 4.13 Proposed sharpness metrics versus subjective MOS

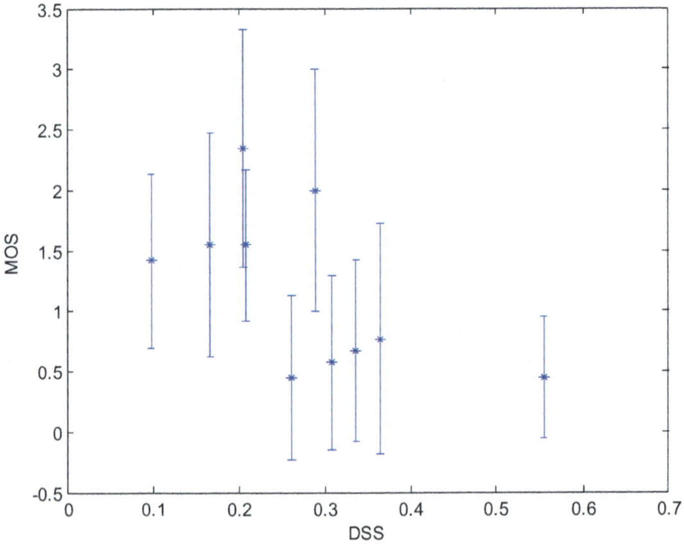

Fig. 4.14 DSS sharpness metrics versus subjective MOS

Our sharpness metrics have a good correlation with subjective perception. Crete's *Asub*-measure has good correlation with the MOS too. DSS is content-dependent and is not appropriate for the case of non-reference estimation. It

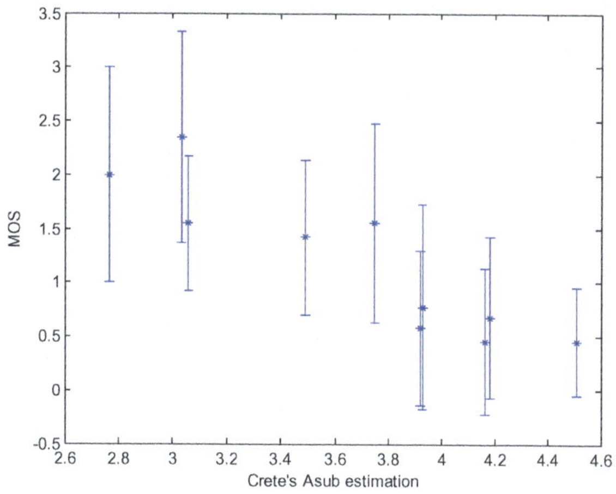

Fig. 4.15 Crete's sharpness metrics versus subjective MOS

Table 4.2 Pearson correlation coefficient between sharpness measures and MOS

Metrics	Correlation coefficient
DSS	0.5894
Crete's	0.9067
Proposed	0.9244
Feature	
F1	0.7089
F2	0.6229
F3	0.7456

is worth noting that image features applied to estimate sharpness have a modest correlation with subjective evaluation, but the AdaBoost ensemble increases the correlation significantly (see Table 4.2).

4.7.3 Evaluation of Sharpening Quality

Figures 4.16 and 4.17 illustrate the responses to the first two questions. There were no hesitations when the photos were blurred (high scores): the corrected photos were marked as sharper and more pleasant. For sharp photos, the response "I do not see any difference" occurred frequently. There is a strong correlation between sharpness and quality. Observers, especially for blurred photos, evaluated the results of processing by our technique positively. According to the opinions of the observers, the corrected photos look sharper than the originals of all samples under examination. Figures 4.18 and 4.19 show examples of processing.

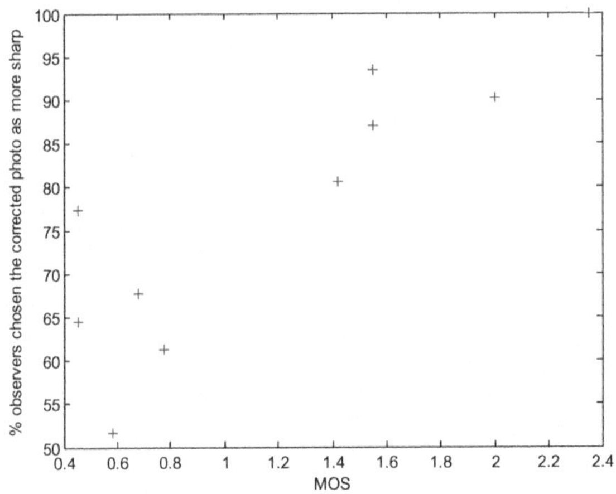

Fig. 4.16 Percentage of observers who chose the corrected photo as the sharpest of a pair versus the subjective MOS

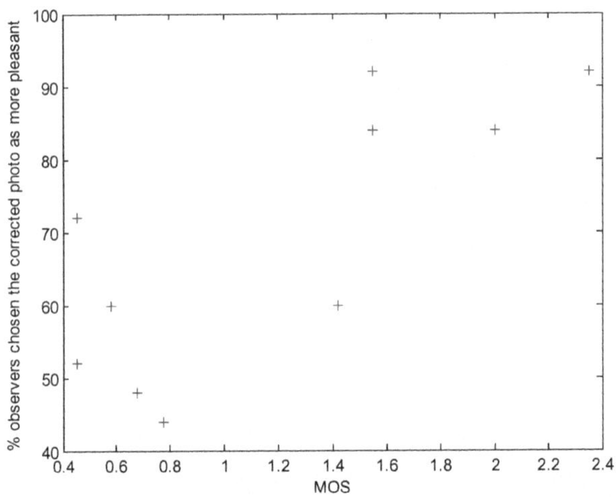

Fig. 4.17 Percentage of observers who chose the corrected photo as the most visually pleasant of a pair versus the subjective MOS

(a)

(b)

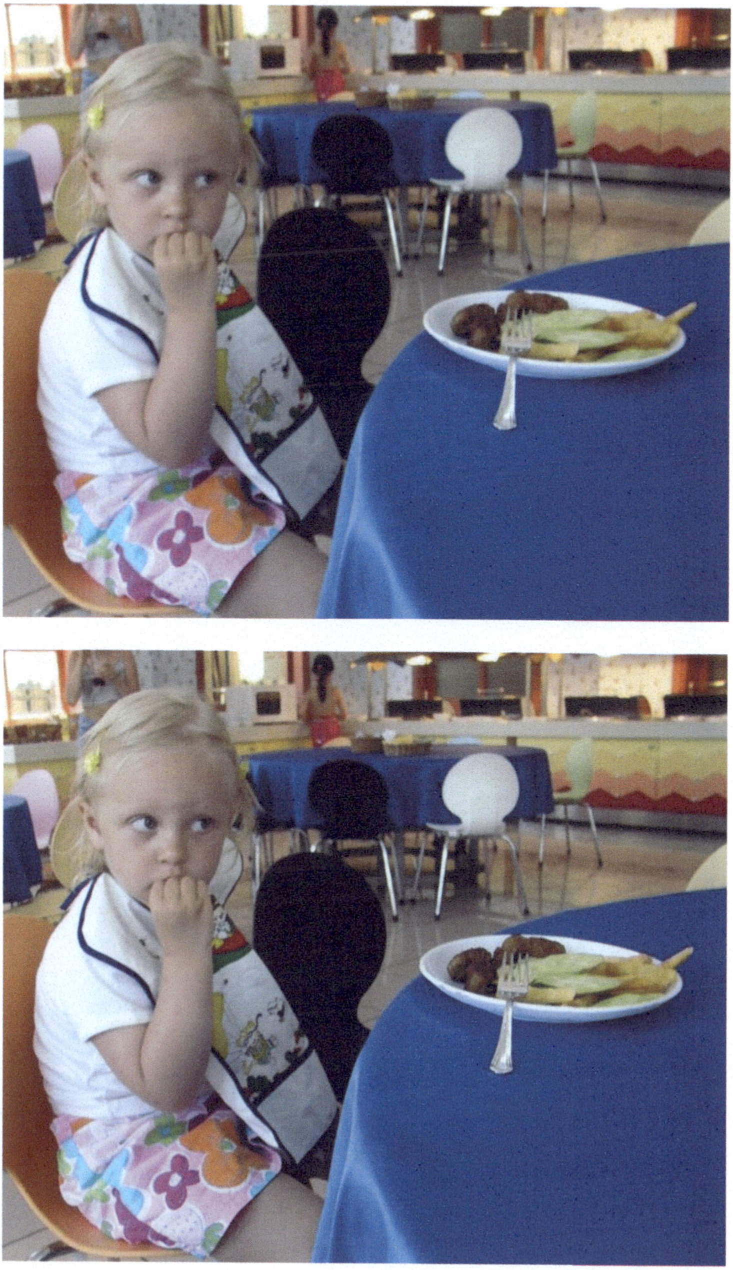

Fig. 4.18 Example 1 of sharpening by our algorithm: **a** initial, **b** corrected

(a)

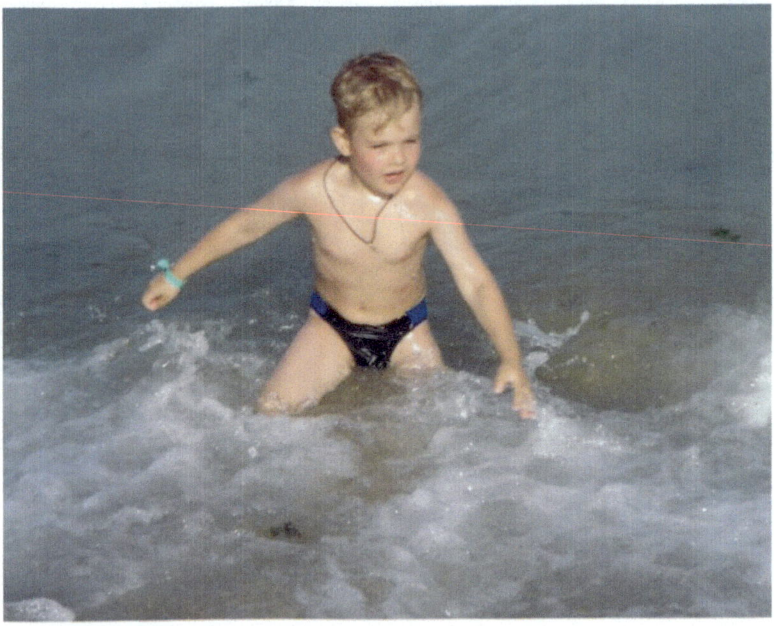

(b)

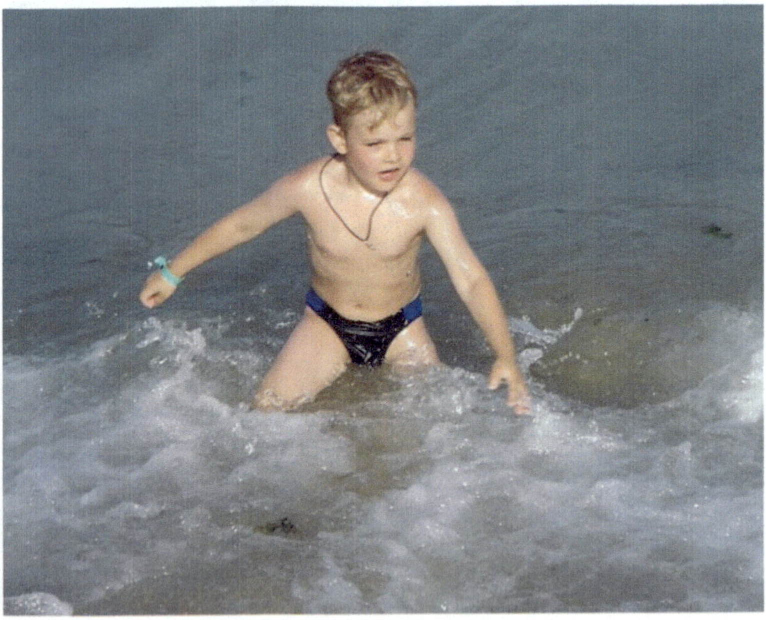

Fig. 4.19 Example 2 of sharpening by our algorithm: **a** initial, **b** corrected

Table 4.3 Outcomes of detection of out-of-focus photos

Algorithm	Number of sharp photos	Number of blurred photos	Number of FPs	Number of FNs	Error rate (%)
Lim's	2650	350	265	35	10
Proposed	1865	339	**17**	**111**	**6**

Bold is used for emphasizing of the highest outcome.

4.7.4 Detection of Out-of-Focus Photos

The described method makes it possible to detect blurred photos and to prevent printing of defective ones. We tested our algorithm on a huge test set for detection of out-of-focus photos and compared the outcomes with those described in Lim et al. (2005). The paper reported a 90% detection rate and 10% false positives from a database of 3000 photos, where 350 images were out of focus. This means 35 out-of-focus photos were not detected and 265 well-focused photos were detected erroneously. Our test set contains 2204 photos including 339 affected by strong blurring. Our outcomes outperform the results of Lim et al. (see Table 4.3). Algorithm parameters were intentionally selected in such a way as to reduce the number of FPs, that is, cases when a good image is classified as a defective one. Obviously, in this case the number of FNs, that is, the number of cases in which a defective image is classified as a good one, increases. Most FPs have a reasonable explanation and occur in cases of portraits or macro-shots that contain a blurred background with a big area. Sometimes our algorithm cannot detect blur originating from shaking of the camera.

References

Crete, F., Dolmire, T., Ladret, P., Nicolas, M.: The blur effect: perception and estimation with a new no-reference perceptual blur metrics. Proc. Electron. Imaging (2007)

Keelan, B.W.: Handbook of image quality: characterization and prediction. Marcel Dekker, Inc (2002)

Kim, S.H., Allebach, J.P.: Optimal unsharp mask for image sharpening and noise removal. J. Electron. Imaging **14**(2) (2005)

Kotera, H., Wang H.: Multiscale image sharpening adaptive to edge profile. J. Electron. Imaging **14**(1) (2005)

Lim, S.H., Yen, J., Wu, P.: Detection of out-of-focus digital photographs. HP Labs Technical Report (2005)

Luong, H.Q., Philips, W.: Sharp image interpolation by mapping level curves. In: Proceedings of Visual Communications and Image Processing Conference (2005)

Polesel, A., Ramponi, G., Mathews, V.J.: Image enhancement via adaptive unsharp masking. IEEE Trans. Image Process. **9**(3), 505–510 (2000)

Safonov, I.V., Rychagov, M.N., Kang, K.M., Kim, S.H.: Adaptive sharpening of photos. In: Proceedings of IS&T/SPIE Electronic Imaging, **6807** (2008)

Schapire, R., Singer, Y.: Improved boosting algorithms using confidence-rated predic-tions. Machine Learning, **37**(3), pp. 297–336 (1999)

Schavemaker, J.G., Reinders, M.J., Gerbrands, J.J., Backer, E.: Image sharpening by morphological filtering. Pattern Recogn. **33**(6), 997–1012 (2000)

Tomasi, C., Manduchi, R.: Bilateral filtering for gray and color images. In: Proceedings Sixth IEEE International Conference on Computer Vision, pp. 839–846 (1998)

Wang, Q., Ward, R., Zou, J.: Contrast enhancement for enlarged images based on edge sharpening. In: Proceedings of IEEE International Conference on Image Processing, vol. 2, pp. II–762 (2005)

Zhang, B., Allebach, J.P., Pizlo, Z.: An investigation of perceived sharpness and sharpness metrics. Proc IS&T/SPIE Electron Imaging **5668**, 99 (2005)

Chapter 5
Removal of JPEG Artefacts

5.1 Introduction

5.1.1 JPEG Compression Pipeline

A technique known as JPEG is widely used for the compression of digital data representing photographic still images. Usually, to achieve a notable compression rate, non-reversible ("lossy") compression is applied: the set of data recovered at the "destination" is not identical to the original set of data. Thus the image represented by the recovered data is not identical to the original image. The hope is that the discrepancy in the destination image, as perceived by the human viewer, will not be noticeable, or in any event will not interfere to an unacceptable degree with the purpose to which the image is to be put.

- Colour Space Conversion. The image first undergoes a colour space conversion where it is remapped from RGB colour space to YCbCr colour space. This colour space conversion assists in the use of different quantization tables (one for luminance, the other for chrominance).

$$
\begin{aligned}
\mathrm{Y} &= 0.299\mathrm{R} + 0.587\mathrm{G} + 0.114\mathrm{B},\\
C_r &= 0.713(\mathrm{R}\ \text{-}\ \mathrm{Y}) + 128,\\
C_b &= 0.564(\mathrm{B}\ \text{-}\ \mathrm{Y}) + 128,\\
\mathrm{R} &= 1.403(C_r - 128) + \mathrm{Y},\\
\mathrm{G} &= \mathrm{Y} - 0.714(C_r - 128) - 0.344(C_b - 128),\\
\mathrm{B} &= 1.773(C_b - 128) + \mathrm{Y}.
\end{aligned}
$$

I.V. Safonov et al., *Adaptive Image Processing Algorithms for Printing*,
Signals and Communication Technology,
https://doi.org/10.1007/978-981-10-6931-4_5

- Segmentation into blocks. The raw image data are divided into 8 × 8 pixel blocks (these blocks are the minimum coded unit). This means that the JPEG compression algorithm depends heavily on the position and alignment of these boundaries.
- Discrete Cosine Transformation (DCT). The image is transformed from a spatial domain representation to a frequency domain representation.

$$F_{ij}(u,v) = \frac{C(u)C(v)}{4}\left[\sum_{n=0}^{7}\sum_{m=0}^{7} f_{ij}(n,m)\cos\frac{(2n+1)u\pi}{16}\cos\frac{(2m+1)v\pi}{16}\right],$$

$$\text{u,v} = 0,\ldots,7, \quad \text{i} = 0,\ldots,\frac{N}{8}-1, \quad \text{j} = 0,\ldots,\frac{M}{8}-1.$$

$F_{ij}(u, v)$ are DCT coefficients of block B_{ij}, $J_{ij}(uv)$ is the luminance value of the pixel (n, m) of the block B_{ij}, $N \times M$ is the dimensions of the image, and

$$C(u) = \begin{cases} \frac{1}{\sqrt{2}}, & \textit{if } u = 0 \\ 1, & \textit{if } u \neq 0 \end{cases}.$$

The result of 2D DCT of each block is an 8 × 8 matrix of DCT coefficients.

- Quantization. Quantization is essentially the lossy part of the algorithm. The quantization table is an 8 × 8 matrix of integer values. DCT coefficients, computed in the previous step, are divided by a corresponding value in a quantization table and then rounded to the nearest integer. Components that have either a small coefficient or a large divisor in the quantization table will likely round to zero. The lower the quality setting, the greater the divisor, causing a greater chance of a zero result. Conversely, all of the quantization table values for the highest quality setting would be 1 s, meaning that all of the original DCT data are preserved. It is widely known that humans are more critical of errors in low-frequency information than in high-frequency information. The JPEG algorithm discards many of these high-frequency (noise-like) details and preserves the slowly changing image information.

Figure 5.1 presents an example of a quantization table:

$$\begin{bmatrix} 52 & 55 & 61 & 66 & 70 & 61 & 64 & 73 \\ 63 & 59 & 55 & 90 & 109 & 85 & 69 & 72 \\ 62 & 59 & 68 & 113 & 144 & 104 & 66 & 73 \\ 63 & 58 & 71 & 122 & 154 & 106 & 70 & 69 \\ 67 & 61 & 68 & 104 & 126 & 88 & 68 & 70 \\ 79 & 65 & 60 & 70 & 77 & 68 & 58 & 75 \\ 85 & 71 & 64 & 59 & 55 & 61 & 65 & 83 \\ 87 & 79 & 69 & 68 & 65 & 76 & 78 & 94 \end{bmatrix}$$

Fig. 5.1 Example of a quantization table

An important point to realize is that the quantization table used for this step differs among nearly all digital cameras and software packages. Since this is the most significant contributor to compression or recompression "error", image will always suffer from degradation when resaving from different compressors or sources. Camera manufacturers independently choose an arbitrary "image quality" name (or level) to assign to the 64-value quantization matrix that they devise, and so the terms cannot be compared between makes or even models by the same manufacturer (i.e. Canon's "fine" versus Nikon's "fine").

- Zigzag scan. The resulting matrix after quantization will contain many zeros. The lower the quality setting, the more zeros will exist in the matrix. By re-ordering the matrix from the top-left corner into a 64-element vector in a zig-zag pattern, the matrix is essentially sorted from low-frequency components to high-frequency ones (see Fig. 5.2). As the high-frequency components are the most likely to round to zero, one will typically end up with a run of zeros at the end of the 64-entry vector. This is important for the next step.
- Differential Pulse Code Modulation on DC component (the first component is zig-zag scan). On a block-by-block basis, the difference in the average value (across the entire block, the DC component) is encoded as a change from the previous block's value. This is known as differential pulse code modulation.
- Run Length Encoding (RLE) on AC components. For the individual entries in the 64-element vector (the AC components), RLE stores each value along with the number of zeros preceding it. As the 1×64 vector contains a lot of zeros, it is more efficient to save the non-zero values and then count the number of zeros between these non-zero values. The RLE stores a skip and a value, where the skip is the number of zeros before this component and the value is the next non-zero component.
- Entropy coding/Huffman coding. A dictionary that represents commonly used strings of values with a shorter code is created. More common strings or patterns use shorter codes (encoded in only a few bits), while less frequently used strings use longer codes. So long as the dictionary (Huffman table) is stored in the file, it is an easy matter to look up the encoded bit string to recover the original values.

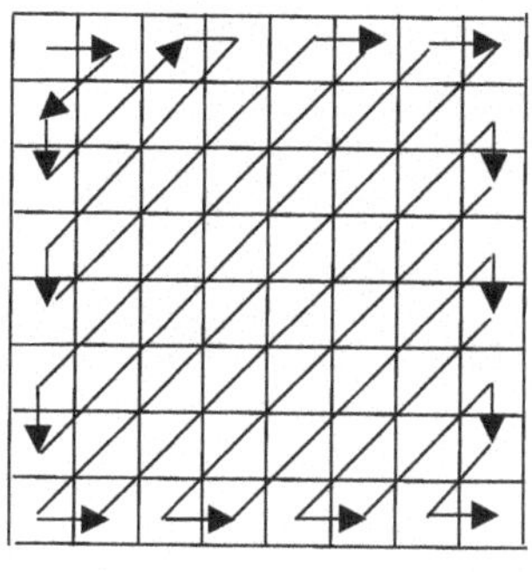

Fig. 5.2 Zig-zag scan pattern

5.1.2 JPEG Quality

When trying to resave a digital photo, one is always faced with a decision as to what "quality settings" (level of compression) to use. The JPEG file format allows one to select an appropriate trade-off between file size and image quality. Different quality levels that can be chosen affect the quantization table, but the quality factor itself is not stored in the image file.

Usually each graphics program tends to use its own compression quality scale and quantization tables, and therefore one cannot simply transfer quality settings from one application to another. This means that the severity of compression artefacts will vary significantly in different graphical applications, even in cases with equal quality factors. This can be seen in Fig. 5.3, where the left-hand image was saved in Photoshop using the "save for web" option while the right-hand image was saved in Matlab, both with 70% quality.

This means that the analysis must take into account the image content and only reliable metadata, such as the quantization table. The file size is less reliable, because of the frequent presence of other metadata, such as EXIF information or the image thumbnail, which can affect the "compression ratio", whose computation involves the file size.

(a) (b)

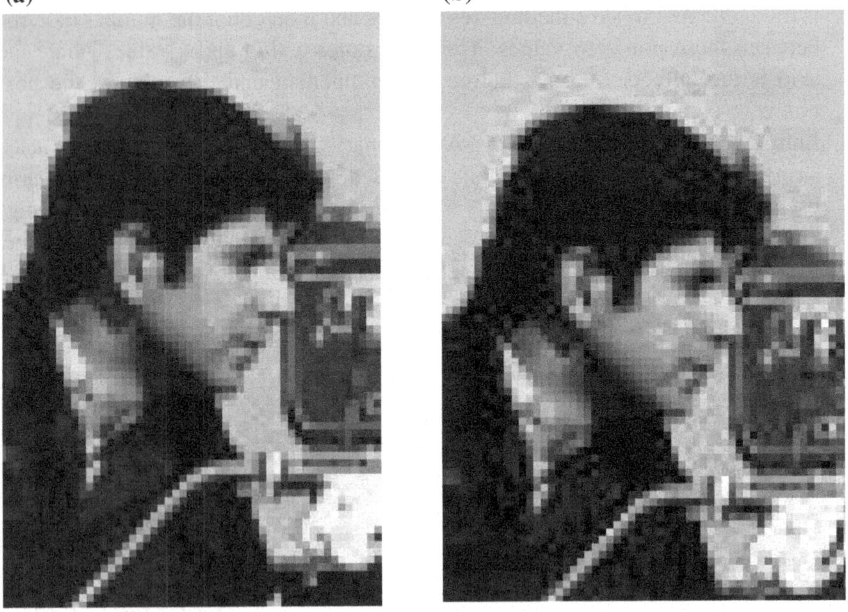

Fig. 5.3 Compressed image with equal quality in different editors: **a** Image saved using Photoshop's "save for web" option with 70% quality and a file size of 17.7 kB; **b** Image saved in Matlab with 70% quality and a file size of 9.5 kB

5.1.3 Blocking Artefacts

When a transform coded image is reconstructed, the image may include visible artefacts of the compression process. One common artefact of block-based transform coding is the "blocking" effect or "grid noise". The blocking effect is the result of the approximation of the DC transform coefficient during quantization. In quantizing the transform coefficients, the encoder selects one of the available quantization parameters closest in value to the coefficient to represent the coefficient. This "rounding off" can produce pixels of visually different colours in adjacent blocks even though the corresponding pixels of the original image were nearly identical. As a result, blocks of pixels created for the transform coding process may be visible in the reconstructed image (see Fig. 5.4). The blocking effect becomes more pronounced as the quantization steps are coarsened to reduce the data rate. The blocking effect is a particular problem along the edges or colour boundaries in the image or in areas of the image that are of relatively uniform colour.

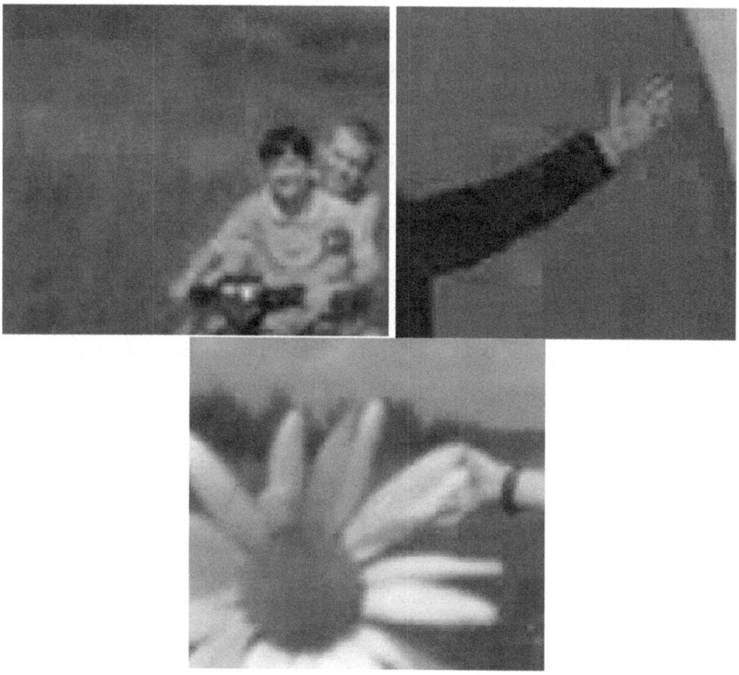

Fig. 5.4 Examples of blocking artefacts

5.1.4 Ringing Artefacts

Ringing artefacts are the result of quantization error (Fig. 5.5 illustrates this effect in the 1D case: when coefficients corresponding to higher frequencies are zeroed out, some fuzziness occurs in the recovered signal), particularly for coefficients representing higher spatial frequencies. Ringing is most likely to occur along strong edges in an image and is common in text because of the prevalence of strong edges. However, ringing artefacts generally manifest as fuzzy grey lines in the vicinity of an edge and are detectable only if the surrounding pixels have relatively high luminance (see Fig. 5.6). If the pixels in the area of the ringing artefact are of low luminance, the lines will be difficult to detect. Further, texture in the area of the ringing artefact tends to hide the artefact. Ringing noise in video is visible as local flickering near the edges. While the blocking effect is the predominant compression artefact in severely compressed images, ringing is typically the most visible artefact at lower compression ratios.

Another type of "ringing" artefact, that is, artefacts resulting from zeroing out of weighting coefficients of high-frequency components of an image block, is the appearance of a special pattern. In some areas, where the texture has some structured patches with the preferred edge direction, quantization can severely degrade areas where the edge direction is not near to 0°, 45°, or 90° while exaggerating the rest. Next, Fig. 5.7 illustrates this effect.

Figure 5.8 also demonstrates that a strong edge is not a necessary condition for ringing artefacts.

However, ringing artefacts near edges are more disturbing, which is why they are the first priority for correction.

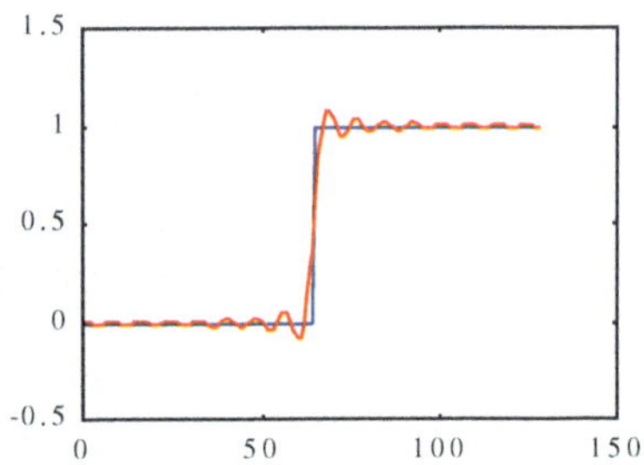

Fig. 5.5 Truncation of DCT coefficients from 128 down to 32 leads to ringing artefacts: the original is a step function. With only 32 low-frequency coefficients left, ringing occurs

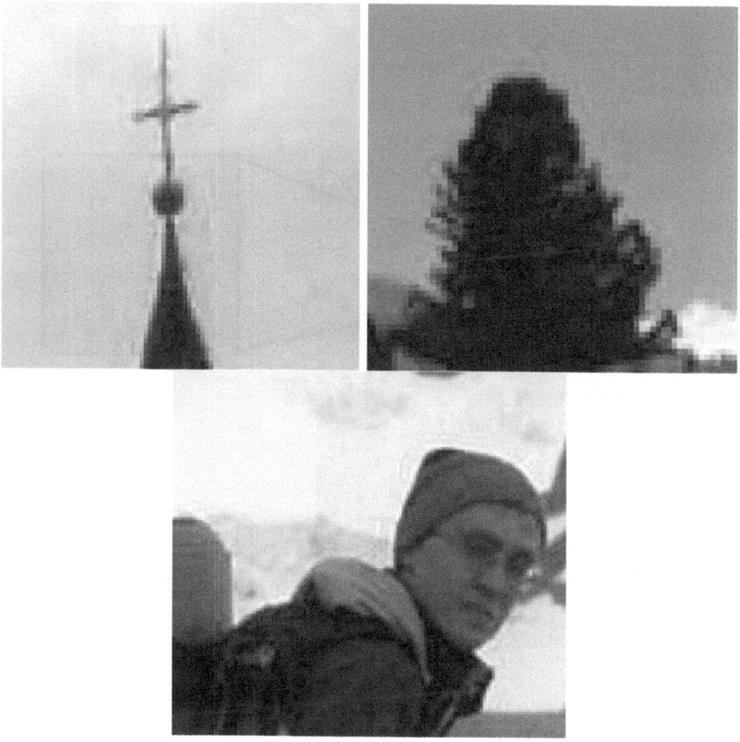

Fig. 5.6 Examples of ringing artefacts near sharp edges

Fig. 5.7 Example of compressed image fragment with regular high-frequency pattern

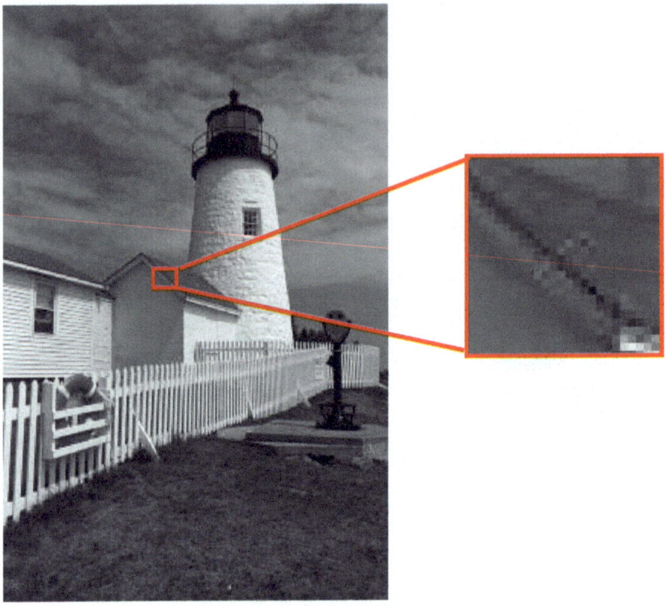

Fig. 5.8 Ringing artefacts can appear near weak edges too

5.1.5 Main Challenges

The main challenge of image quality prediction is to construct an adequate methodology that allows quality estimation of an arbitrary JPEG-coded image without an "ideal" reference image.

The main challenge of DCT block compression artefacts detection is to determine whether the present image discontinuity is due to compression or due to non-degraded image details. This classification makes it possible to concentrate solely on degraded blocks during the correction procedure, thus avoiding image over-blurring in "good" regions.

In the case of ringing artefacts, it is difficult to distinguish between textures and patterns resulting from quantization of high-frequency components.

5.2 Proposed Method of Detection of Artefacts

To reduce the effect of artefacts caused by some information loss, a variety of different algorithms and methods have been suggested. The simplest way to enhance a degraded image is to apply low-pass filtering. However, this approach will cause blurring of the original edges of the image. Therefore, many methods have been proposed to accommodate the adaptability of enhancement techniques.

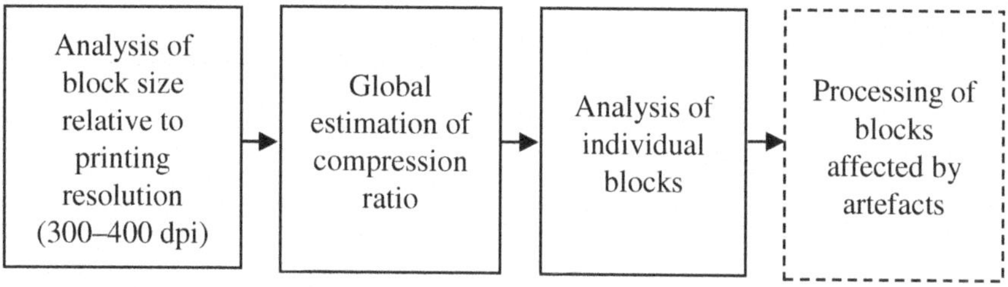

Fig. 5.9 General artefacts detection pipeline

It seems that the most reliable method is to apply some rough global estimation of the image at first and then to classify each block individually for blocking and ringing artefacts. The general workflow will follow the pipeline below (Fig. 5.9):

Since the block size in JPEG coded images is equal to 8 × 8 pixels, artefacts will not be visible at a high printing resolution. It is possible to infer some estimation of minimal printing resolution (also depending on the type of printing device) such that artefacts will not be perceptible.

Our first goal was to derive a global estimate of the compression ratio which would later be used for compressed image classification on "good" images without perceptible artefacts and "bad" images with artefacts such as blocking and ringing.

The process of finding the classifier consists of the following steps.

1. Choosing a database of images compressed with various ratios by various image editors or digital cameras.
2. Manual labelling of images into two classes ("good" and "bad").
3. Computing features related to the compression ratio and presence of artefacts (explained below in detail).
4. Computing of the parameters of the hyperplane separating two classes.

For this study, a database consisting of 600 images was used.

A rough global estimate of the compression ratio is based on the value of bpp (bits-per-pixel), which is computed as the file size (in bytes) × 8 divided by the number of pixels (width × height):

$$\text{bpp} = \text{file\_size} \times 8/(\text{M} \times \text{N})$$

As mentioned previously, the file size is not a very accurate measure of the compression ratio because of the frequent presence of other metadata in the image file. Therefore, it is more appropriate to estimate the volume of coded DCT coefficients, which are stored in a file.

Fan and de Queiroz (2003) propose to construct special kinds of histograms for computing global characteristics of the image compression ratio.

For every image block, the following values are computed:

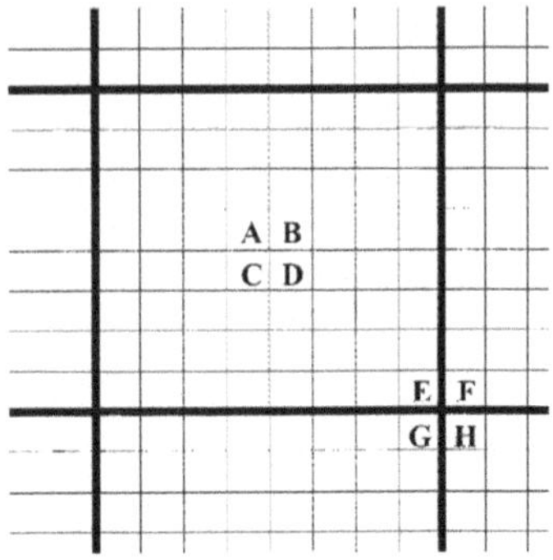

Fig. 5.10 Pixels in the middle of the block and near the block boundary

$$Z' = |A - B - C + D|,$$
$$Z'' = |E - F - G + H|,$$

where the block boundary lies between the pixels E, F, G, and H, while A, B, C, and D are taken from the block interior (see Fig. 5.10). In the next step, normalized histograms of Z' and Z'' (H' and H'' respectively) are computed, and the characteristic value for the image compression ratio is defined as the difference between these histograms:

$$K = \sum_n |H'(n) - H''(n)|.$$

Figure 5.11 presents the absolute difference between histograms H' and H''. From this figure, it can be seen that this difference becomes larger if the image has been compressed.

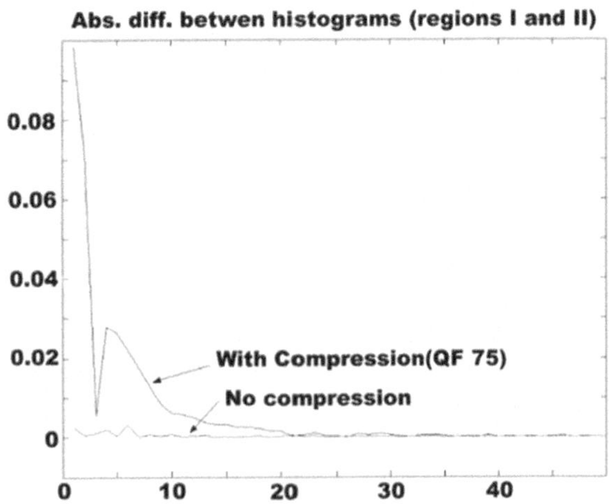

Fig. 5.11 Absolute difference between histograms H' and H''

Having computed K, Fan and de Queiroz (2003) propose to compare it to a predefined threshold value. If K is greater than the threshold, this means that some artefacts are present and the image is classified as "good"; otherwise there are no artefacts and the image is "bad".

In the current work, it is proposed to enhance the global image quality estimation. Instead of computing just one characteristic K and comparing it with a threshold, which seems to be a very rough estimation, here the modified value of K is computed and plotted against another characteristic value. After that, it is possible to construct a hyperplane separating "good" images and "bad" images.

The current work proposes to introduce two values, Z_1 and Z_2, by analogy with Z.

$$Z_1 = \min\left(255, \left[\frac{1}{8}\sum_{k=1}^{8}\left|B_{ij}(7,k) - B_{ij}(8,k)\right| + \frac{1}{8}\sum_{k=1}^{8}\left|B_{ij}(9,k) - B_{ij}(10,k)\right|\right]\right),$$

$$Z_2 = \min\left(255, \left[\frac{1}{8}\sum_{k=1}^{8}\left|B_{ij}(8,k) - B_{ij}(9,k)\right|\right]\right),$$

where B_{ij} is the image block with coordinates i, j: $B_{ij} = I(8i + k, 8j + l)$, $k, l = 1\ldots8$. This means that when k or l is greater than 8, there are pixels in the neighbouring block. These two sums, Z_1 and Z_2, are mean differences between pixels: Z_2 is the mean difference between pixels along the block boundary, and Z_1 is the average absolute difference between pixels adjacent to the block boundary.

Next, the value of K is computed as the difference between values of Z_1 and Z_2 in normalized histograms H_1 and H_2:

$$K = \frac{64}{MN}\sum_{n}|H_1(n) - H_2(n)|.$$

Next Fig. 5.12 shows a plot of values of K normalized by the number of blocks against values of bpp. Here, red circles indicate "bad" images with visible compression artefacts while blue crosses indicate "good" images without perceptible artefacts.

We can see from Fig. 5.12 that the computed features show good separability and it is possible to construct a separation line between "good" and "bad" images. Here, it seems to be more important to avoid type II errors, that is, False Negatives (FNs). Here, by FN we mean a situation in which a "bad" image is erroneously marked as a "good" one.

To improve the separability of classes, a modified version of the value K named Kb was computed, taking into account the maximal difference between standard deviations of the current block (Block 1) and a block shifted by four pixels (Block 2) (see Fig. 5.13).

Histogram $H^{\Delta\sigma}$ was computed using the absolute differences between standard deviations in Blocks 1 and 2 (see Fig. 5.13).

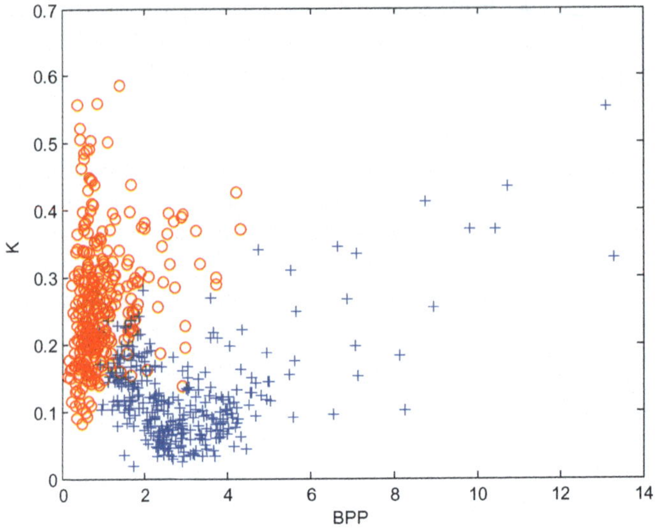

Fig. 5.12 Plot of values of modified *K* normalized by the number of blocks against values of bpp. Red circles indicate "bad" images with visible artefacts of compression and blue crosses indicate "good" images without perceptible artefacts

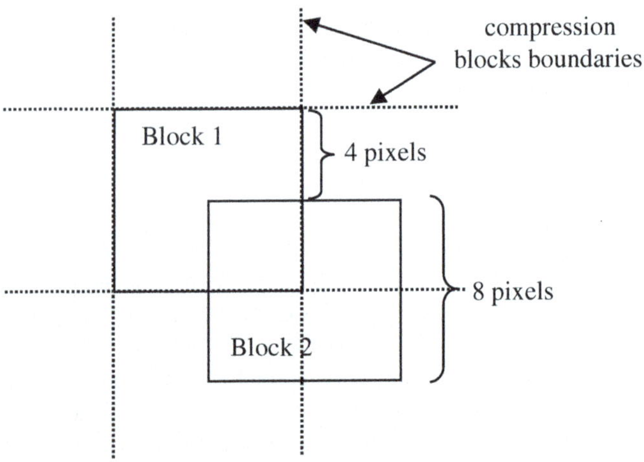

Fig. 5.13 Blocks for estimation of the difference between variances

$$\Delta\sigma = \left|Var(Block1_{ij}) - Var(Block2_{ij})\right|,$$

Figure 5.14 presents a plot of Kb against bpp, where Kb is computed according to the following:

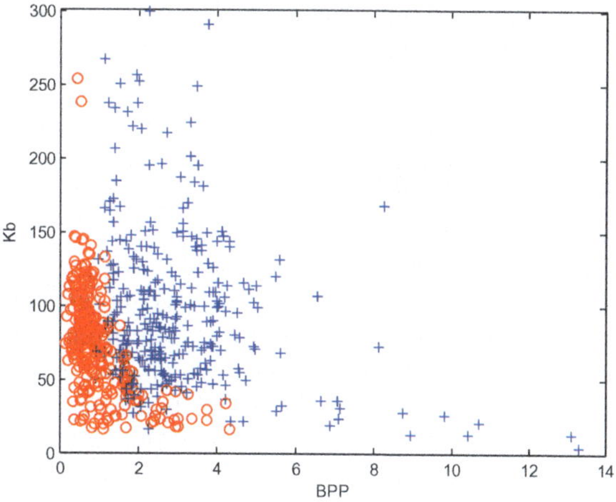

Fig. 5.14 A plot of values of *Kb* against values of bpp. Red circles indicate "bad" images with visible artefacts of compression and blue crosses indicate "good" images without perceptible artefacts

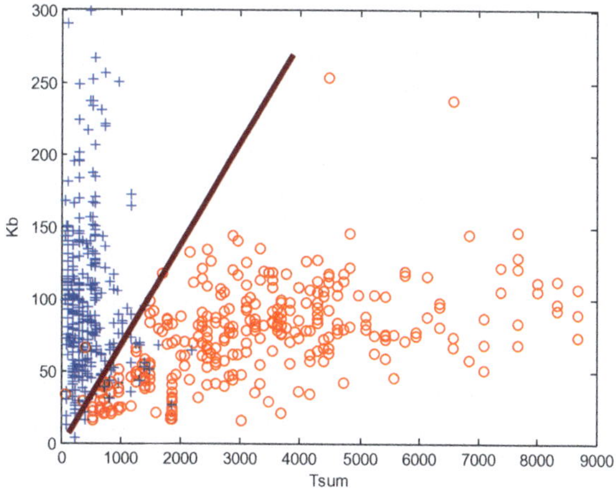

Fig. 5.15 Plot of values of *Kb* against the norm of the quantization table. Red circles indicate "bad" images with visible artefacts of compression and blue crosses indicate "good" images without perceptible artefacts

$$K_b = \frac{64}{MN}\sum_n \left[|H_1(n) - H_2(n)|\left(\max(H^{\Delta\sigma}) - H^{\Delta\sigma}\right)\right].$$

To avoid the use of unreliable file size values, Fig. 5.15 provides a plot of *Kb* against the norm of the corresponding quantization table computed according to the following formula:

$$\|Q\| = \sum_{i,j} Q_{ij}.$$

The last plot seems to be the most separable: it is possible to separate "good" and "bad" images. Here it seems to be more important to avoid type II errors, that is, FNs. Here, an FN means a situation in which a "bad" image is erroneously marked as a "good" one. That is why it is easiest to separate all the "good" images from the "bad" ones in the last image.

5.2.1 Blocking Artefacts

The second step in artefact detection is to analyse each individual block and to mark the blocks affected by perceptible blocking artefacts. The main challenge of this step is to distinguish compression noise from lightness variations in the target image. Almost all prior art methods propose to detect blocking artefacts in the spatial domain based on differences between neighbouring pixels. Others propose to count the number of non-zero DCT coefficients in the frequency domain.

In this work, it is proposed to design a method based on a learning framework, construct a binary classifier, and use it for no-reference estimation of present (or absent) blocking artefacts (Tolstaya et al., 2010). The classification system is a two-class classifier. For the case of separating the set of training vectors belonging to two classes, let $\{x_1; y_1\}; \ldots ; \{x_m; y_m\}$ denote a set of training data, where $x_i \in R^N$ is a feature vector and $y_i \in \{-1, +1\}$ is its class label. Classification is the application of Real AdaBoost based on boosting and weighted voting of an elementary classifier committee (Friedman et al. 2000). The decision function is constructed as a weighted sum of elementary classifiers:

$$F(x) = sign\left[\sum_{m=1}^{M} w_m h_m(x)\right].$$

The elementary classifier here consists of a comparison of the corresponding feature vector component with some threshold value, computed in advance at the training stage. Function $F(x)$ assigns the input test feature vector x to one or another class.

There are several AdaBoost algorithms that differ in terms of their approaches to the optimization of weights w_i. In some realizations of these algorithms it is possible to adjust the parameters of simple weak learners, in particular to optimize thresholds. Real AdaBoost is the generalization of a basic AdaBoost algorithm first introduced by Freund and Schapire (1996).

Recent developments in the field of machine learning have demonstrated that boosting-based methods may have a satisfactory performance by combining weak learners. Furthermore, the boosting procedure can also be regarded as a feature selection process if the weak learner uses a single feature in each stage.

For example, Hanghang et al. (2005) proposed a boosting method, which makes it possible to compute a global estimate of an input image quality. The authors propose to select a training base, build a binary classifier using the AdaBoost method, and then calibrate the classifier's output so that it will give quality metrics consistent with metrics obtained from human observers. The results demonstrate a good classification rate with a linear correlation value of over 90% between the prediction results and the mean human score. This work is a continuation of work done by Wang et al. (2002), who propose to compute three features: the average differences across block boundaries, average absolute difference between in-block image samples, and zero-crossing rate. After that the features are combined by a simple formula:

$$S = \alpha + \beta B^{\gamma_1} A^{\gamma_2} Z^{\gamma_3},$$

and the unknowns, $\alpha, \beta, \gamma_1, \gamma_2$ and γ_3, are estimated from data obtained from human observers, while B, A, and Z are the blockiness estimate, activity, and zero-crossings of the gradient, correspondingly.

Although Tong et al. (2005) mention that the feature selection step is crucial in constructing an efficient classifier, in fact they propose very simple features. For blocking and ringing artefacts, the features consist of image pixel values taken from a target image block and arranged in a vector. Further, in the current work, we propose to extend the application of AdaBoost by:

- individual classification of each block instead of classification of the whole image;
- using more complex features rather than just pixel intensities.

For this work, a set of six JPEG images compressed with various quantization tables was used. Regions containing strong blocking artefacts were labelled manually. In order to avoid erroneous classification, we used images that were different from the "bad" ones as the "good" images. Unlike Hanghang et al. (2005), who used the same images for creating "good" and "bad" versions, in our study the "good" images were taken from a totally different set of images compressed with a low compression ratio.

Feature vectors were extracted from every marked block of "bad" images and labelled with "1" and arbitrary image blocks from "good" images were labelled with "−1". The AdaBoost classifier was trained on a training set of over 6300 feature vectors using 10-fold cross-validation. The details of feature computation are given below.

Let us consider the matrix of basis functions of 8×8 DCT. Figure 5.16 demonstrates the 64 basis functions $p_{i,j}$ used for JPEG encoding.

For example, the basis function $p_{2,2}(k, l)$ is presented in Fig. 5.17.

The perceptibility of blocking artefacts varies greatly with the DCT frequency. From the basis functions above, an a priori matrix of boundary discontinuities was constructed. This matrix has a size of 64×64 and every element is the boundary discontinuity between two basis functions. One such matrix L is constructed for left boundaries (where every block's left boundary is compared with every block's right boundary), while another matrix B is constructed for bottom boundaries (where every block's bottom boundary is compared with every block's top boundary). Boundaries are compared by computing the norm of the difference as the sum of squares of differences between basis functions:

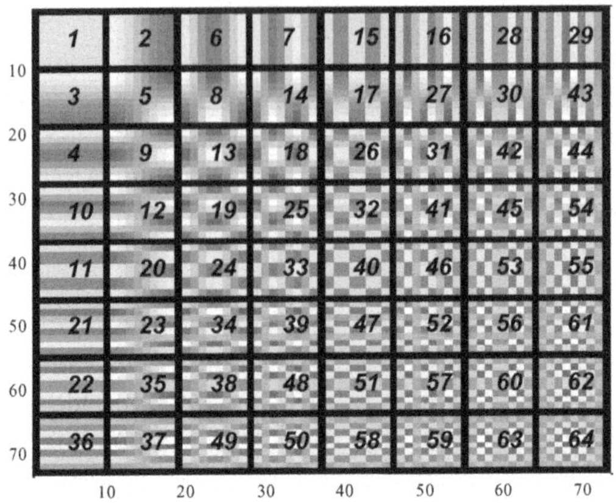

Fig. 5.16 DCT basis functions. A zig-zag scan index is overprinted on each basis function, depending on the position of its non-zero DCT coefficient in frequency space

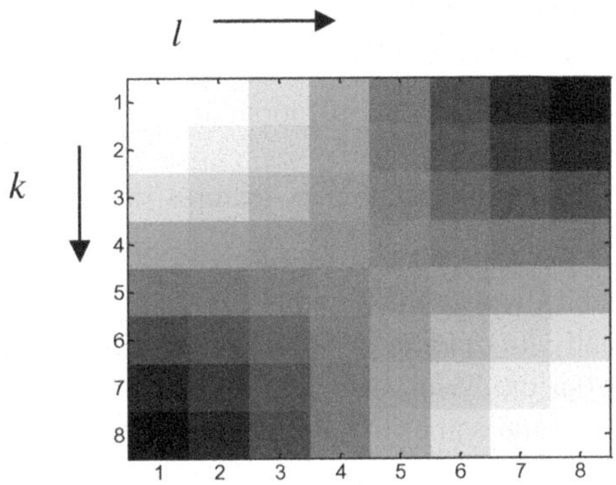

Fig. 5.17 Basis function $p_{2,2}(k, l)$

$$L_{8i+j,8m+n} = \sqrt{\sum_{k=1}^{8} \left(p_{i,j}(k,8) - p_{m,n}(k,1)\right)^2},$$

$$B_{8i+j,8m+n} = \sqrt{\sum_{k=1}^{8} \left(p_{i,j}(8,k) - p_{m,n}(1,k)\right)^2}.$$

The feature vector is extracted from every image and its components are computed according to the following. First of all, two adjacent image blocks $T_{i,j}$ and $T_{i+1,j}$ (or $T_{i,j}$ and $T_{i,j+1}$) are considered in order to check whether a perceptible boundary exists between them:

$$\Delta T_{i,j} = \frac{1}{8}\sqrt{\sum_{k=1}^{8}\left(T_{i,j}(k,8) - T_{i,j+1}(k,1)\right)^2}, \text{ for the left boundary;}$$

$$\Delta T_{i,j} = \frac{1}{8}\sqrt{\sum_{k=1}^{8}\left(T_{i,j}(8,k) - T_{i+1,j}(1,k)\right)^2}, \text{ for the bottom boundary.}$$

If the number of non-zero DCT coefficients in adjacent blocks is equal to 1, then $\Delta T_{i,j}$ is computed between these blocks. If the mean difference $\Delta T_{i,j}$ is smaller than some threshold (for example, 3 in an eight-bit image) it is assumed that no blocking artefacts (and no perceptible boundary) exist.

If $\Delta T_{i,j}$ is greater than a threshold and the number of non-zero DCT coefficients is greater than 1, a more complex classifier based on boosting is applied. A procedure of feature vector computation is explained below. Let $t_{i,j}$ be a block of corresponding DCT coefficients of image block $T_{i,j}$. Then

$$F^{\text{hor}}_{64(8a+b)+8c+d} = \left|\frac{t_{i,j}(a,b)}{Q_{a,b}} - \frac{t_{i,j}(c,d)}{Q_{c,d}}\right|\left(L_{\max} - L_{8a+b,8c+d}\right),$$

$$F^{\text{vert}}_{64(8a+b)+8c+d} = \left|\frac{t_{i,j}(a,b)}{Q_{a,b}} - \frac{t_{i,j}(c,d)}{Q_{c,d}}\right|\left(B_{\max} - B_{8a+b,8c+d}\right)$$

for features of left and bottom boundaries, where a, b, c, and d = 1… 8. In this case, the length of the feature vector is equal to 64 * 64. To simplify it, it is proposed to compare only the coefficients of basis functions which will potentially create more annoying discontinuities. These basis functions are more likely to have a non-zero DCT coefficient. Figure 5.18 shows the blocks that we propose to compare (marked in grey):

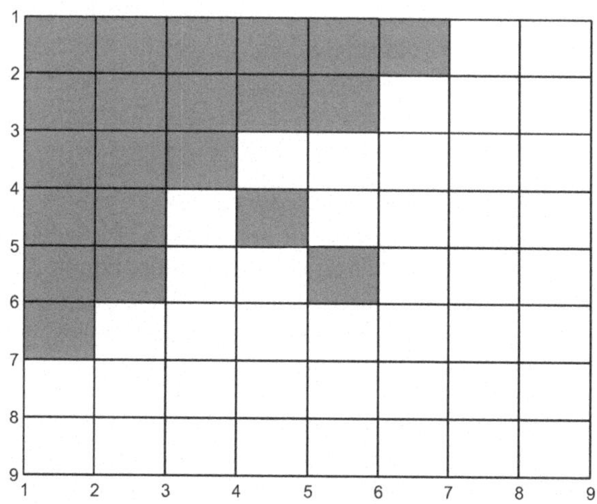

Fig. 5.18 Blocks that we propose to compare (marked in grey)

Table 5.1 Percentage of detected blocks affected by blocking artefacts, depending on the number of bits-per-pixel

Photoshop JPEG quality	8	20	40	60	80	90	99
Bits-per-pixel	0.46	0.67	0.99	1.83	3.57	5.87	9.56
Percentage of blocks affected by blocking artefacts	99	98	94	68	9	<1	<1

This approach saves a considerable amount of memory and computation. Actually, the length of the feature vector that we used is equal to 441 elements (21 elements compared with each other).

The AdaBoost classifier was trained on the database of 6300 labelled feature vectors. Training shows a good convergence rate, with 3% errors on ten-fold cross-validation. Table 5.1 demonstrates that number of detected blocks, affected by blocking artifacts, decreases with growing bits-per-pixel value for compressed 24 bpp colour image.

5.2.2 *Ringing Artefacts*

Since ringing artefacts do not manifest across the boundaries of the blocks used in compressing the image, the blocks used for ringing prediction are most conveniently the same as those used for compression.

However, ringing artefacts are most notable when a high-dynamics block (especially an edge block) has an adjacent smooth block. In this case, fuzzy grey lines appear in the vicinity of an edge and this fuzziness belongs to the same block containing the edge.

Figure 5.19 illustrates the ringing effect and corresponding non-zero DCT coefficients. The left panel presents image fragments and the right panel presents the corresponding DCT matrix, where white squares correspond to zero DCT coefficients and black squares to non-zero DCT coefficients. As can be seen, ringing blocks have smooth neighbours, and smaller DCT coefficients represent blocks with strong edges, so the ringing artefact is more perceptible.

The pattern of the ringing artefacts that appear may be characterized by the following rules.

- The current block (affected by ringing) has high activity; that is, it is represented by a sufficient number of non-zero DCT coefficients.
- At least one of its four neighbours (top, bottom, left, or right) is a smooth block (with a small number of non-zero DCT coefficients).
- The current block is compressed and the cause of artefacts is essentially quantization, which means that the number of zero DCT coefficients is also noticeable.

The following is a fragment of the image "Lighthouse" (see enlarged portion in Fig. 5.8) and its non-zero DCT coefficients. We can see that the top left block, which shows ringing, has a low-frequency neighbour, the top right block (5.20).

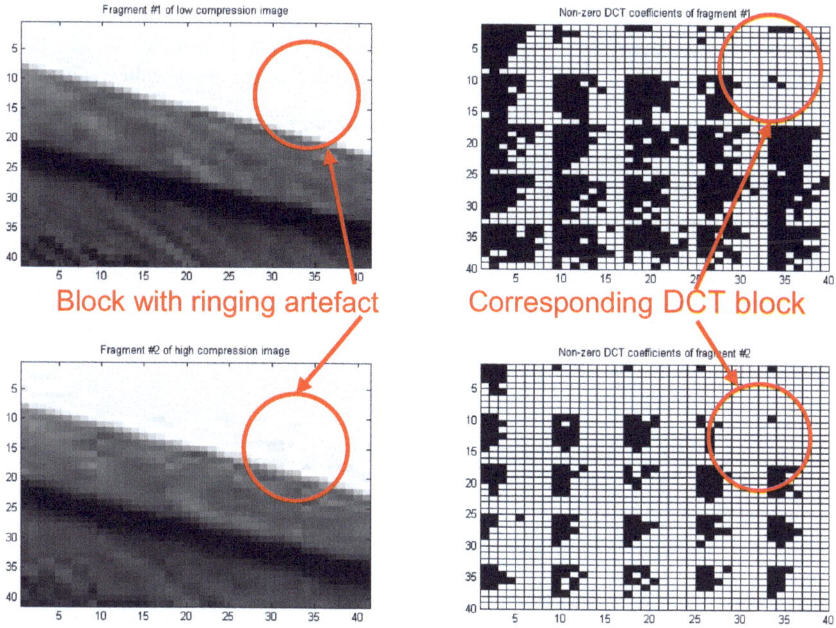

Fig. 5.19 Image fragments with ringing effect and corresponding non-zero DCT coefficients. The left panel presents image fragments; the right panel shows the corresponding DCT matrix, where white squares correspond to zero DCT coefficients and black squares correspond to non-zero DCT coefficients

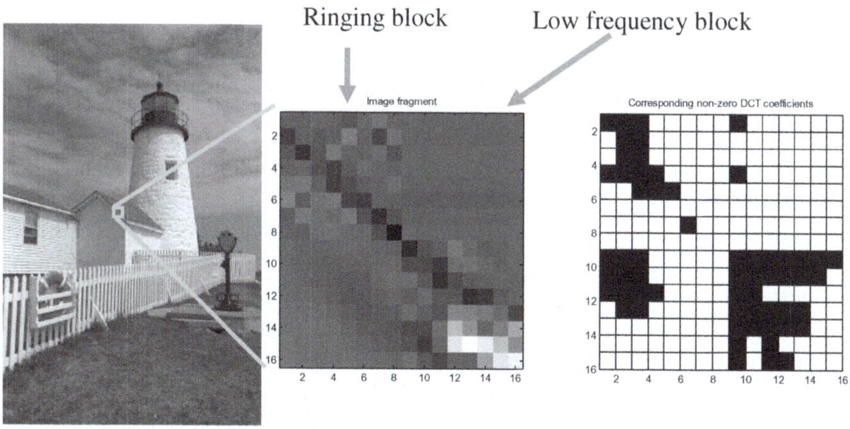

Fig. 5.20 Fragment of the image "Lighthouse" (see enlarged portion in Fig. 5.8) and its non-zero DCT coefficients

Based on all these assumptions, first of all, "suspicious" blocks in the image are revealed. Below, some details of "ringing probability" are presented.

For each 8 × 8 block of an image being processed, the following values are found.

1. The maximal index of the non-zero DCT coefficient if the block is arranged with a zig-zag order in a vector, i_{max}.
2. The number of non-zero DCT coefficients N_{DCT}.
3. The same values for its four neighbours: the neighbouring top, bottom, left, and right 8 × 8 blocks ($i^k_{\max} N^k_{DCT}$, k = "top", "bottom", "left", "right").

The simple "ringing probability", RP, is computed using the following formulas:

$$\mathrm{G}_1 = \min_{\substack{k=\text{"top"}\\ \text{"bottom"}\\ \text{"left"}\\ \text{"right"}}} \left\{N^k_{DCT} - N_{DCT}\right\},$$

$$G_2 = \min_{\substack{k=\text{"top"}\\ \text{"bottom"}\\ \text{"left"}\\ \text{"right"}}} \left\{i^k_{\max} - i_{\max}\right\},$$

$$\mathrm{RP} = \frac{\sqrt{|\mathrm{G}_1\mathrm{G}_2|}(64 - N_{DCT})}{64} \times \theta(\mathrm{N_{DCT}} > 4) \times \theta(\mathrm{G}_1 < 0) \times \theta(\mathrm{G}_2 < 0) \times \theta(\mathrm{i}_{\max} \geq 5),$$

where $\theta(\mathrm{a} > b) = \begin{cases} 1, & \text{if a} > \text{b}, \\ 0, & \text{if a} \leq \text{b} \end{cases}$; $(64 - N_{DCT})$ is the number of zero DCT coefficients in the current block. The higher the number, the higher the compression; $\theta(G_1 < 0)$ ensures that the block under examination has the greatest number of non-zero DCT coefficients among its neighbours; $\theta(G_2 < 0)$ ensures that the block under examination has the greatest index of the last non-zero DCT coefficient among its neighbours; $\theta(i_{max} \geq 5)$ is the condition that the frequency of ringing artefacts is quite high.

From Fig. 5.16, the basis function index, which contributes to the most disturbing artefacts, can be estimated.

Figures 5.21 and 5.22 demonstrate several examples of RP It is additionally assumed that if RP is smaller then 0.1 * max(RP), the RP will be equal to 0.

The presented measure detects the probable localization of ringing artefacts very effectively (without a high-pass filter). However, its serious disadvantage is that it gives many false positives (FPs). The next step is to measure ringing severity. Various metrics have been proposed in the literature, mostly based on edge detection and the estimation of oscillations near major edges.

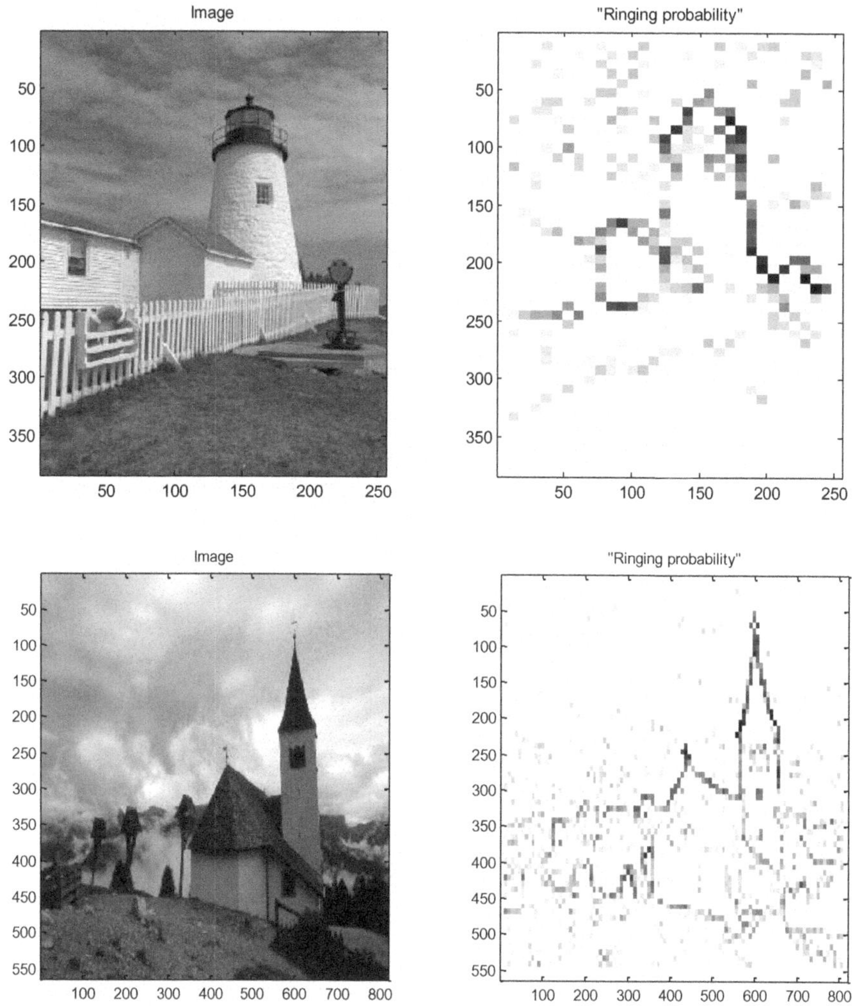

Fig. 5.21 Examples 1 and 2 of "ringing probability"

Feng and Allebach (2006) proposed to detect false ringing blocks by analysing the average edge gradient in a block and also adopt the concept of just noticeable differences proposed by Chou and Li (1995). Ringing metrics are based on measuring local activity as the average absolute value of luminance gradients.

Oguz et al. (1998) proposed to use the so-called Visible Ringing Measure (VRM), based on the average local variance around major edges. First, a VRM mask is constructed by applying edge detection and morphological operations under the edge mask. Then the local average variance around the edges is constructed, avoiding long step edges.

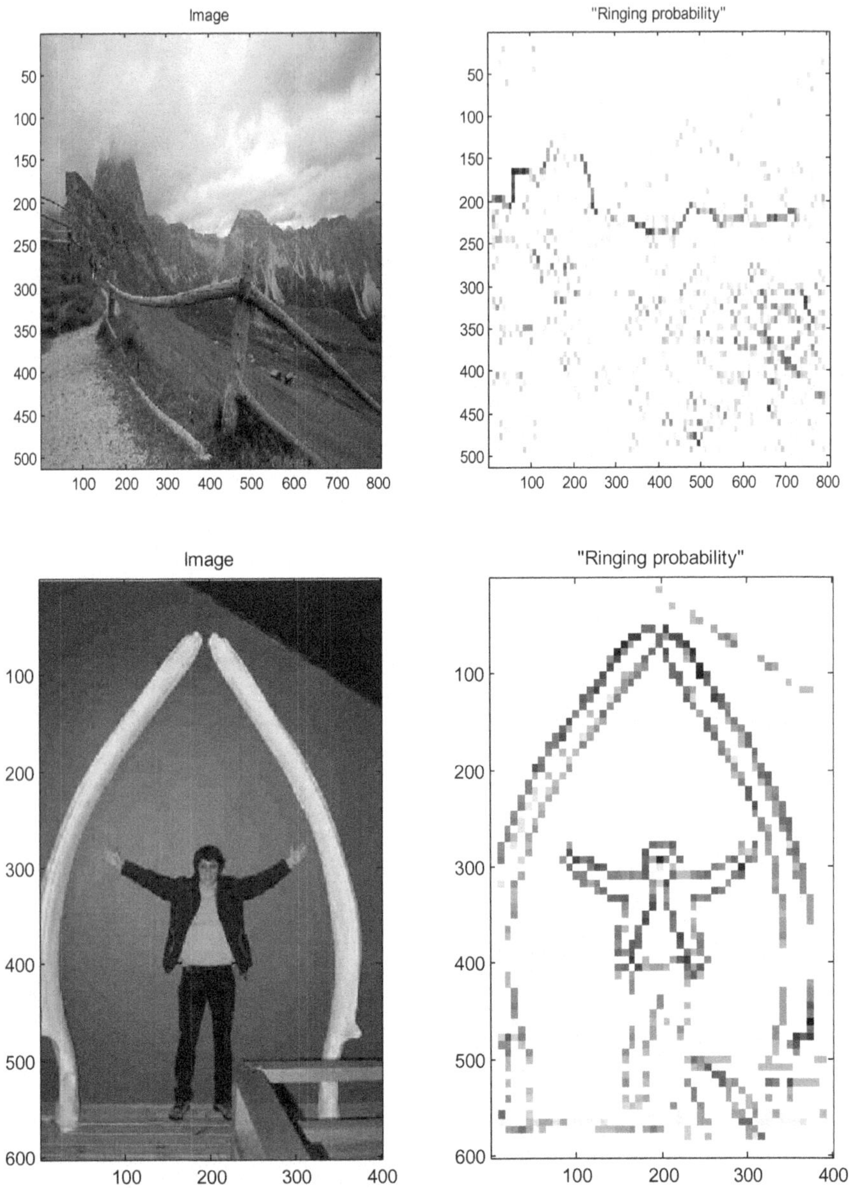

Fig. 5.22 Examples 3 and 4 of "ringing probability"

Work (Yang et al. 2001) continues (Oguz et al. 1998) by proposing a restoration scheme based on maximum a posteriori (MAP) estimation.

In Chang and Chen (2006), the image is partitioned into smooth, edge, or textured areas, adopting a quad-tree method. Here, the de-ringing method is based on the application of morphological operations. Voting is applied to choose the optimal structuring element for the morphological operations. This method is based on computing the difference between the reference (true) and coded images. It is proposed that the ringing metric be defined for each important vertical edge. First the vertical edges in the original image are found (weak edges and noise are again discarded by means of thresholding) and the difference between the processed image and the reference is calculated. Then each row in the processed image is scanned and the ringing around each edge is measured. All of the cited methods have quite high computational complexity, involving an edge-detection step, or need a reference (uncompressed) image.

Construction of a ringing measure can pursue different goals: one is the measurement of artefact severity for image quality comparisons (a posteriori estimate), while another is estimation of the parameters needed for ringing correction. The following part describes the estimation of parameters, which will be further applied to ringing artefact correction.

First of all, in this work, it is proposed that the number of potential ringing regions be reduced by thresholding the "ringing probability". The next important step is to estimate the local variance of artefacts for subsequent correction. This estimate will allow the application of some adaptive blurring filter to reduce ringing perceptibility.

It is not appropriate to only compute the variance of a potential ringing region, because it will most probably include a sharp edge. The goal is to measure the variance of fuzziness in the vicinity of an edge. For this purpose, pixel intensities are clusterized in a potential ringing block. Figure 5.23 shows the decoded image block, which exhibits ringing artefacts and contains a sharp edge. In such a block, pixels are clusterized into two classes using a k-means algorithm. The mean of two variances of two classes is computed, avoiding edge transition areas. The variance of a set of pixel intensities $\{x_1 \ldots x_N\}$ is computed according to a standard formula ($<x>$ is the mean value):

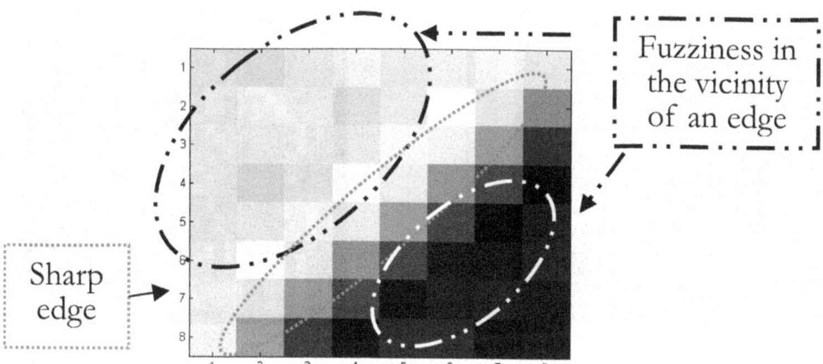

Fig. 5.23 Decoded image block exhibiting the ringing artefact and containing a sharp edge

$$\sigma^2 = \frac{1}{N-1}\sum_{i=1}^{N}(x_i - \langle x \rangle)^2.$$

In Yang et al. (2001), the authors apply k-means clustering to construct a flat surface model to replace a rippled surface with a flat one in order to estimate the true (uncompressed) image intensities. They assume that the true image estimate is a montage of flat surfaces. In order to manage a broad class of images, a flat surface model is applied locally to small regions of the image. First of all, a number of classes is estimated for different patches of the image (for a texture patch and for a patch containing only sharp edges). Then the authors reduce their model to three-class clustering for the sake of computational simplicity.

In this work two-, three- and four-class clustering is performed, and the clustering with the smallest error is chosen. This seems to be a sufficiently simple and effective procedure. Some spurious variance outbreaks mark regions where four classes is insufficient but these outbreaks are easily detected and removed, using the estimate of clustering error.

As a result, an estimate of fuzziness introduced by ringing is given as the variance in potential ringing regions. The variance between actual pixel intensities and "flat-surface estimates" obtained by clustering is computed. The estimated

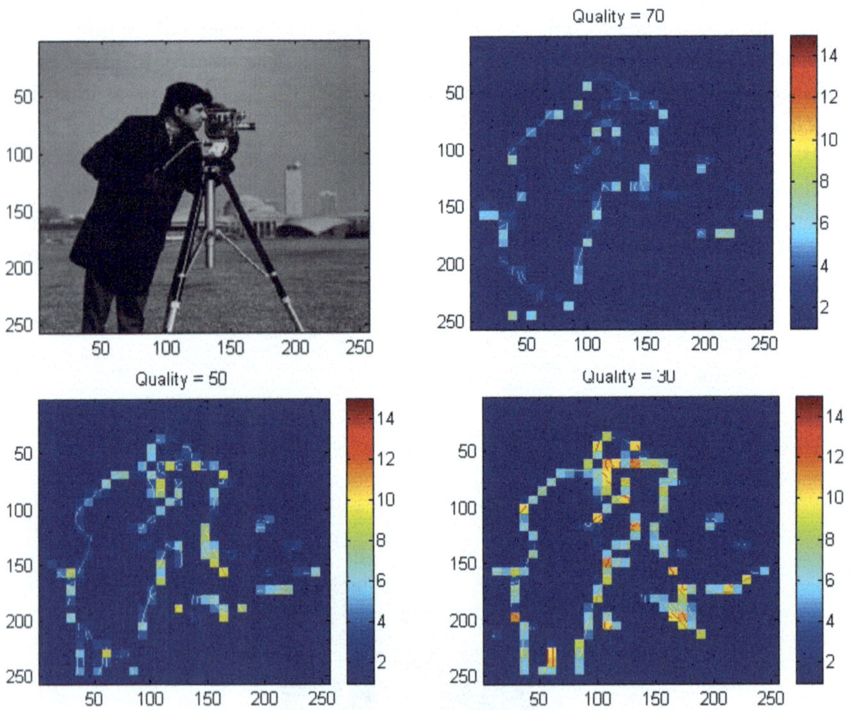

Fig. 5.24 "Cameraman" image saved in Photoshop with different quality options and the associated fuzziness (lightness variance in the vicinity of edges) in potential ringing regions

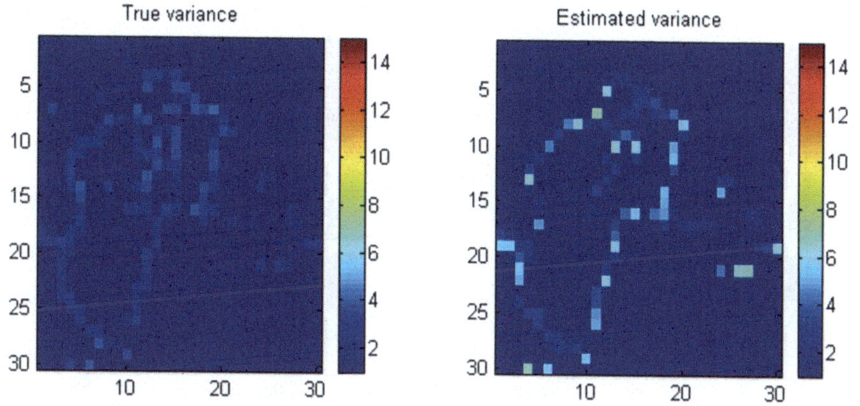

Fig. 5.25 True and estimated fuzziness for Q = 70 (Photoshop "save for web" option)

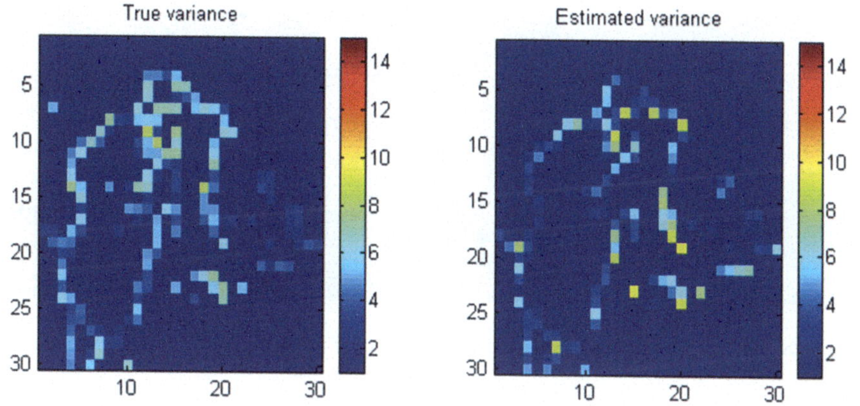

Fig. 5.26 True and estimated fuzziness for Q = 50 (Photoshop "save for web" option)

variance can also be regarded as a measure of the ringing artefact, taking into account, however, the fact that in natural uncompressed images, sharp edges will not be ideal step edges and the variance estimate will always be greater than 0.

It is worth noting here that the proposed technique is computationally efficient, since the "ringing probability" is computed very fast and clustering is performed not for all image blocks but only for potential ringing regions.

Figures 5.24, 5.25, 5.26 and 5.27 illustrate potential ringing regions, with colour representing ringing fuzziness. The image "Cameraman" was saved in Photoshop using the "save for web" option using quality levels of 70, 50, and 30. As we can see from Fig. 5.24, fuzziness in the potential ringing regions grows with decreasing quality.

To compare the estimated fuzziness with the "true" fuzziness, the difference between the true (uncompressed) image and images coded with quality levels of 30,

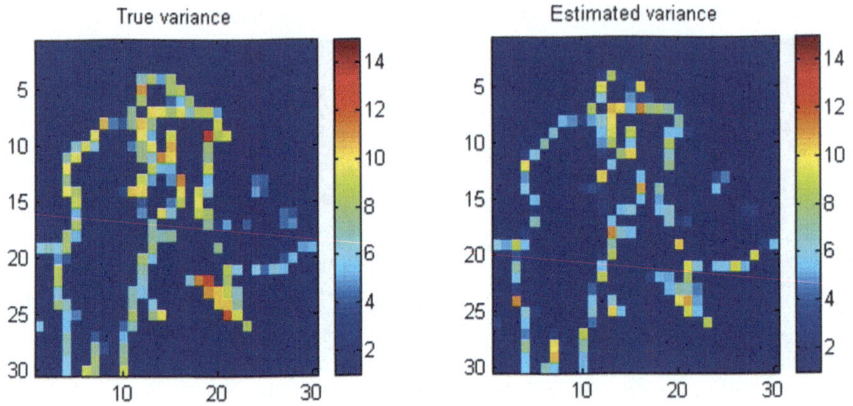

Fig. 5.27 True and estimated fuzziness for Q = 30 (Photoshop "save for web" option)

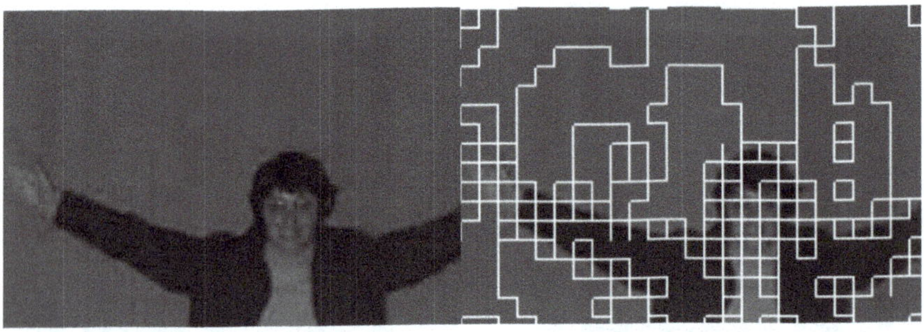

Fig. 5.28 Results of blocking artefact detection

50, and 70 was analysed. In potential ringing blocks, the variance of the difference was computed using the same formula. The following figures present the estimated fuzziness and true fuzziness obtained with the true (uncompressed) image for different quality settings.

As we can see from Figs. 5.25, 5.26 and 5.27, the estimation precision is quite high for high quality images and decreases with decreasing image quality. On the other hand, the variance is often underestimated so that adaptive blurring for image restoration will not over-blur an image.

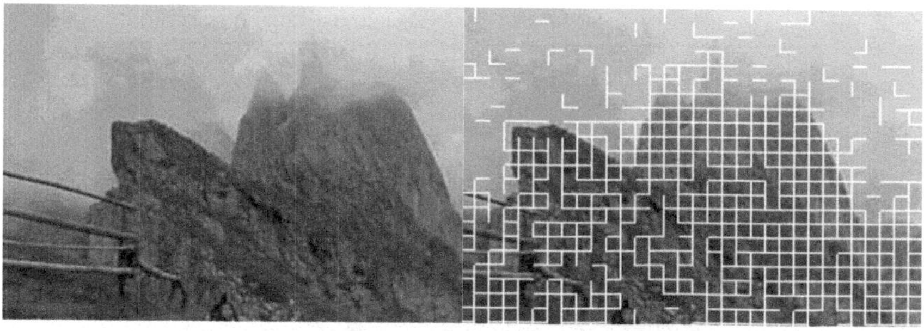

Fig. 5.29 Results of blocking artefact detection (the enlarged region is in Fig. 5.28)

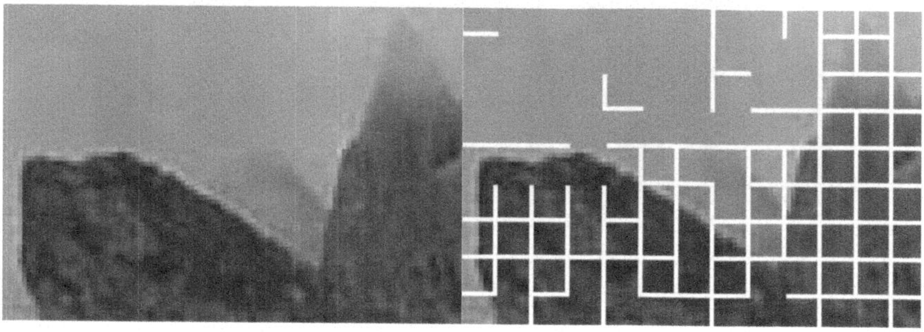

Fig. 5.30 Results of blocking artefact detection (enlarged)

5.2.3 Results

5.2.3.1 Blocking

The results show that the blocking detection algorithm is able to detect more than 95% of visible block boundaries. On good images, about 5% of good boundaries are marked as "bad", but from the point of view of artefact removal, it is more appropriate to mark the "good" boundaries than to skip the "bad" ones. Figures 5.28, 5.29 and 5.30 demonstrate the detected block boundaries.

The proposed method has the following advantages: it is fast and computationally efficient, it allows block boundaries to be detected even in highly detailed areas, and it depends on the quantization table but not on the compression ratio. Moreover, each block is analysed individually and a procedure for classification of each block can save time in further processing and help avoid over-blurring as a result of processing blocks that are not affected by blocking artefacts.

Among the disadvantages is that some erroneously detected artefacts may exist and some artefacts may not be detected at all (FNs and FPs).

Among the possible additions to the method may be an analysis of colour features, since colour artefacts are notably visible, and very often colour artefacts

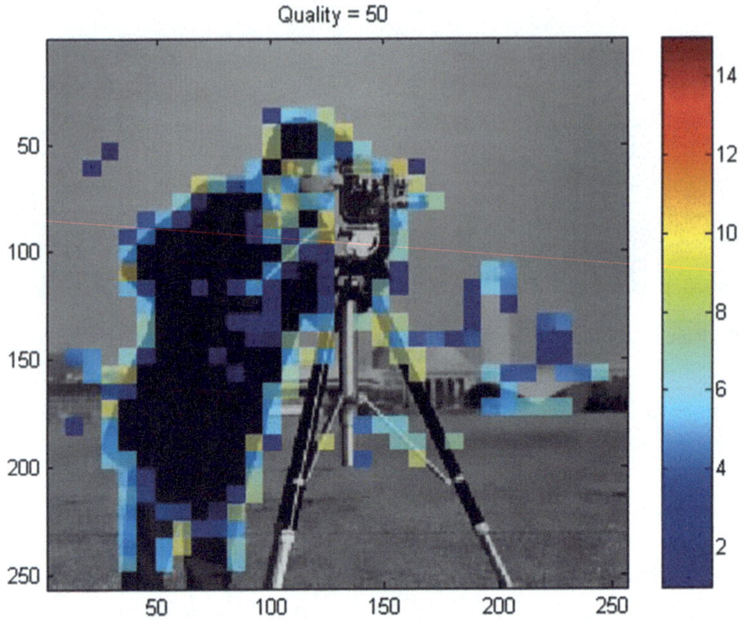

Fig. 5.31 The image "Cameraman" with a quality level of 50, with colour representing ringing fuzziness

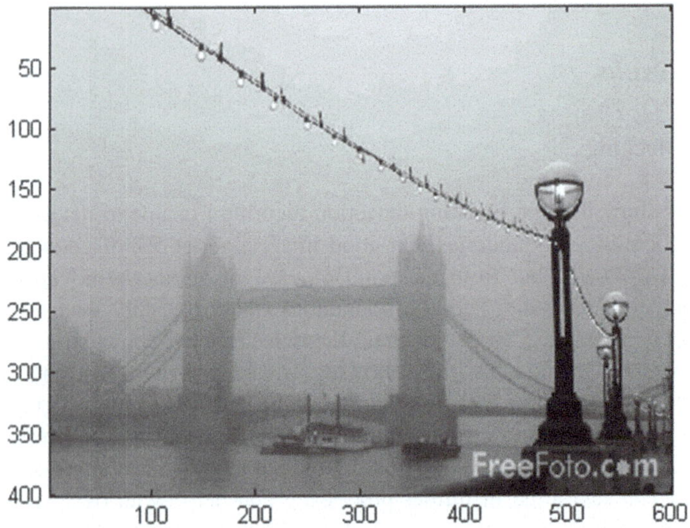

Fig. 5.32 "Tower Bridge" image (at www.FreePhoto.com)

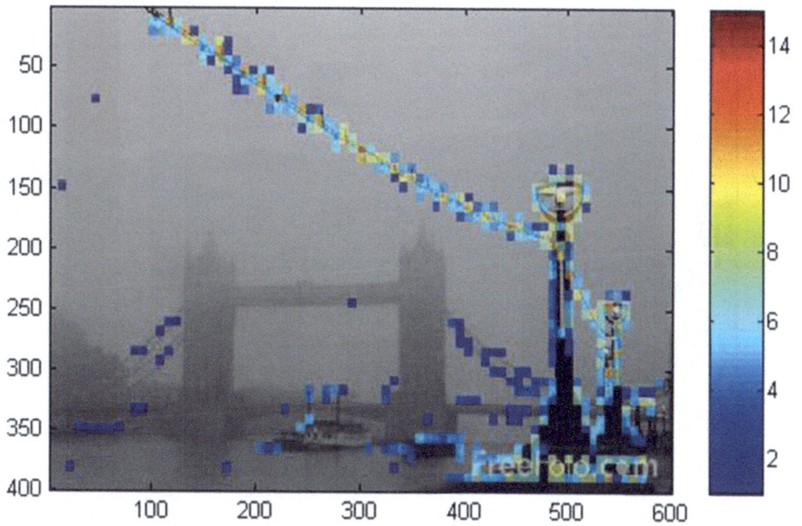

Fig. 5.33 "Tower Bridge" image with colour representing ringing fuzziness

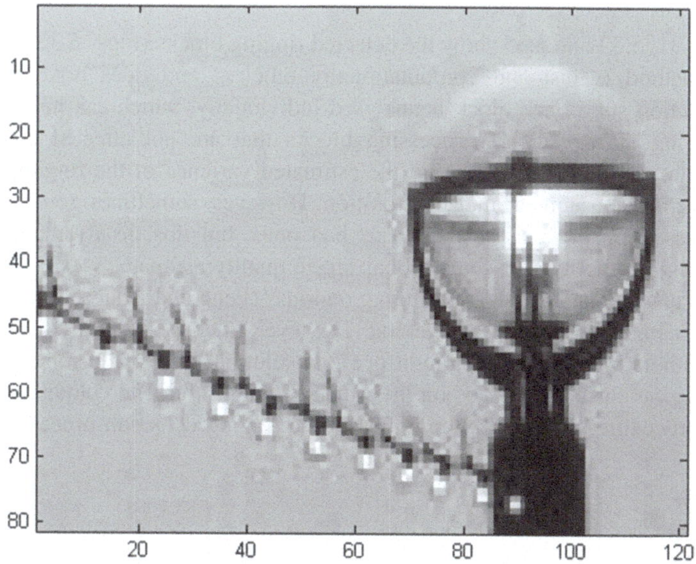

Fig. 5.34 Fragment of "Tower Bridge" image

also appear on the block boundaries. Also, it is worth investigating the minimal resolution at which artefacts are perceptible more accurately (on screen, for example, artefacts become visible at 300–400% zoom) and conducting principal components analysis (PCA) to infer which features are more significant in order to reduce the computation time.

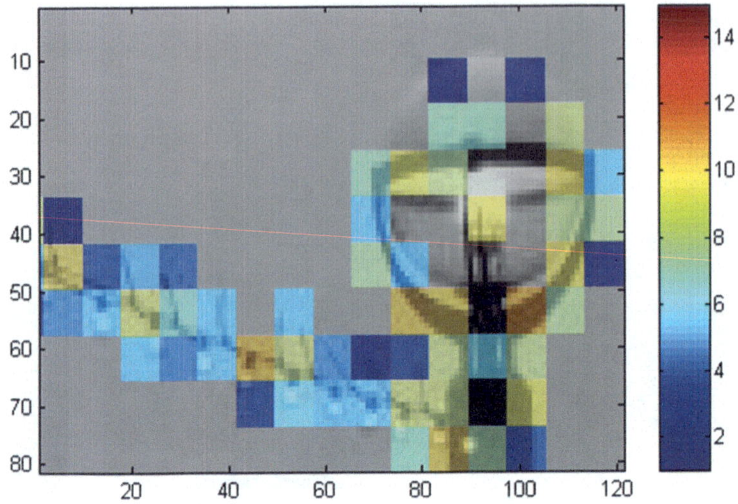

Fig. 5.35 Fragment of "Tower Bridge" image with colour representing ringing fuzziness

5.2.3.2 Ringing

Figures 5.31, 5.33 and 5.35 show the detected ringing blocks (Figs. 5.32 and 5.34).

The method is fast and computationally efficient and does not include an edge-detection step. Each block is analysed individually, which can help to avoid over-blurring as a result of processing blocks that are not affected by ringing artefacts and save time as well, and the estimated variance of the ringing artefact can be used in subsequent image restoration. However, sometimes good blocks in uncompressed images can be treated as bad ones, but this disadvantage can be eliminated with the help of some global image quality estimate.

Analysis of so-called "isolated ringing regions" (Feng and Allebach 2006) could possibly improve the proposed method. However, such a situation is quite rare. These authors proposed a straightforward solution: the inclusion of image smooth-region clustering (based on the watershed algorithm and colour similarity), but this procedure significantly complicates the ringing detection procedure.

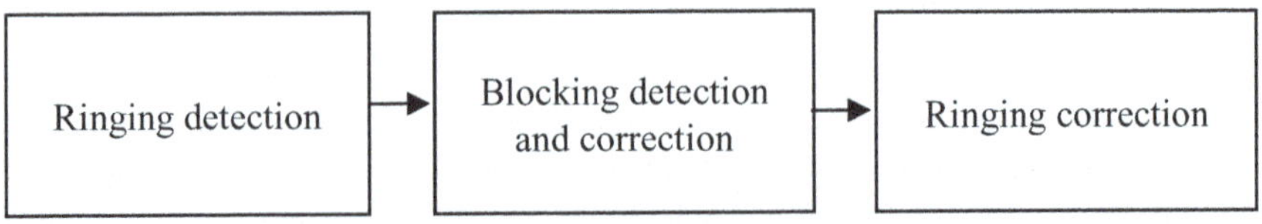

Fig. 5.36 Overall procedure for removal of JPEG artefacts

5.3 Correction of Artefacts

A sophisticated procedure for artefact detection allows very delicate image modification and preservation of image details. Since the detection step requires non-modified image data (which means that detection should precede correction), the overall procedure for the removal of JPEG artefacts consists of the steps depicted in Fig. 5.36.

After the ringing detection step, the ringing variance matrix is stored in memory and blocking detection is performed one by one following correction of the current block if needed.

5.3.1 Blocking Artefacts

Sharpness preservation is a challenging problem in the correction of blocking artefacts. The main problem is the need to distinguish between real edges and false edges generated by blocking artefacts. In this work, it is assumed that real edges have a significantly greater drop in lightness. A very simple approach for the correction of blockiness is proposed. Let us consider two adjacent 8 by 8 pixel blocks (see Fig. 5.37).

$$
\begin{aligned}
k_1 &= f(|\mathrm{a}_1 - \mathrm{a}_0|),\\
k_2 &= f(|\mathrm{a}_0 - \mathrm{b}_0|),\\
k_3 &= f(|\mathrm{b}_1 - \mathrm{b}_0|),
\end{aligned}
$$

where $f(x) = \frac{1}{2\sigma_r} e^{-x^2}$.

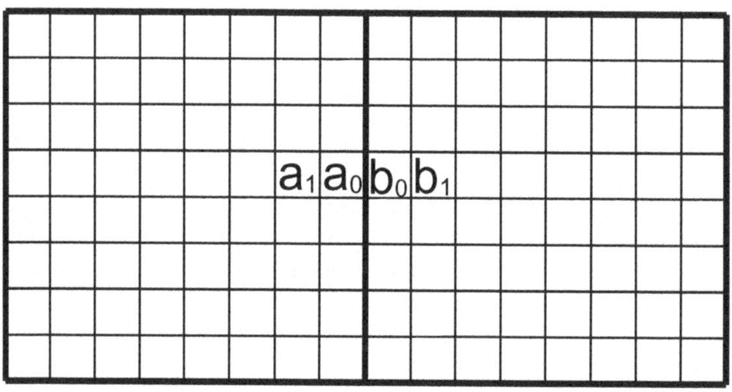

Fig. 5.37 Pixels near boundary between adjacent 8 × 8 pixel blocks

After the coefficients have been computed, pixels a_0 and b_0 are replaced by c_0 and d_0, respectively:

$$\begin{aligned} c_0 &= a_0 + k_1 a_1 + k_2 b_0, \\ d_0 &= b_0 + k_2 a_0 + k_3 b_1. \end{aligned}$$

This means that if the difference between a_0 and b_0 is big, the edge is true, k_2 is small, the new value of c_0 depends on a_1, while d_0 depends on b_1, and the edge is preserved. The larger the value of σ_r, the more intense the smoothing of the block boundaries. σ_r varies from 0.05 to 0.5. In a particular application, σ_r is fixed at a value of 0.1.

5.3.2 Ringing Artefacts

In the case of ringing artefacts, fuzziness in the vicinity of a strong edge can be effectively removed by application of a bilateral filter (Tomasi and Manduchi 1998), as mentioned in Sect. 1.4.3. Ringing correction is performed only in blocks where the ringing probability is greater than zero. The image block is filtered with a bilateral filter. The bilateral filter is implemented in a separable manner: the block of image I is filtered first row-wise and then column-wise to speed-up filtering.

The bilateral filter helps smooth small variations of lightness occurring as ringing artefacts, and at the same time this operation preserves true image edges. Moderate application only in blocks affected by artefacts helps to avoid over-blurring of the image.

Domain filtering helps to avoid inter-block discontinuity. Parameter σ_d is set to 0.1 or can vary in some interval.

Range filtering is adaptive: σ_r depends on the estimated ringing variance. The greater the variance, the greater the value of σ_r. However, for the sake of saving computational resources, adaptivity can be omitted, since it does not provide drastic improvements. When σ_r is set to 0.1, the results are still quite acceptable.

5.3.3 Results

Traditionally, the Peak Signal to Noise Ratio (PSNR) is employed as a numeric criterion for estimation of the quality of JPEG-compressed image correction. In general, the PSNR makes it possible to estimate a positive or negative correction effect. However this metric is quite coarse: results for extremely high PSNRs sometimes look a little over-blurred. PSNR is computed according to the following formulas:

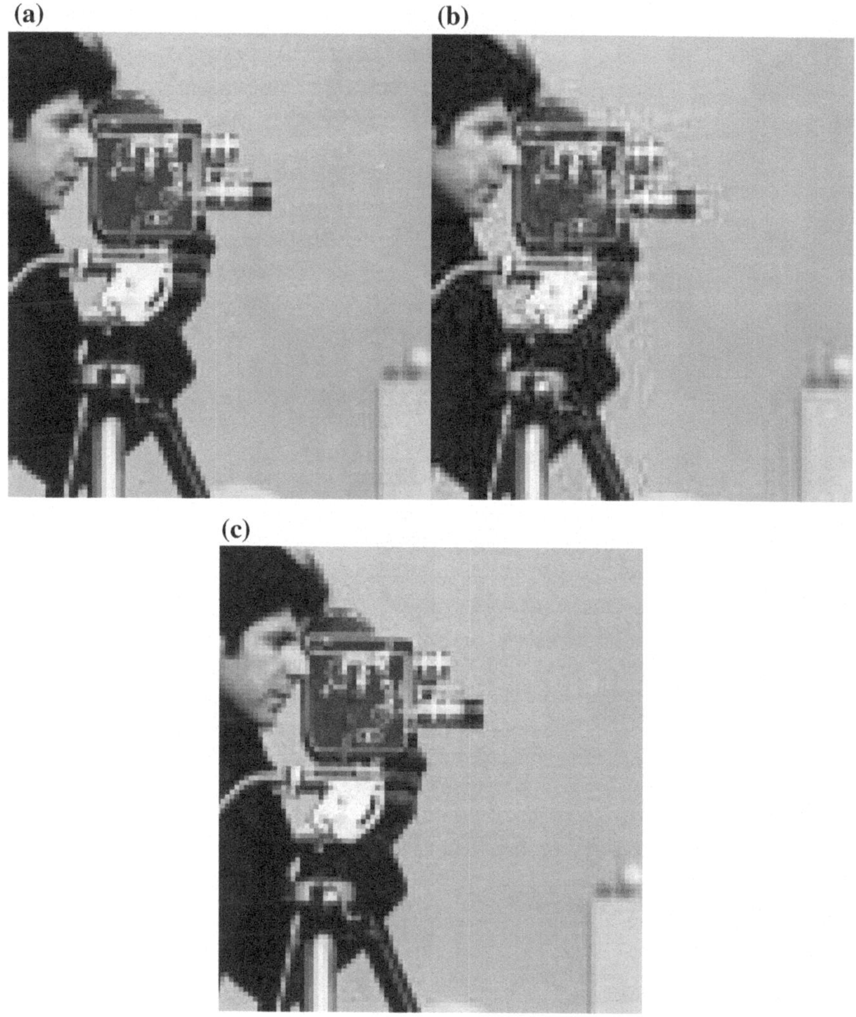

Fig. 5.38 "Cameraman" image: **a** True image (without compression); **b** 20% quality in Photoshop "save for web"; **c** 80% quality in Photoshop "save for web"

$$MSE = \frac{1}{MN}\sum_{i=1}^{M}\sum_{j=1}^{N}(I - I_{true})^2,$$

$$PSNR = 10\log_{10}\left(\frac{255^2}{\mathrm{MSE}}\right).$$

We compared the correction abilities of our solutions with existing utilities intended for the correction of JPEG-compressed images. We tested three

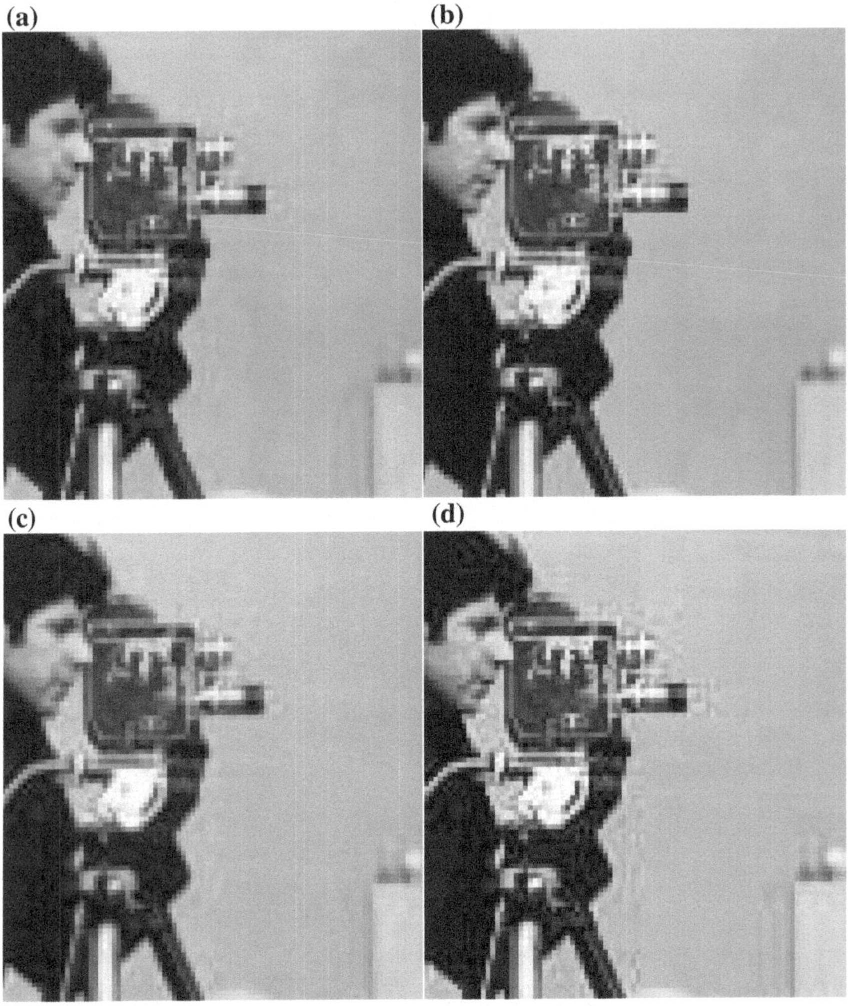

Fig. 5.39 Results of enhancing an image with 20% quality in Photoshop "save for web" using : **a** The proposed method; **b** Shape-adaptive DCT; **c** JPEG Enhancer; **d** DeJPG

well-known solutions: shape-adaptive DCT (SA-DCT) (Foi et al. 2007), Topaz Labs DeJPEG, and JPEG Enhancer. For manual or semi-automatic tools, we used the default settings (Figs. 5.38, 5.39 and 5.40).

Figure 5.41 demonstrates the PNSRs for the images "Lena" and "Cameraman" for various bpp values. JPEG images were created using the Photoshop CS2 "save for web" option at quality levels of 8, 20, 40, 60, 80, 90, and 99%.

The SA-DCT algorithm achieves state-of-the-art performance in terms of PSNR. The results of SA-DCT look good for the low-quality image (high compression ratio) but outcomes for the middle quality image look over-blurred and textures are

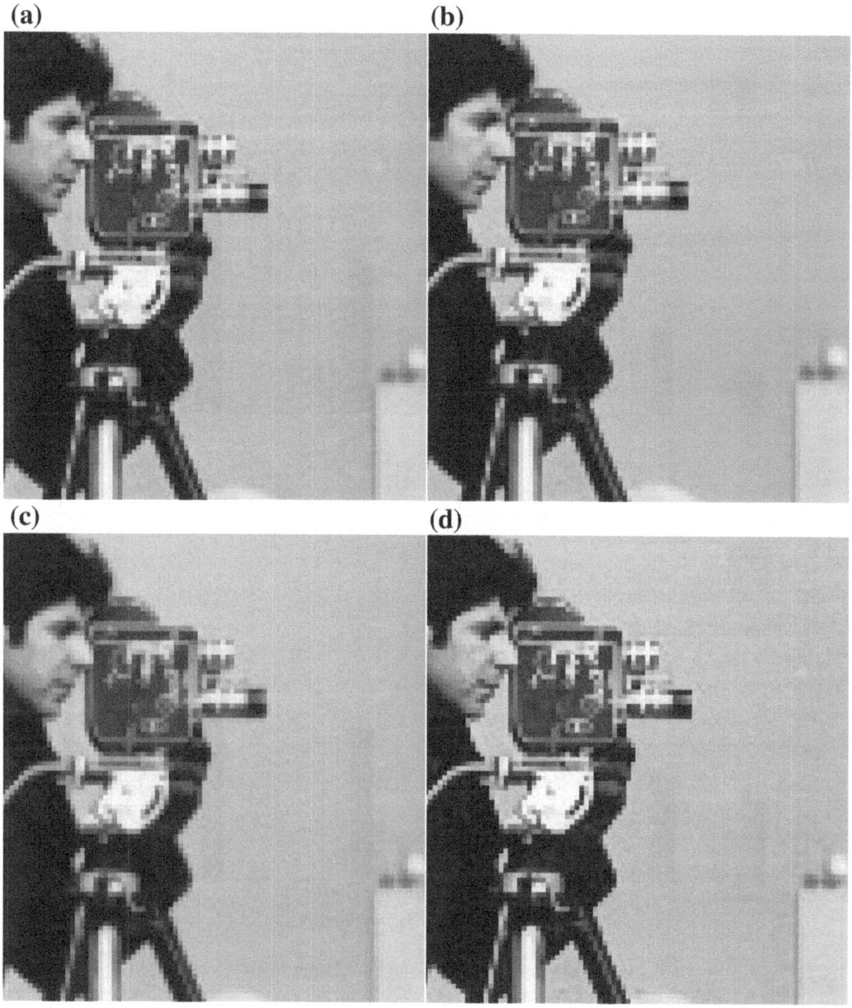

Fig. 5.40 Results of enhancing the image with 80% quality in Photoshop "save for web" using: **a** The proposed method; **b** Shape-adaptive DCT; **c** JPEG Enhancer; **d** DeJPG

deteriorated. For low-quality images, the proposed method suppresses ringing artefacts and attenuates blocking artefacts but parts of blocks are visible. For medium-quality images, the outcomes of the proposed method look perfect: ringing and blocking artefacts are suppressed completely, while texture is not deteriorated. Figures 5.42 and 5.43 show one more example.

Table 5.2 shows the processing time for an image with a size of 2480 × 3507 pixels. It can be seen that the proposed method outperforms its competitors in terms of speed as well.

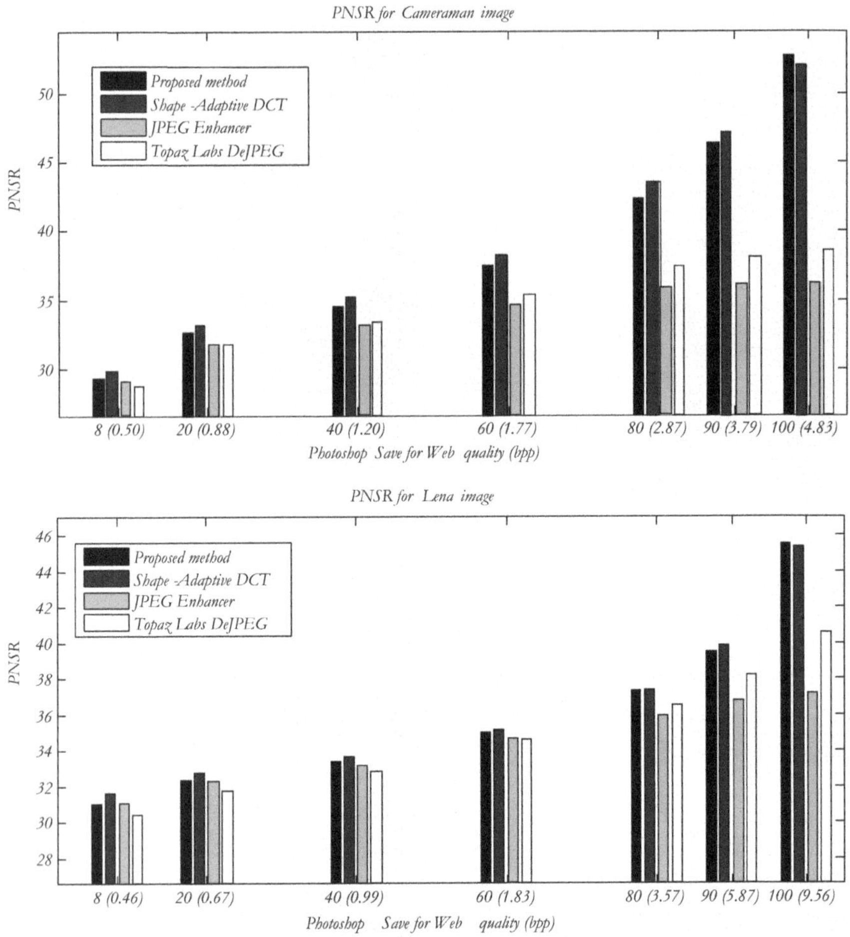

Fig. 5.41 Results of PNSR computation for the images "Cameraman" (top) and "Lena" (bottom) for various bits-per-pixel values

The proposed detection method is fast, accurate, and computationally efficient, helping to reduce the amount of computation during correction, and each block is processed individually, which can help avoid over-blurring as a result of processing blocks that are not affected by ringing artefacts and can save time as well. The

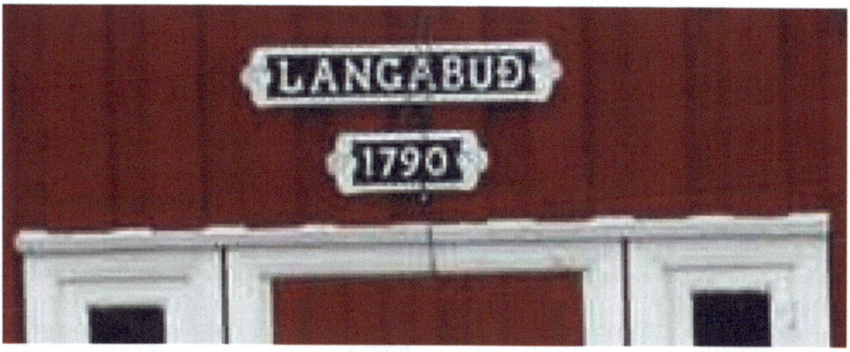

Fig. 5.42 Source image affected by JPEG artefacts

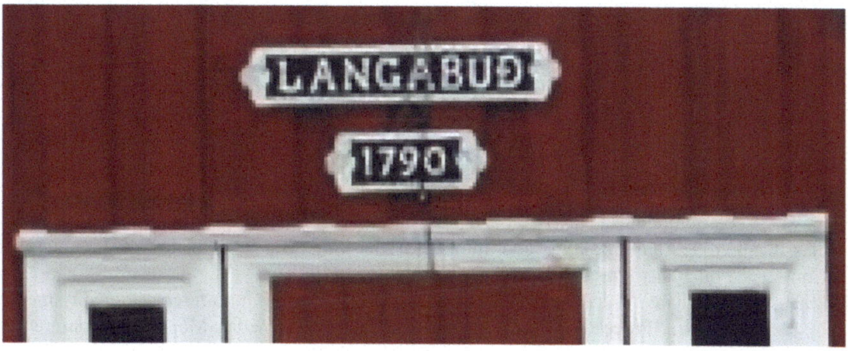

Fig. 5.43 Resulting image

Table 5.2 Processing time for an image of size 2480 × 3507 pixels

Solution	Proposed method	UnJPEG	DeJPEG	JPEG Enhancer	SA-DCT (PS plug-in)
Time (s)	9	26	37	25	811

estimated variance of the ringing artefact allows the application of an adaptive approach to de-ringing.

Classification of image areas on texture, edges, and smooth regions for the de-blocking procedure could improve the proposed method by allowing the differentiation of the correction procedure depending on block characteristics. Meanwhile, for classification, the work of Tong and Venetsanopoulos (1998) can be applied, for fast classification in the frequency domain using DCT coefficients of the image coding block.

References

Chang, Y.W., Chen, Y.Y.: Alleviating-ringing-artefact filter using voting scheme. ICGST Int. J. Graph. Vis. Image Process. Spec. Issue Appl. Image Process. Tech. **6**, 9–14 (2006)

Chou, C., Li, Y.: A perceptually tuned sub-band image coder based on the measure of just-noticeable distortion profile. IEEE Trans. Circ. Syst. Video Technol. **5**(6), 467–476 (1995)

Fan, Z., de Queiroz, R.L.: Identification of bitmap compression history: JPEG detection and quantizer estimation. IEEE Trans. Image Process. **12**(2) (2003)

Feng, X., Allebach, J.P.: Measurement of ringing artefacts in JPEG images. In: Allebach J.P., Chao H. (eds.) Proceedings of IS&T/SPIE Electronic Imaging, vol. 6076, pp. 74–83. Digital Publishing (2006)

Foi, A., Katkovnik, V., Egiazarian, K.: Pointwise shape-adaptive DCT for high-quality denoising and deblocking of grayscale and color images. IEEE Trans. Image Process. **16**(5), 1395–1411 (2007)

Freund, Y., Schapire, R.: Experiments with a new boosting algorithm. In: International Conference on Machine Learning, vol. 96, pp. 148–156 (1996)

Friedman, J., Hastie, T., Tibshirani, R.: Additive logistic regression: a statistical view of boosting. Ann. Stat. **38**(2), 337–374 (2000)

Oguz, S.H., Hu, Y.H., Nguyen, T.Q.: Image coding ringing artefact reduction using morphological post-filtering. In: IEEE Second Workshop Multimedia Signal Processing, pp. 628–633 (1998)

Tolstaya, E.V., Rychagov, M.N., Kim, S.H., Choi, D.C.: Removal of blocking and ringing artefacts in JPEG-coded images. In: Proceedings of IS&T/SPIE Electronic Imaging (2010)

Tomasi, C., Manduchi, R.: Bilateral filtering for gray and color images. In: Proceedings of IEEE Conference on Computer Vision, pp. 839–846 (1998)

Tong, H., Venetsanopoulos, A.N.: A perceptual model for JPEG applications based on block classification, texture masking, and luminance masking. In: Proceedings of International Conference on Image Processing, vol. 3, pp. 428–432 (1998)

Tong, H., Li, M., Zhang, H-J., Zhang, C., He, J., Ma, W-Y.: Learning no-reference quality metric by examples. In: Proceedings of 11th International Multimedia Modelling Conference, pp. 247–254 (2005)

Wang, Z., Sheikh, H.R., Bovik, A.C.: No-reference perceptual quality assessment of JPEG compressed images. In: Proceedings IEEE International Conference on Image Processing, vol. 1, pp. I–1 (2002)

Yang, S., Hu, Y., Nguyen, T.Q., Tull, D.L.: Maximum-likelihood parameter estimation for image ringing-artefact removal. IEEE Trans. Circ. Syst. Video Technol. **11**(8), 963–973 (2001)

Chapter 6
Descreening of Scanned Images

6.1 Introduction

Modern laser or ink-jet printers and offset machines print using just several paints. Different shades of colour or grey are simulated by special ordering of binary speckles or dots, called the halftoning procedure. Conventionally halftones can be separated on two parts: *Amplitude Modulation* (AM) and to *Frequency Modulation* (FM) halftone. AM halftone is characterized by varying printed dots size for visualization of different shades. In AM colour halftoning, CMYK (Cyan, Magenta, Yellow, Black) ink dots of variable size are printed in overlapping grids. The grids are placed at different angles for each of the ink colours as shown in Fig. 6.1. Smaller halftone dots absorb less light and as a result of increasing the amount of reflected light, the apparent density decreases, and the object looks lighter.

Figure 6.2 demonstrates examples of scanned AM-halftoned images produced by electrophotographic and offset machines. FM halftone is based on changing of relative density of dots on a page. The most popular FM halftoning method is error diffusion dithering, where the quantization error for each pixel is distributed among its neighbourhoods.

Printed halftone images look almost natural on paper; however, their representation significantly degrades when halftoned images are scanned or copied. One of well-known unpleasant effects is periodical Moiré pattern. Such a distortion appears due to the overlapping of two regular structures, for example, grid of halftone cells and lattice of a scanner.

The common approach for quality improvement after scanning of printed documents and prevention of Moiré patterns is called descreening or the inverse halftoning technique. The descreening procedure is oriented on the recovery of continuous-tone (contone) image structure. A typical way this is achieved is with the smoothing of a scanned image. However, it is necessary to emphasize, any descreening algorithm faces the following dilemma: low-frequency smoothing of a halftone image inevitably leads to blurring of sharp edges of text and graphics

I.V. Safonov et al., *Adaptive Image Processing Algorithms for Printing*,
Signals and Communication Technology,
https://doi.org/10.1007/978-981-10-6931-4_6

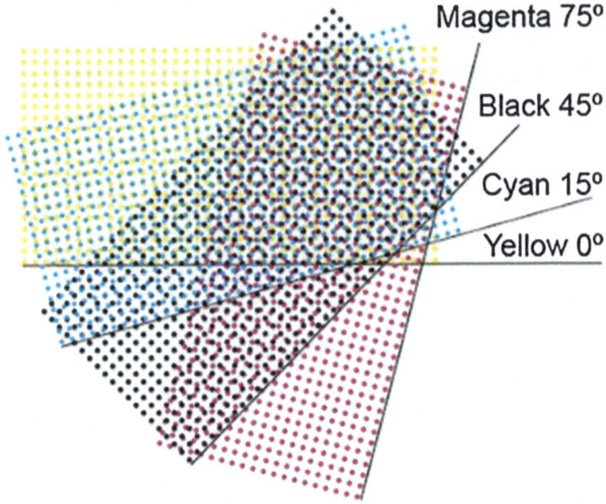

Fig. 6.1 Illustration of colour AM-halftoning

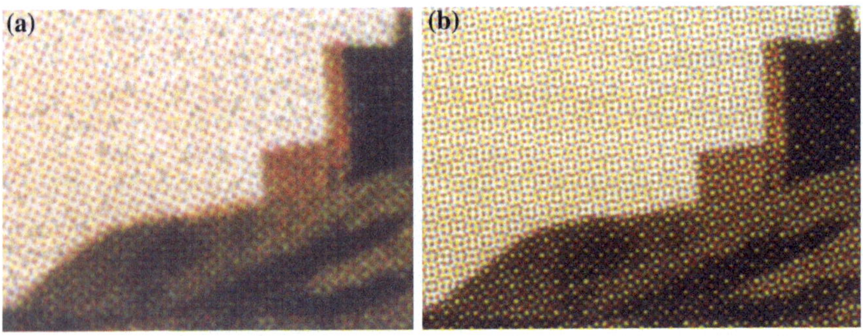

Fig. 6.2 Enlarged fragments of halftone image: **a** printed on a laser printer, and **b** printed on an offset machine and scanned with 600 dpi resolution

primitives. On the other hand, insufficient smoothing keeps the screen in halftoned areas.

Summarizing, it is possible to formulate the following requirements for a descreening algorithm where:

- a sufficient halftone suppression prevents the appearance of Moiré patterns during copying;
- high-frequency details on images such as sharp edges and textures are preserved;
- the smoothing elements on images without halftone such as text, charts and similar graphic objects filled by solid colour are avoided;
- robust processing of images uses halftone types and parameters;
- colour distortions are avoided.

6.2 Existing Approaches for Inverse Halftoning

There are plenty of works related to descreening. Some of them are oriented on particular halftone types such as AM or FM, but from the practical point of view the most interesting universal approaches are suitable for any halftone method. Existing descreening, algorithmic solutions are based on processing in frequency or spatial domains as well as on combined approaches.

One of the most evident and popular approaches for inverse halftoning is the detection in frequency domain of amplitude peaks, which correspond to halftone grid frequencies, and suppression of the peaks via notch filtration. Such solutions are oriented on descreening of AM-halftoned grayscale images, inasmuch as for colour images the inter-channel frequency mixing significantly complicates localization of spectral peaks for colour channels. Kuo et al. (2001a, b, c) describes the algorithm for the detection of halftoned regions in the wavelet domain and filtration of the regions by a quasi-optimal selective filter. Aizenberg and Butakoff (2002) describes filters for suppression of periodical and quasi-periodical noises including AM halftone. The filter is based on detection and suppression of spectral peaks, which correspond to periodical noise. For that purpose, the image spectrum is processed by median filter. If ratio of a pixel in the spectrum to median value exceeds threshold, then the sample is recognized as the one belonging to halftone frequency and is substituted by the median. Luo et al. (1998) describes a combined algorithm, where recovering of contone images is performed by means of filtration in frequency and spatial domains sequentially. Processing in a spatial domain is intended for removing of remaining artefacts after filtration in a frequency domain as well as for sharpening. Detection and removal of the halftone pattern in a frequency domain is impractical for hardware implementation due to huge memory requirements. For colour image, frequency of halftone cells of C, M, Y and K channels lie in R, G and B channels simultaneously. In this case, it is too hard to detect frequencies of halftone cells properly.

Spatial approaches are more suitable for implementation in printing scanning devices. Kuo et al. (2001a, b, c) depicts descreening based on a two-stage Sigma-filter. At the first stage an image is smoothed. Then, the smoothed image is used for selection of neighbourhood pixels for a filter mask. However, such piecewise smoothing often leads to a cartoon effect. Siddiqui et al. (2008) describes the method for inverse halftoning based on anisotropic diffusion. In the approach, a feature vector is calculated for edge preservation and for the estimation of the smoothing ratio of halftone oscillations. Extracted features characterize edge strength and orientation to vicinity. The method has low memory consumption and is highly parallelable.

Shou and Lin (2004) describe an approach based on texture classification using neural network. The method includes preliminary network training with manually labelled fragments of images. Smoothing ratio varies depending on an outcome of texture classifications.

Siddiqui and Bouman (2007a, b) describe an algorithm that combines two image processing techniques: modified SUSAN filtering and Resolution Synthesis-based Denoising (RSD) predictor. The RSD predictor is based on the classification of a vicinity around a current pixel, and the application of a linear, low-pass filter, adjusted for the detected classes. The resulting image of the RSD predictor is treated as a first order estimation and is used for the adjustment of smoothing ratio for the SUSAN filter. A disadvantage of the approach is that it is quite a complex procedure which necessitates preliminary training. The solution is also oriented on a specific halftone model as well.

Nishida (2004) considers a method for adaptive inverse halftoning. The solution includes two stages. The first stage is intended for analysis of a downsampled copy of the original scanned image and it performs separation into two general areas: a slow varied background and foreground with sharp edges such as text, lines, etc. The background pixels are smoothed for halftone suppression and combined with the original image. On the second stage, non-processed pixels are checked in relation to the background area and are smoothed in the original resolution. However, the approach produces visual artefacts around contours.

6.3 Adaptive Descreening

6.3.1 General Workflow

We propose a descreening algorithm that meets the above-mentioned requirements for halftoned image reconstruction. Our approach is applicable for both grayscale and colour images. In the book, we discuss the processing of photos in RGB colour system only; handling of images in other colour systems and grayscale images is similar; and experienced persons can easily modify our approach for other image types.

Figure 6.3 shows a block diagram of basic processing stages of the method. The descreening algorithm has five stages. The first stage is the initial decrease of the edge transition slope length via local tone mapping to restore edges, which were blurred during scanning. The second stage is adaptive low-pass filtering for suppression of halftone pattern. The main idea of the adaptive filtering is smoothing of pixels, which correspond to flat regions on contone image before halftoning, i.e. the regions where halftone pattern changes insignificantly; and modest smoothing of pixels corresponding to edge regions before halftoning, i.e. the regions where halftone pattern changes significantly.

The next two stages are intended for compensation of edge smoothing via sharpening. For sharpening we apply the same approaches as described in Chap. 4, but some details are different. The third step is decreasing of edge transition slope length via local tone mapping. Further, local contrast on ends of slope is increased via the modified unsharp mask filter. The final stage is a global contrast

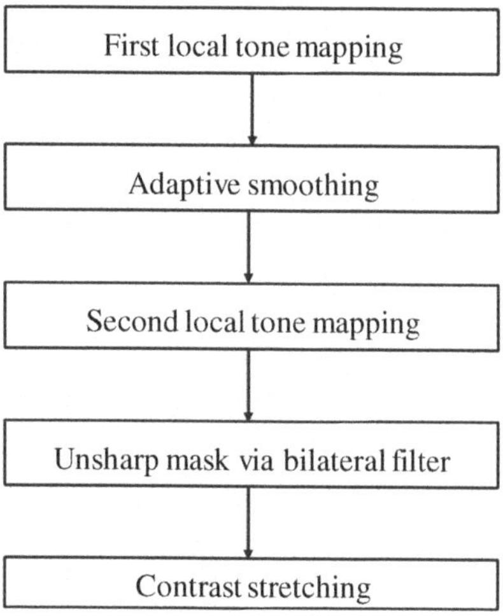

Fig. 6.3 General pipeline of the descreening algorithm

enhancement to compensate for contrast decrease due to low-pass filtering. Figure 6.4 shows enlarged fragment of image after processing stages.

6.3.2 First Local Tone Mapping

Some blurring of scanned image occurs during scanning. Sharpening via decreasing of edge transition slope allows better preservation of edges on sequent adaptive low-pass filtering. Edge transition slope is decreased in brightness channel by application of local tone mapping. We compute brightness channel for RGB colour image as follows:

$$Y(i,j) = \max[R(i,j), G(i,j), B(i,j)],$$

where i and j are coordinates of a pixel.

Algorithm for local tone mapping for sharpening is described in Sect. 4.4. To prevent appearing of cartoon-effect it is worthwhile to make a blending of initial brightness channel Y with outcome of local tone mapping Ye:

$$Ys(i,j) = Y(i,j) \times (1 - g(H - L)) + Ye(i,j) \times g(H - L),$$

where alpha-channel is function like shifted Gamma-correction:

$$g(x) = 1 - e^{-(x - Tltm)/50},$$

where x is scaled to [0, 255] range; parameters L, H and $Tltm$ are described in Sect. 4.4.

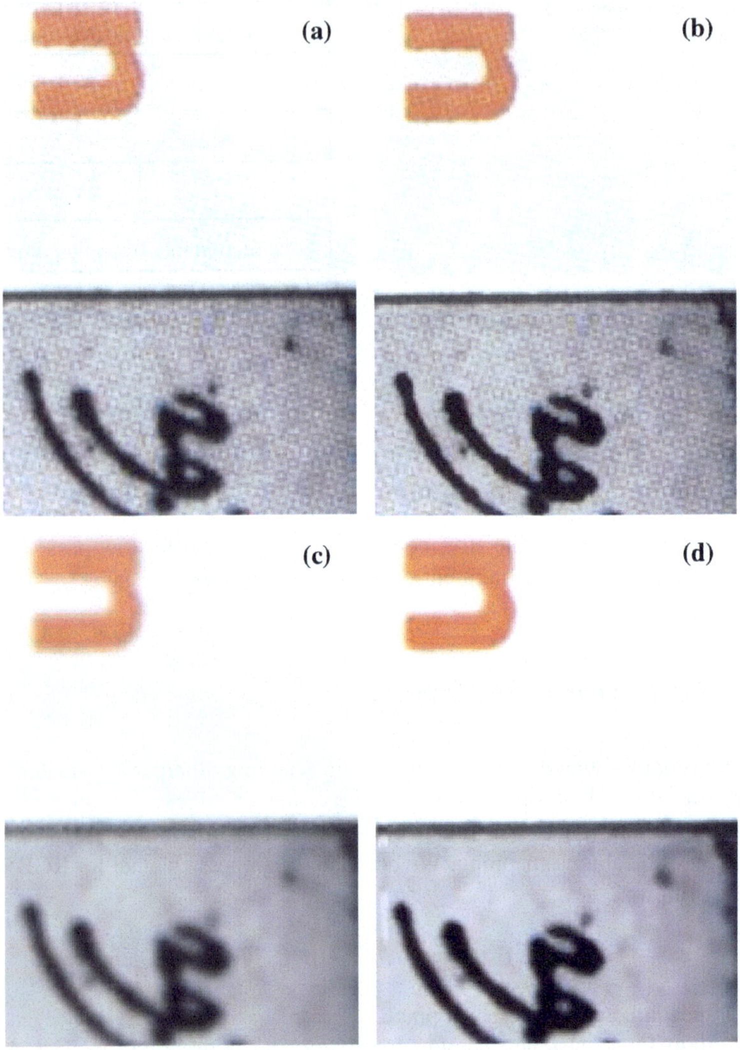

Fig. 6.4 Processing steps: **a** fragment of initial image, **b** image after first local tone mapping, **c** image after adaptive smoothing, **d** final processing outcome

Figure 6.5 demonstrates plot of function $g(x)$. Further, we take into account alterations in brightness channel and modify RGB channels as it is described in Sect. 1.3.

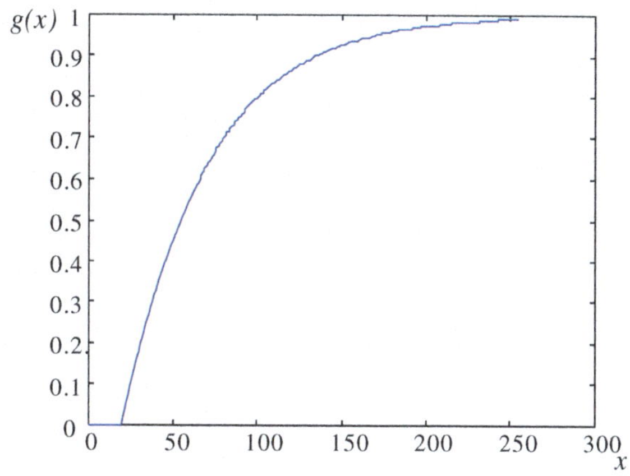

Fig. 6.5 Plot of alpha-channel $g(x)$

6.3.3 *Adaptive Smoothing*

The main idea of this method is based on a simple assumption: to involve only the surrounding pixels belonging to the same area as the processed pixel. Flat regions have local blocks similar to each other. On the contrary blocks across edges differ strongly. This fact can be used for edge detection on screened images. The conventional way is to compare image blocks for all directions. However, the computational complexity of directional edge, detection filters are high. We propose to calculate similarity between the central block of window and different adjacent block which is selected randomly. If similarity is high then pixel belongs to flat region, otherwise pixel belongs to edge region. The approach was inspired partly by reductive NL-means filter from Nvidia CUDA SDK, Kharlamov and Podlozhnyuk (2007). The filter has low noise suppression abilities in comparison with traditional NL-means by Buades et al. (2005) but it is much faster.

Our adaptive smoothing algorithm consists of the following steps. A square sliding window moves throughout an image and the image is filtered non-recursively. The size of the window is $Kb \times Kb$, size of blocks is $Bn \times Bn$. Figure 6.6 illustrates blocks in local window for window size 7×7 and block size 3×3. One of adjacent blocks is selected randomly: n and m are random values form the set

$$\{-Kb/2, -Kb/2+1, \ldots -1, 1, \ldots Kb/2-1, Kb/2\}.$$

Similarity M between central block of window and block selected randomly is calculated according to the formula:

$$M[D(i,j)] = \frac{1}{\sqrt{1+\left|\frac{D(i,j)}{\sigma_R^2}\right|}},$$

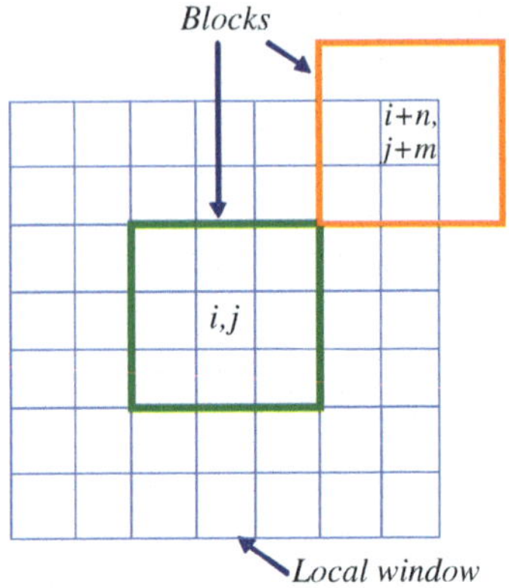

Fig. 6.6 Blocks in local window

where σ_R is parameter for edge detection control, *D(i, j)* is squared Euclidean distance between blocks for current pixel,

$$D(i,j) = \sum_{k=-Bk/2}^{Bk/2} \sum_{l=-Bl/2}^{Bl/2} \Big[(R(i+n+k,j+m+l) - R(i+k,j+l))^2 + (G(i+n+k,j+m+l) - G(i+k,j+l))^2 + (B(i+n+k,j+m+l) - B(i+k,j+l))^2 \Big].$$

Further, if similarity *M* is greater than *T* it means that blocks are similar; and current *(i, j)* of R, G and B colour channels is strongly smoothed by convolution with *Hs* kernel; otherwise the colour channels are smoothed modestly by convolution with *Hm* kernel:

$$F\{x\} = \begin{cases} 1, & x > T \\ 0, & x \leq T \end{cases},$$

$$Rs = R * Hs,$$

$$Gs = G * Hs,$$

$$Bs = B * Hs,$$

$$Rm = R * Hm,$$

$$Gm = G * Hm,$$

$$Bm = B * Hm,$$

$$R'(i,j) = Rs(i,j) \times F\{M(D(i,j))\} + Rm(i,j) \times [1 - F\{M(D(i,j))\}],$$

$$G'(i,j) = Gs(i,j) \times F\{M(D(i,j))\} + Gm(i,j) \times [1 - F\{M(D(i,j))\}],$$

$$B'(i,j) = Bs(i,j) \times F\{M(D(i,j))\} + Bm(i,j) \times [1 - F\{M(D(i,j))\}].$$

All the parameters depend on a scanning resolution. For example, for 600 dpi $Kb = 7$, $Bn = 3$, $\sigma_R = 8$, $T = 0$.

6.3.4 Sharpening and Global Contrast Enhancement

The next step of the descreening method is the sharpening of the brightness channel of the smoothed image. The brightness channel is extracted from R′ G′ B′. The second decreasing of edge transition slope length is performed identically to the described in Sect. 4.4 except of a calculation of the parameters L and H: L is minimal brightness in the vicinity and H is maximal brightness. An increasing of local contrast near edges enhances image and is positively estimated by observers. We apply Unsharp mask via bilateral filter (see Sect. 4.5) for sharpness enhancement.

Final stage of the method is global contrast stretching. We increase contrast of brightness channel Ye' by 5% to compensate contrast degradation due to smoothing. After sharpening and contrast enhancement it is necessary to reflect changes in brightness channel on colour R, G and B channels as it is described in Sect. 1.3.

6.4 Bilateral Descreening

6.4.1 General Concept

We suggest one more adaptive smoothing approach is working in an anisotropic manner on the basis of the bilateral filtration concept, as well as edge detection and classifying. Figure 6.7 shows flow chart of the approach.

The first step is directional edge detection and estimation of edge orientation. It is applied for each colour channel independently. Based on the outcome of the edge detection filter, each pixel is classified as either lying in *flat area* or *edge region*. We perform Gaussian blurring for suppression of halftone patterns in *flat* areas. We distinguish *roof* and *ramp* edges based on analysis profile of intensity across the edge. For smoothing of edge regions, we apply a modification of bilateral filtering. Parameters for smoothing of edge regions depend on the type of edge.

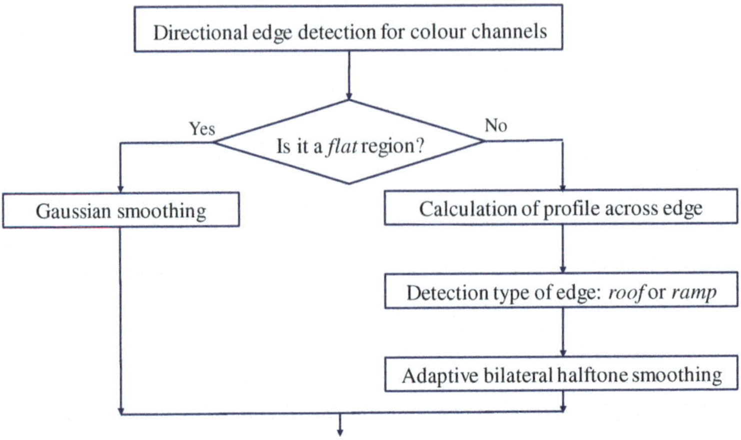

Fig. 6.7 Flow chart of Bilateral descreening algorithm

6.4.2 *Edge Detection*

We apply classification of each pixel of halftoned images into two categories: *flat* and *edge* regions. First class corresponds to areas which are characterized by monotonic, intensity variation. The second class corresponds to pixels, which is grouped on either edges, or near edges in areas with abrupt changing of intensities (Fig. 6.8).

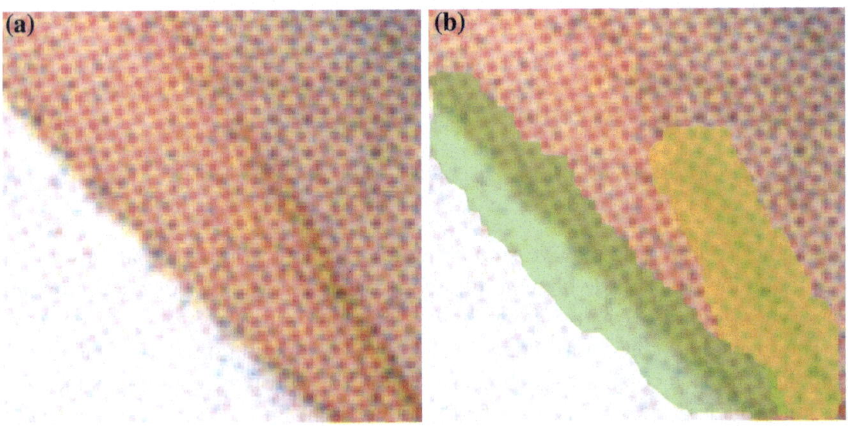

Fig. 6.8 **a** Fragment of an image, **b** fragment of image with marked *edge* and *near to edge* areas by yellow and green colours, all remaining pixels are classified as *flat*

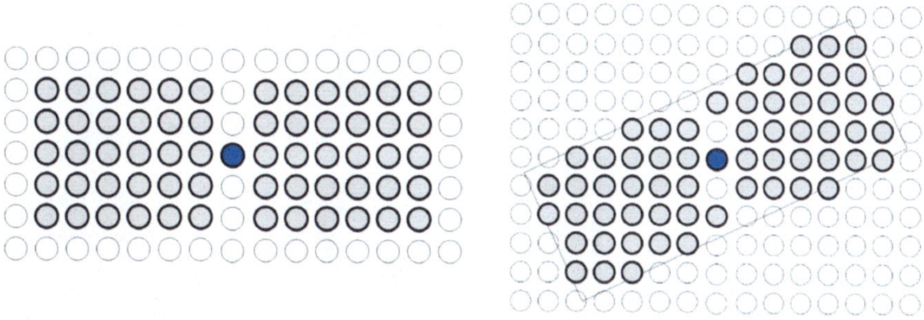

Fig. 6.9 An illustration of two positions of a rectangular mask

Edge detection for pixels of halftoned images differ noticeably from the similar task for continuous-tone images. Conventional edge detection via high-pass filters does not work in case of halftoned images, because edges of halftone cells relate to high frequencies as well. For edge detection, we propose the calculation of the difference between the averaged values of the left and right part of a rectangular mask (see Fig. 6.9) instead of the difference between separate pixels. Printing and scanning resolutions should be considered for proper selection of the height of mask. Height of the mask should be larger than size of a halftone cell. To detect edges of various orientations it is necessary to consider several rotated masks. $G^n(i,j) = |mV_1 - mV_2|$ is the difference between mean values of halves of the mask for a pixel with coordinates (i,j) in the direction n. $G^n_R(i,j)$, $G^n_G(i,j)$, $G^n_B(i,j)$ are calculated independently for R, G and B colour channels of an image. Aggregated estimation of gradient is calculated as:

$$G^n(i,j) = \sqrt{[G^n_R(i,j)]^2 + \left[G^n_G(i,j)\right]^2 + [G^n_B(i,j)]^2}.$$

After calculation of the estimations of gradient for all orientations we obtain the vector of gradients $\{G^n(i,j)\}$, $n = 0, \ldots N$, where N is the number of orientations. Then, we select direction $n_{\max}$ corresponding to maximal gradient in the vector. We suppose that following the rule for edge detection: a colour edge exists if the gradient value in direction $n_{\max}$ for at least one of the colour components exceeds a predetermined threshold *eTh*. Hence, if threshold is exceeded, then the current pixel classified as belonged to *edge* region, and as belonged to *flat* area otherwise.

Figure 6.10 shows examples of *edge* and *near to edge* areas detection outcome. Black pixels designate *edge* and *near to edge* areas, white pixels correspond to *flat* regions.

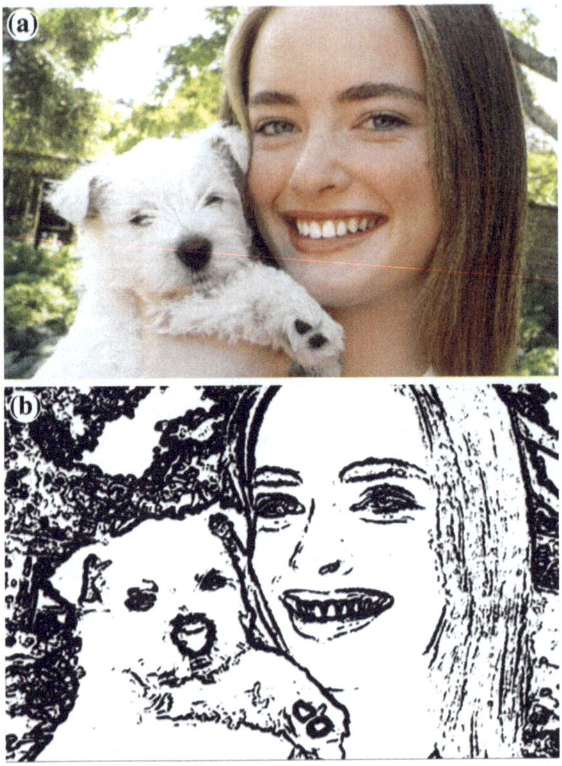

Fig. 6.10 **a** Initial image, **b** edge detection outcomes

6.4.3 *Smoothing*

Any low-pass filter can be applied for smoothing of flat regions. We recommend the use of Gaussian blur. However, excessive smoothing can lead to the loss of sharp details in an image. In an ideal case, size of the convolution kernel should be greater than the size of a halftone cell in a scanned image.

Smoothing of *edge* and *near to edge* areas is based on several heuristics. First, projections L_c on line in direction $n_{\max}$ for each colour components are calculated. The projections L_c can be considered as the analogues of continuous-tone profiles of edge slope in direction with maximal gradient. Figure 6.11 demonstrates example of profile across edge.

Then, we classify edge profile types. There are two types of profiles: *ramp edge* and *roof edge*. Figure 6.12 shows ideal profiles for *ramp* and *roof edge*. If a derivative of the profile changes sign once, then the current pixel belongs to *roof edge* type, otherwise it belongs to *ramp edge*.

Next step is an estimation of parameters conditionally named as *high level* (*HL*) and *low level* (*LL*). An estimation of the levels for *ramp edge* is based on the assumption that the midpoint of L_c separates two regions. The maximal value is

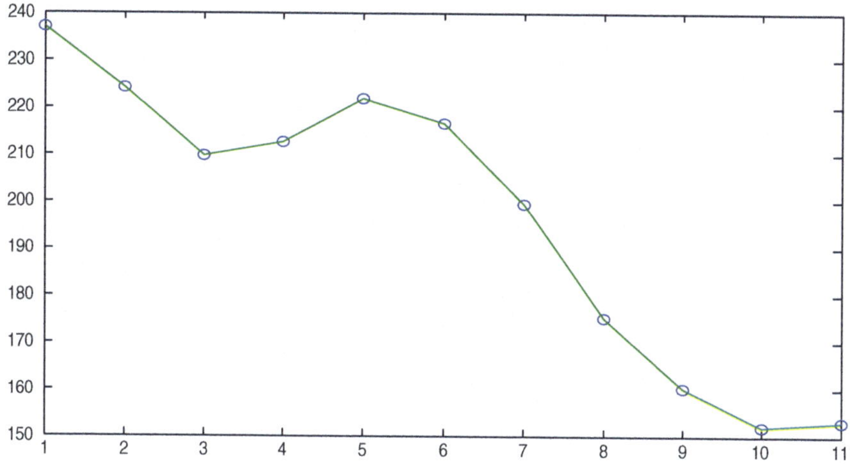

Fig. 6.11 An example of estimation of a profile across an edge

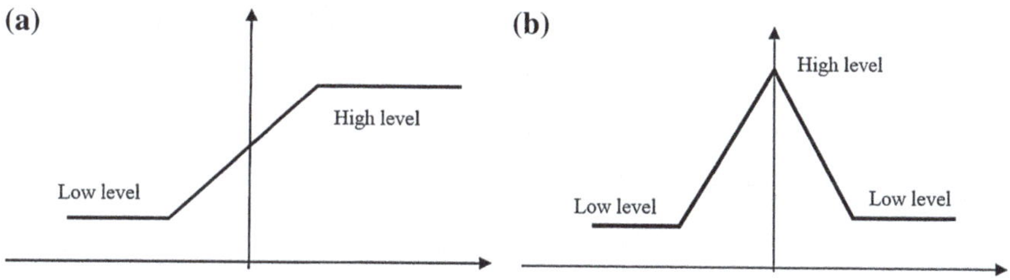

Fig. 6.12 **a** Ideal profile of *ramp edge*, **b** ideal profile of *roof edge*

calculated for half of the profile where the average amplitude grows from midpoint with index $k = (P-1)/2$ to higher values, P is the length of a profile. The maximal value corresponds to HL. The other half of the profile is processed for the estimation of a minimal value that corresponds to LL parameter. HL and LL values for *roof edge* are calculated by looking for global maximal and minimal values in the edge profile $L(k)$.

If $\tilde{x}(i,j) \geq (HL + LL)/2$, then pixel $x(i,j)$ relates to *high level plateau*, otherwise $x(i,j)$. relates to *low level plateau*, where $\tilde{x}(i,j)$ is:

$$\tilde{x}(i,j) = \frac{L[(P-1)/2-1] + L[(P-1)/2] + L[(P-1)/2+1]}{3}.$$

The assignment of all pixels to HL and LL leads to a cartoon effect. To avoid it we propose to use weighted averaging like bilateral filtering (Tomasi and Manduchi 1998). We called the following filter *Bilateral halftone smoothing*:

$$y_e(i,j) = \frac{\sum_{n=-\frac{N-1}{2}}^{\frac{N-1}{2}} \sum_{m=-\frac{M-1}{2}}^{\frac{M-1}{2}} \left[\frac{1}{2\pi\sigma_p\sigma_s} e^{-\frac{\tilde{x}(i-n,j-m)-\mu^2}{2\cdot\sigma_p^2}} e^{-\frac{(i-n)^2+(j-m)^2}{2\cdot\sigma_s^2}} \right]}{\sum_{n=-\frac{N-1}{2}}^{\frac{N-1}{2}} \sum_{m=-\frac{M-1}{2}}^{\frac{M-1}{2}} \left[\frac{x(i-n,j-m)}{2\pi\sigma_p\sigma_s} e^{-\frac{\tilde{x}(i-n,j-m)-\mu^2}{2\cdot\sigma_p^2}} e^{-\frac{(i-n)^2+(j-m)^2}{2\cdot\sigma_s^2}} \right]},$$

where N and M are sizes of mask; σ_s^2 is variance of spatial Gaussian;

$$\begin{cases} \mu = HL|\tilde{x}(i,j) \geq (HL+LL)/2 \\ \mu = LL| \text{otherwise} \end{cases},$$

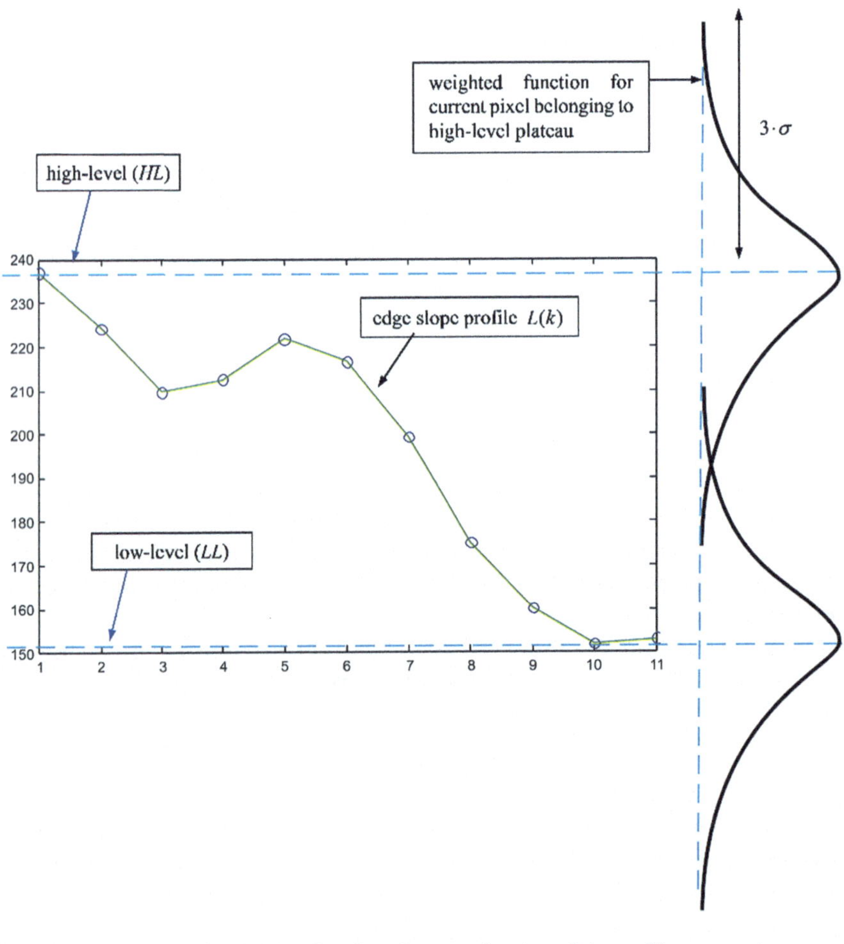

Fig. 6.13 Illustration of weighting functions for an estimation of the profile across edges

$$\sigma_p = \begin{cases} 10, & \text{if } \frac{HL-LL}{k_s} < 10 \\ 150, & \text{if } \frac{HL-LL}{k_s} > 150 \\ \frac{HL-LL}{k_s}, & \text{otherwise} \end{cases},$$

where coefficient k_s depends on scanning resolution, $k_s = 2$ for 600 dpi and $k_s = 2.5$ for 300 dpi.

Figure 6.13 shows an example of the edge profile with indicated HL and LL values and weighting functions.

6.5 Results

6.5.1 Approaches for Descreening Quality Evaluation

It is hard to estimate the quality of inverse halftoning solutions objectively. A majority of publications devoted to descreening demonstrate qualitative examples only. We consider four approaches for the evaluation of quality for inverse halftoning techniques:

- subjective assessment by conducting a survey about visual quality of processed images;
- elaboration of a test chart, and numerical estimation of specific regions of the chart;
- re-training of blind BRISQUE metric (Mittal et al. 2012) for assessment of descreened images;
- a simulation of scanned halftoned images and application of reference structural similarity index (SSIM) metric (Wang et al. 2004).

The following existing solutions for inverse halftoning are evaluated: *SilverFast HDR Studio* in automatic mode, *Descreen 4.0* plug-in for Adobe PhotoShop with default settings, *AKVIS Noise Buster* and our implementation of hardware friendly descreening (HFD) filter described by Siddiqui et al. (2008). The paper considers processing of grayscale images only. We processed R, G, and B channels by HFD filter independently. In general, the technique provides a good descreening outcome with a small colour distortion. Figures 6.14 and 6.15 illustrate the enlarged examples of processing results of suggested in *Adaptive* and *Bilateral descreening* approaches in comparison with enumerated solutions.

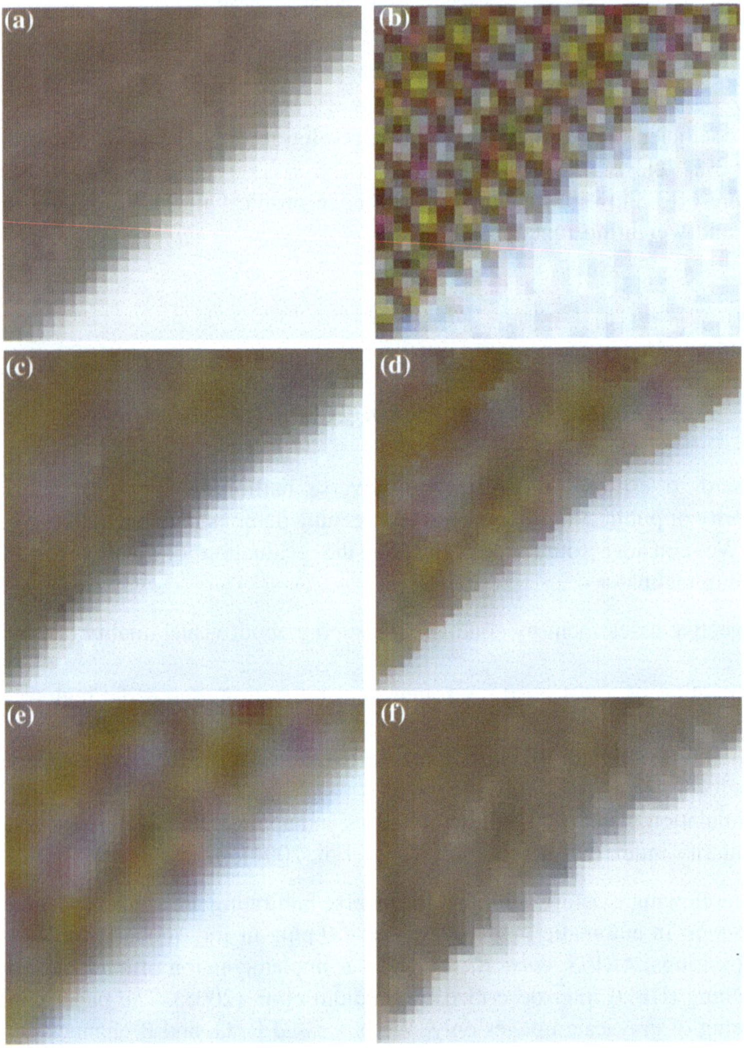

Fig. 6.14 Illustration of processing: **a** original continuous-tone image, **b** halftoned image, scanned with 600 dpi resolution, **c** result of adaptive descreening, **d** outcome of bilateral descreening, **e** *Descreen 4.0* utility, **f** *AKVIS Noise Buster*

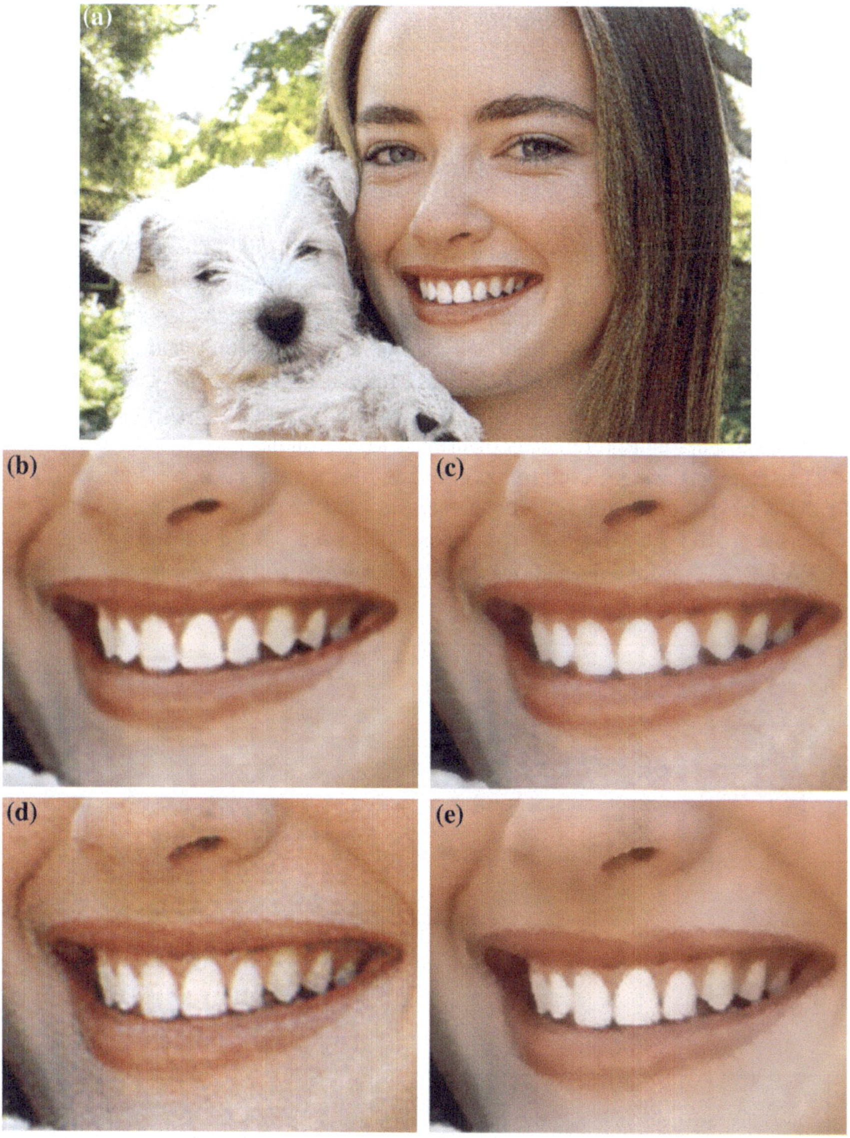

Fig. 6.15 Illustration of descreening: **a** original halftoned image, scanned with 600 dpi resolution, **b** result of adaptive descreening, **c** outcome of bilateral descreening, **d** *Descreen 4.0* utility, **e** *AKVIS Noise Buster*

6.5.2 Subjective Assessments

We conducted a survey to collect user opinions for quality evaluation of processing of real-world scanned images. Pages of magazines were scanned with 600 dpi resolution, halftone structure smoothing and then printed again by MFP Samsung CLX-6240. We demonstrated the resulting hardcopies with descreened images to 30 observers for quality evaluation. Test hardcopies contained three sets of 5 images, corresponding to 4 different descreening solutions that we mentioned above and the original scanned image without any inverse halftoning. First set contains outcomes of descreening solutions for photo with a red car (see Fig. 6.16, left panel); second set contains results for page combining text and pictures (see Fig. 6.17, left panel); third set contains outcomes for MFP advertisement (see Fig. 6.18, left panel). Every person was asked to rank 5 images in each set according to personal preferences of printed image quality.

The score was assigned the following way: for the first place, we added 5 points for the corresponding solution, for the second place 4 points, and so on.

In Figs. 6.16, 6.17 and 6.18 (right panels) we present the mean opinion score (MOS) for the considered solutions. Per the survey outcome, *Adaptive descreening*

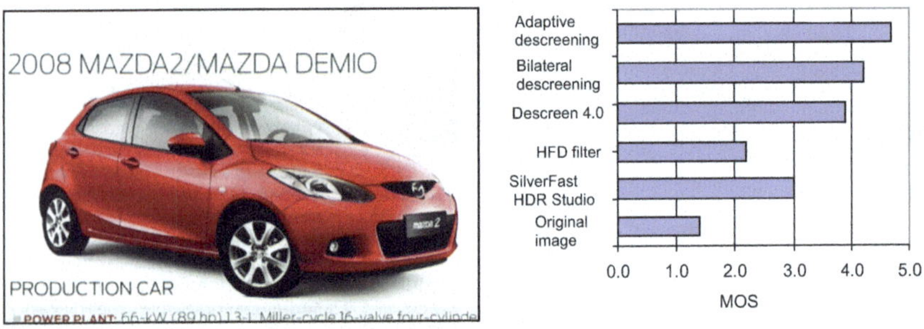

Fig. 6.16 Test image #1 and survey results for it

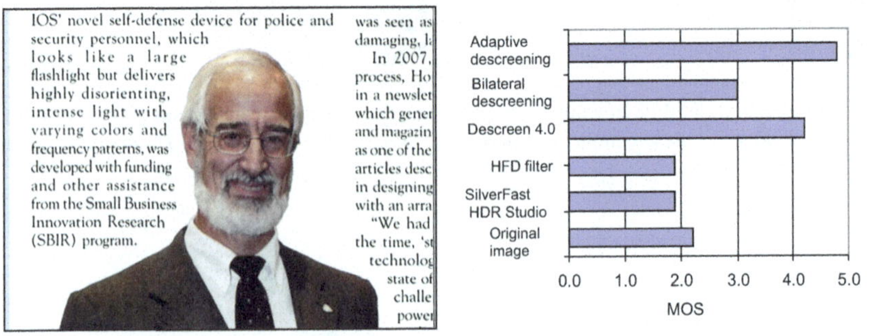

Fig. 6.17 Test image #2 and survey results for it

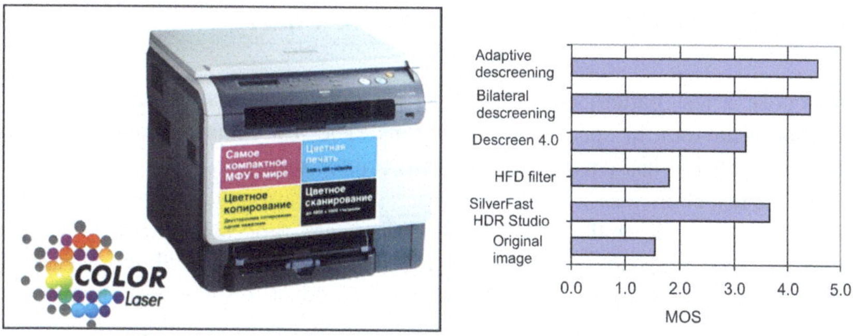

Fig. 6.18 Test image #3 and survey results for it

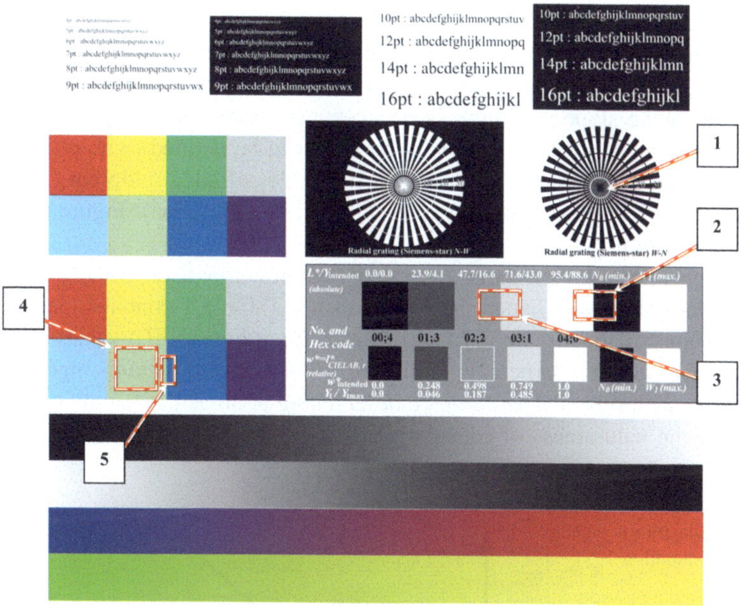

Fig. 6.19 Test chart

demonstrated the highest score results for all tested images among other analysed solutions. *Bilateral descreening* is less effective for highly textured areas like as hairs and small details, but provides excellent halftone pattern suppression and edges sharpening for images with graphics.

6.5.3 Numerical Estimations for a Test Chart

We propose a test pattern (see Fig. 6.19) for numerical analysis of descreening algorithm behaviour in specific local regions. The test image comprises several parts from ISO/IEC test chart. The chart was printed using high quality equipment. Further, the test chart was scanned and processed by various techniques for inverse halftoning. Grayscale and colour gradient bars are processed properly without cartoon effect or other artefacts resulting from all the tested algorithms. Areas with black text on white background and white text on black background are intended for evaluations of text preserving abilities during descreen procedures.

Our solutions provide better text preservation in comparison with competitive approaches especially for text size 10 pt and greater. Smaller text has visually significant degradations, but looks better than after processing by other descreening methods nevertheless.

To estimate sharpness, we suggest the analysis of the edge profile inclination across two colour patches (see regions 3 and 5 in the test chart, Fig. 6.19). We assume, the edge slope with high slope steepness corresponds to good sharpening techniques. Hence, the edge sharpness measure can be defined as an estimation of slope steepness for two nearest pixels on both sides of edge midpoint (*Edge Line* inclination) and for next pixels (*Next to Edge Line* inclination). Figure 6.20 illustrates the criterion.

The higher value of the *Edge Line* steepness in comparison with *Next to Edge Line* steepness indicates better sharpening result. Table 6.1 contains the measurement outcomes of *Edge Line* (EL) and *Next to Edge Line* (NEL) values. The table demonstrates strong sharpening ability of *Bilateral descreening* algorithms in comparison with others. The *adaptive descreening* approach is mainly oriented on preserving the naturalness of edges.

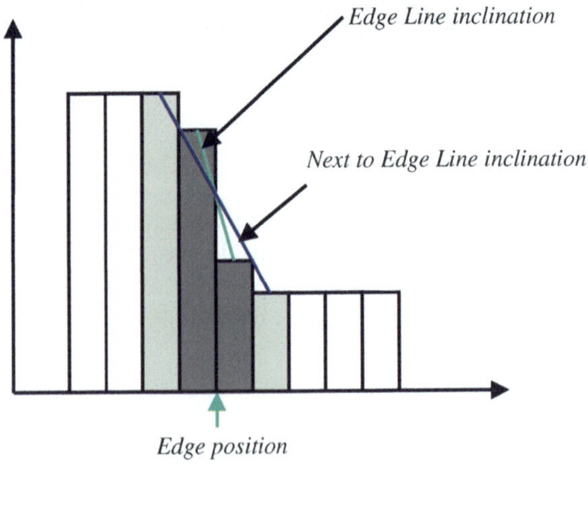

Fig. 6.20 Illustration to edge sharpness estimation

Table 6.1 Estimation of *Edge Line* (EL) and *Next to Edge Line* (NEL) values (region #5, Fig. 6.19)

Scan resolution		Input image		Descreen 4.0		AKVIS		HFD filter		Adaptive descreen		Bilateral descreen	
		EL	NEL	EL	NEL	EL	NEL	EL	NEL	EL	NEL	EL	NEL
600 dpi	*R*	51	46	28	37	38	41	29	38	33	43.5	56	38.5
	G	45	54	27	36	38	41	29	37.5	39	43.5	53	38
	B	9	10	2	1.5	14	2	3	4	21	18.5	3	1.5
300 dpi	*R*	71	41	33	42	44	44	26	22	35	49	67	44.5
	G	58	38.5	32	40.5	44	44.5	24	23	34	49	67	44.5
	B	18	6	3	0.5	16	2	11	1	12	17	11	8

Table 6.2 Estimation of the smoothing ability for flat areas (region #4, Fig. 6.19)

Scan resolution		Input image	Descreen 4.0	AKVIS	HFD filter	Adaptive descreen	Bilateral descreen
600 dpi	*R*	16.8	5.42	6.18	2.84	3.82	3.99
	G	13.9	4.11	4.83	2.49	2.95	2.97
	B	12.7	4.6	4.41	2.3	2.73	3.15
300 dpi	*R*	13.7	3.98	5.19	2.34	2.78	2.91
	G	9.96	3.11	3.87	2.12	2.34	2.3
	B	10.0	3.61	4.6	2.31	2.78	2.7

Table 6.3 Estimation of ability for preserving of high frequencies (region #1, Fig. 6.19)

Item	SilverFast HDR	HFD filter	Descreen 4.0	Adaptive descreen	Bilateral descreen
Black star, *pixels*	94	49	**48**	53	59
White star, *pixels*	89	48	**46**	51	53

Bold is used for emphasizing of the best outcomes.

Next quality criterion was based on the calculation of standard deviations for R, G and B channels within colour square (region 4 in the test chart, Fig. 6.19). Smaller standard deviation means better ability for suppression of halftone structure in the descreening result. Table 6.2 contains standard deviations for analysed solutions. The HFD filter provides a smaller standard deviation. However, strong smoothing leads to losing of fine details.

To estimate the preservation of high frequency details, we used area 1 in Fig. 6.19. Quality metric is based on the measuring of diameter of a completely blurred region. Table 6.3 contains measurement results. Smaller values designate to better results. Descreen 4.0 demonstrates the best outcomes. Our algorithms demonstrate ability for good high-frequency details preservation.

6.5.4 Modified BRISQUE Metric

The main idea of BRISQUE (*blind/referenceless image spatial quality evaluator*) metric is based on the assumption that the distribution of spatial features for processed images should correlate with distributions of the features in natural images. Initially, BRISQUE was trained on images from the LIVE dataset (Wang et al. 2004). The dataset contains 29 pristine images with 779 of their copies distorted by JPEG 2000, JPEG, white noise, Gaussian blur and a Rayleigh fast-fading channel simulation with various degrees of severity of the distortions. Each of the distorted images has an associated difference mean opinion score (DMOS) which represents the subjective quality of the image. Outcome of the training is a regression model constructed by Support Vector Machine (SVM). The regression model allows us to assign a quality metric for an image without a reference.

Since original BRISQUE metric was not trained for halftoned images, it cannot be used for quality estimation of inverse halftoning algorithms. To overcome this, we added to training set 58, scanned halftoned images processed by different descreening methods with various parameters. Totally, 232 images with corresponding subjective DMOS estimations were added to training set. After that, the regression model was re-trained. As a result, we have modified BRISQUE metric that is capable of estimating quality of halftoned and enhanced halftoned images.

Table 6.4 contains BRISQUE measures for *Adaptive* and *Bilateral descreening* algorithms in comparison with Gaussian blur and *Bilateral texture filter* (BTF) by Cho et al. (2014). The table demonstrates, *Bilateral descreening* outperforms other approaches according to BRISQUE metric.

Table 6.4 Quality assessment by re-trained BRISQUE

Image	Gaussian blurring	BTF	Adaptive descreening	Bilateral descreening
1	51.3	40.1	45.8	**19.9**
2	50.2	32.2	39.1	**16.3**
3	53.7	34.3	44.9	**19.3**
4	54.8	35.4	43.7	**26.7**
5	52.4	38.6	40.9	**17.5**

Bold is used for emphasizing of the best outcomes.

6.5.5 *Assessment via Structural Similarity Index*

The structural similarity index (SSIM) is a full reference metric that is the measurement of image quality. It is based on an initial distortion-free image as reference or etalon. SSIM was designed to improve on traditional reference metrics such as peak signal-to-noise ratio (PSNR) and mean squared error (MSE). There are plenty publications which demonstrate SSIM has a good correlation with subjective visual quality evaluation. However, for a lot of real-world distortions obtaining of pair from etalon and distorted images is a challenging task. In our case, problem is: how to simulate scanned halftoned colour image from initial contone image?

We discovered a way for the simulation of colour scanned halftoned images by means of Adobe Photoshop. The approach includes seven steps.

1. Upsizing of initial contone image 8 times and then bilinear interpolation is applied.
2. Applying of *Pixelate/Color Halftone* filter with radius 20.

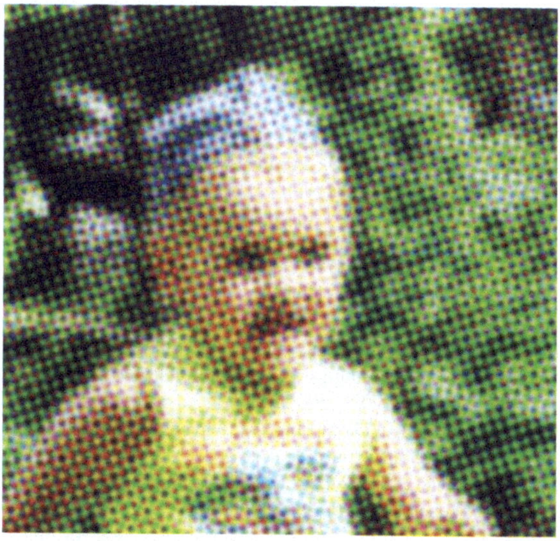

Fig. 6.21 A fragment of simulated scanned halftoned image

Table 6.5 Quality assessment by SSIM

Image	Gaussian blurring	BTF	Adaptive descreen	Bilateral descreen
1	0.51	0.48	**0.67**	0.52
2	0.45	0.44	**0.62**	0.47
3	0.47	0.46	**0.60**	0.49
4	0.54	0.49	**0.71**	0.55
5	0.37	0.35	**0.50**	0.38

Bold is used for emphasizing of the best outcomes.

3. Gaussian blurring with radius 20. It simulates of toner blending in printing process.
4. The addition of 7% of noise simulates noise in a printing process.
5. Downsizing of an image 8 times where bilinear interpolation is used.
6. Gaussian blurring with a radius of 0.4 helps to simulate low-pass filtering in scanning process.
7. Addition of 3% of noise. It simulates noise in the scanning process.

Figure 6.21 demonstrates a fragment of simulated scanned halftoned images. One can compare the image with examples of real halftoned scans in Fig. 6.2. Outcomes of simulation look like the real halftoned images, which are scanned with resolution 300 dpi.

Given simulation technique allows to create pairs from etalon images and simulated scanned halftoned images. It opens door for application of SSIM for quality assessment of descreening methods. Table 6.5 contains SSIM measures for *Adaptive* and *Bilateral descreening* algorithms in comparison with Gaussian blur and *Bilateral texture filter*. The table demonstrates, *Adaptive descreening* outperforms other approaches according to SSIM metric.

References

Aizenberg, I., Butakoff, C.: Frequency domain median-like filter for periodic and quasi-periodic noise removal. In: Proceedings of IS&T/SPIE Electronic Imaging, vol. 4667 (2002)

Buades, A., Coll, B., Morel, J.M.: A non local algorithm for image denoising. In: Proceedings IEEE International Conference on Computer Vision and Pattern Recognition, vol. 2, pp. 60–65 (2005)

Cho, H., Lee, H., Kang, H., Lee, S.: Bilateral texture filtering. ACM Trans. Graph. **33**(4), 128 (2014)

Kharlamov, A., Podlozhnyuk, V.: Image Denoising. Tech. Rep. NVIDIA, Inc. (2007)

Kuo, C., Rao, A.R., Thompson, G.: Color halftone descreening based on color sigma filters. Proc. SPIE **4300**, 377–385 (2001a)

Kuo, C., Rao, A.R., Thompson, G.: Wavelet based halftone segmentation and descreening filter design. IEEE Int. Conf. Acoust. Speech Signal Process. **3**, 1573–1576 (2001b)

Kuo, C., Tewfik, A.H., Rao, A.R.: Colour halftone document segmentation and descreening. IEEE Image Process. **2**, 1065–1068 (2001c)

Luo, J., de Queiroz, R., Fan, Z.: A robust technique for image descreening based on the wavelet transform. IEEE Trans. Signal Process. **46**(4), 1179–1184 (1998)

Mittal, A., Moorthy, A.K., Bovik, A.C.: No-reference image quality assessment in the spatial domain. IEEE Trans. Image Process. **21**(2), 4695–4708 (2012)

Nishida, H.: Adaptive inverse halftoning for scanned document images through multiresolution and multiscale analysis. Proc. SPIE **5296**, 192–203 (2004)

Shou, Y.W., Lin, C.T.: Image descreening by GA-CNN-Based texture classification. IEEE Trans. Circuits Syst. **51**(11), 2287–2299 (2004)

Siddiqui, H., Bouman, C.: Training-based algorithm for moire suppression in scanned halftone images. In: Proceedings of IS&T/SPIE Electronic Imaging, vol. 6498 (2007a)

Siddiqui, H., Bouman, C.: Training-based descreening. In: IEEE Transactions on Image Processing, vol.16, pp.789–802 (2007b)

Siddiqui, H., Boutin, M., Bouman, C.A.: Hardware-friendly descreening. IEEE Int. Conf. Image Process. 1740–1743 (2008)

Tomasi, C., Manduchi, R.: Bilateral filtering for grey and colour images. In: Proceedings Sixth IEEE International Conference on Computer Vision, pp. 839–846 (1998)

Wang, Z., Bovik, A.C., Sheikh, H.R., Simoncell, E.P.: Image quality assessment: from error visibility to structural similarity. IEEE Trans. Image Process. **13**(4), 600–612 (2004)

Chapter 7
Automatic Red Eye Correction

7.1 Introduction

The red eye artefact is an annoying shortcoming of photographs. Many consumer photos are affected by the red eye effect, which frequently arises when shooting with a flash. This defect considerably worsens the impression made by a photo. The phenomenon is caused partly by a small angle between the flash and the lens of the camera. This angle has decreased due to the miniaturization of digital cameras. Some instruments have been made for mitigation of the phenomenon. For example, the use of one or several pre-flashes can reduce the size of a human's pupils. Despite such efforts, red eye remains a worrisome defect in consumer photography. Figure 7.1 demonstrates the typical red eye effect.

The severity of the red eye effect varies for different races. Keelan (2002) carried out an in-depth investigation of the origin of red eye and factors affecting its severity. When light impinges on the human eye at an angle not too far from the optical axis, it may propagate through the pupil, be reflected back from the fundus, and exit the pupil at approximately the same angle it entered. The fine blood vessels is the cause the reflected light to be red in colour. The human pupil contains the pigment melanin, which gives it a dark colour. The amount of melanin in the pupil correlates with skin and hair pigmentation, so that darker-skinned and darker-haired individuals typically have a high melanin concentration in their pupils. Melanin attenuates the light propagating through the pupil and a higher melanin content leads to a decrease of the severity of the red eye effect. Although there are some variations of pigmentation within nationalities, the red eye effect is less prominent in races with a darker complexion on average. However, for the Caucasian race, the red eye effect is a large and troublesome problem. Table 7.1 contains the outcomes of experiments for different demographic groups in identical shooting conditions.

At the present time, the general trend is toward image enhancement in the fully automatic regime. Automatic red eye correction is an essential feature for modern photo printers. An ideal algorithm should have a high detection and correction

I.V. Safonov et al., *Adaptive Image Processing Algorithms for Printing*,
Signals and Communication Technology,
https://doi.org/10.1007/978-981-10-6931-4_7

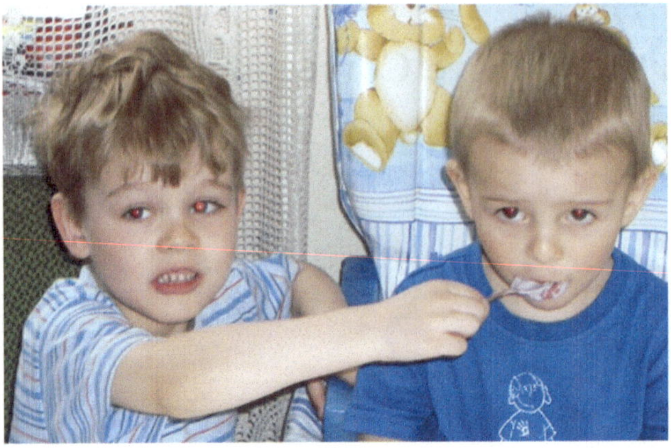

Fig. 7.1 Typical red eye effect

Table 7.1 Percentage occurrence of visually detectable red eye for different demographic groups

Group	Detectable red eyes, %
African-American	9
Asian	15
Latin American	41
Caucasian adult	70
Caucasian youth	82

quality, should operate independently from face orientation, and should have a relatively low computational complexity and memory consumption. Satisfying all these requirements is a great challenge.

7.2 Analysis of Photos with Red Eyes

To investigate the issues involved in red eye detection, we have collected and analysed about 1000 photos. Different photographers using various models of digital still cameras and smartphones captured the images. Several dozen scanned photos were included in our dataset too. The sizes of the images vary from 3 to 15 Mpix. The majority of the photos were stored in JPEG file format with EXIF information.

First, we have marked the positions of red eyes on the images from the dataset. Further, we randomly divided the collection into a training set containing about 950 photos and a test set of 50 photos. The total number of red eyes in the training set is 2738: 1304 paired and 130 single eyes.

By analysing photos, we selected the following main issues in the creation of a detection algorithm.

- The colour tone of red eyes varies significantly: from yellow and orange to magenta and violet.
- The colours of red eyes and skin tone distributions partially overlap.
- Face orientations vary widely.
- Occasionally only part of a face is visible.
- Sometimes only one eye from pair is visible or has colour distortion.

We picked 5000 pieces of skin tones and 5000 samples of colours inside red eye regions. Figure 7.2 shows the distributions of skin tones and the red eyes tints in the a*b* plane of L*a*b* colour space. The colours of the red eyes occupy a wide range. There is a significant overlap between eye tints and skin tones. A mixture of glints, which are specular reflections of flash in the pupil area with a red colour from the eye retina, is one of the causes of the intersection of the distributions. It is worth noticing that the overlay of those distributions in the brightness channel is even more significant in comparison with the overlap in chromaticity channels.

At the present time, many automatic approaches involve detecting faces in a frame and subsequently detecting red eyes within face areas. However, many face detection algorithms are able to detect faces that are oriented in an upright frontal view only. The approaches are unable to detect faces that are rotated in-plane or out-of-plane with respect to the image plane and also cannot detect faces in cases where only part of the face is visible. We counted the percentage of face orientations on photos from our dataset. One can see our outcomes in Table 7.2. There are too many cases in which a preliminary face detection stage does not help to detect red eyes.

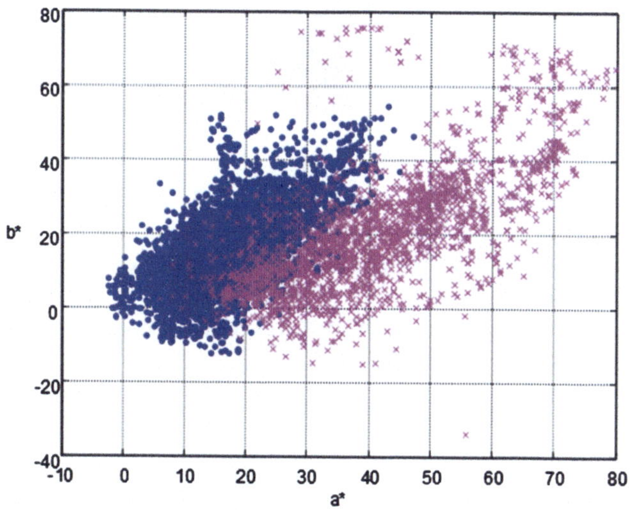

Fig. 7.2 Projection of distributions of red eye tints (magenta crosses) and skin tones (blue points) onto a*b* plane of L*a*b* colour space

Table 7.2 Percentages of various face orientations in the training set

Orientation	Example	%
Upright frontal	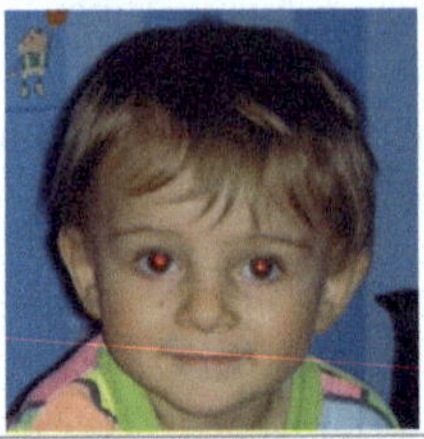	72
Rotated in-plane	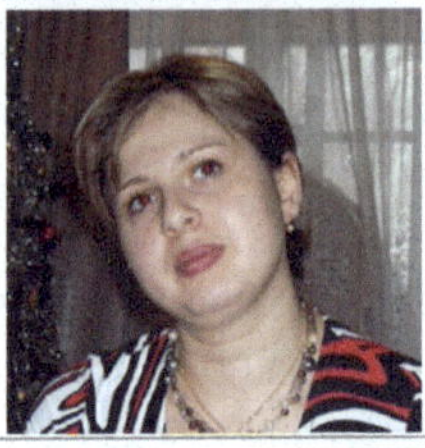	17
Rotated out-of-plane	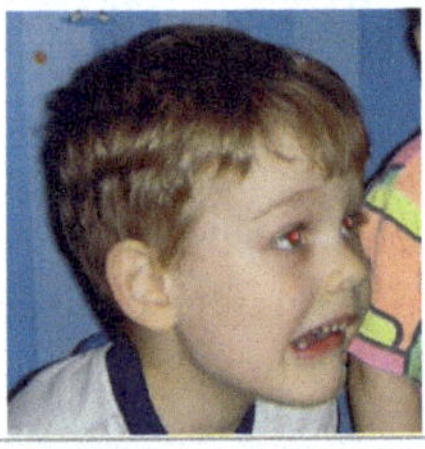	6
Only part of the face is visible	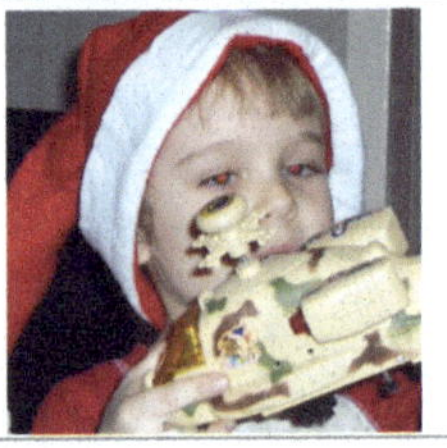	5

We separated red eyes into two groups: relatively large and well distinguishable regions and small areas with low local contrast. We named the first group HQ and the second group LQ. In the training set, 72% of red eyes belonged to the HQ group. It is obvious that the severity differs between HQ and LQ cases. It is necessary to correct the majority of HQ red eyes, whereas correction of LQ red eyes is a desirable action.

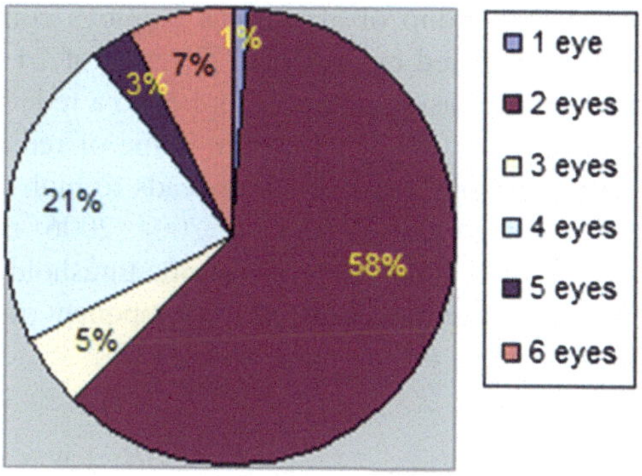

Fig. 7.3 Percentages of numbers of red eyes in a photo

In addition, we counted the number of red eyes in each photo. Figure 7.3 demonstrates a chart of the number of red eyes in a photo. Less than 3% of photos contain more than six red eyes.

7.3 Existing Techniques for Red Eye Correction

There are many papers and patents devoted to semi-automatic and automatic red eye correction. Existing approaches include two phases: red eye area detection and correction/retouching. The detection phase combines stages from the following list: human face and/or eye detection, segmentation and labelling of red regions, glint detection, calculation of features and classification of regions, and pair verification. For the characterization of red eye detection errors, we enter the number of False Positives (FPs) and False Negatives (FNs), where FN is the situation when a red eye region is not detected, and FP is the situation when a non-red-eye region is detected falsely.

The segmentation of red regions is usually done by means of thresholding in some colour space: YCbCr (Patti et al. 1998; Ioffe 2003), Lab (Corcoran et al. 2005), or RGB (Zhang et al. 2004). A technique described by Luo et al. (2008) provides a slightly better result. The colours that are typical for red eyes are bounded by curves in a*b* planes of L*a*b* colour space. The parameters of the curves depend on the brightness (L*) value. One of the disadvantages of the approach is the expensive conversion to L*a*b* colour space.

Often, a special *redness* factor is calculated for pixels of an image. Then segmentation is performed in a redness channel (Hardeberg 2001; Schettini et al. 2004; Gasparini and Schettini 2005). Wu (2010) describes the usage of two different channels of redness to increase the number of detected red eyes. However, due to the reasons mentioned above, in many cases it is impossible to discriminate different red eye pixels and skin pixels using thresholds in any colour or redness channel.

Another group of algorithms employs correlation or matching filters for segmentation of red eye regions. A series of 24 symmetrical annular edge detection filters of increasing radius is applied to a redness image in Czubin et al. (2002) and Smolka et al. (2003). However, some of red eye regions have non-round shapes, and the use of 24 convolutions leads to high computational complexity.

Luo's algorithm (Luo et al. 2004, 2006) demonstrates a combination of these two common approaches: usage of a threshold and a matching filter for the redness channel, where redness is the a* component of L*a*b* colour space. The following filter produces a final redness score as:

$$RS = AR_1 + w \times (\frac{AR_1}{AR_2})^n,$$

where AR_1 and AR_2 are the average values of redness of the inner and outer squares, respectively, according to Fig. 7.4, w is a weight factor, and n is a predefined constant. The first term designates an absolute redness value and the second term represents the redness contrast between the inner square and surrounding margin. Application of this filter to the redness channel followed by segmentation by a threshold produces a binary image. To detect red eyes of different sizes, filters with multiple kernel sizes are used. The outcome is a combination of the filter outputs by logical OR.

Obviously, there are many false positives after the segmentation of red areas. There are several common approaches to decrease the number of FPs. Zhang et al. (2004) describe an application of glint detection. Their method examines only red regions surrounding bright pixels. However, about 10–15% of red eyes from our dataset have no glint. Another popular approach is the application of face detection before segmentation (Held 2002; Gaubatz and Ulichney 2002). The detection of oval skin tone areas or modifications of Viola and Jones's (2001) algorithm, for example the one by Egorova et al. (2009), are frequently applied. Eye outline detection is used sometimes jointly with face detection (Ioffe 2003) and sometimes separately (Luo et al. 2006). Face and eye detection algorithms are orientation dependent and their applicability is limited as mentioned above. Thus, those approaches lead to an increase in the number of FNs for collections of photos with red eyes in arbitrary orientation.

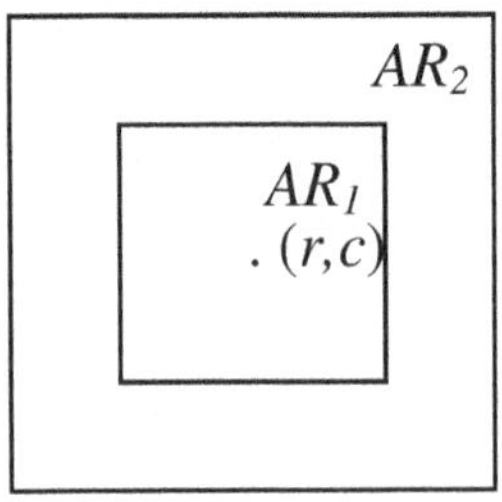

Fig. 7.4 A concentric template of a redness filter in Luo's approach

Corcoran et al. (2005) describe the main characteristics of red eye regions. In some way, all the red eye detection algorithms employ characteristics from the following list:

- a reddish tone;
- a round shape;
- an area that is relatively small with respect to the whole image area;
- compactness;
- location in the vicinity of regions with skin tones;
- the presence of whitish pixels due to sclera of an eye in close proximity;
- a relatively high local contrast near the region.

The careful tuning of parameters has made it possible to create a decision tree for effective rejection of FPs. The general trend is to use more sophisticated machine learning techniques instead of a decision tree. Many red eye correction methods employ the adaptive boosting (AdaBoost) algorithm because it has a high performance and a compact model size (Luo et al. 2008). Boosting (Freund and Schapire 1996; Schapire and Singer 1999) is a family of machine learning algorithms that produce a classifier which is a weighted combination of simple classifiers; as a rule, it is an ensemble of decision trees. In addition, algorithms from the AdaBoost family have the ability to select relevant and more informative features during training. Luo et al. (2004) describe the selection of 25 final features from an initial set containing 178 features by means of the AdaBoost training procedure.

The generalization capability of any learning algorithm depends on the quality of a training set. HP (Ulichney et al. 2003) offered an automatic red eye correction algorithm as a free web service: www.redbot.net. The aims of the researchers were to collect many typical "real-world" photos damaged by the red eye effect and to get feedback from users about correction performance and quality. This is actually a good way of fine-tuning an algorithm. Unfortunately, the service is no longer available.

One more method used to eliminate FPs is pair verification for eyes (Schildkraut and Grey 2002). Moreover, a situation in which only one eye from pair is corrected is unwanted, especially for embedded applications, where it is hard to implement interactive editing. However about 5% of photos contain single red eyes, which happens, for instance, when a face is partially obscured by other objects. Eye pair verification is a useful procedure but single eyes should not be ignored.

A reasonable way of avoiding the need to process photos taken without a flash is by analysis of an EXIF Flash tag (Chen et al. 2005; Oberhardt et al. 2003). It is certainly useful to decrease the processing time and reduce the number of FPs.

The general approach to correction consists of desaturation and darkening of pixels in the red eye area. Many solutions apply various modifications of a seeded region-growing algorithm for merging adjacent reddish pixels. To improve the visual quality, a mask from the segmented pixels is smoothed and used as the alpha channel for blending the initial and desaturated pixels (Gasparini and Schettini 2005; Luo et al. 2004).

Unfortunately, there are no commonly adopted performance metrics or publicly available datasets containing photos with red eyes. Usually, authors characterized their own solutions via the number of FPs and FNs (Volken et al. 2006). Some papers discuss the various types of detection errors and their severity (Willamowski and Csurka 2006). There are no publications devoted to a deep analysis of correction and retouching quality in the case of the red eye effect.

7.4 Red Eye Detection

7.4.1 Segmentation

Taking into account the outcome of our analysis, we propose to use the colour and edge of red regions jointly for the segmentation of red eyes. In contrast to other red eye detection methods, the proposed method does not use thresholding in any colour space or in the redness channel but uses a set of colours typical for red eyes and human skin defined by two 3D tables. Notice that these sets can intersect with each other. Such tables can be constructed for any colour space and can circumscribe sets of an arbitrary form. We carry out the processing of images in YCbCr colour space because the majority of photos are stored in JPEG file format, which utilizes that colour system. It makes it possible to avoid extra conversions between colour spaces. Nevertheless, we can adapt the proposed approach for any colour space. In addition, sets of colours typical for red eyes and human skin may be defined by a set of analytic functions. Gomez and Morales (2002) demonstrate an example of generation of such functions for detection of skin tone pixels.

The mentioned 3D tables contain *typicalness levels* that characterize a given colour as the colour of human skin and red eyes correspondingly. The typicalness level is an integer value ranging from 0 to 3, where 3 means that a given colour is very typical for human skin or red eyes and 0 means that the given colour is not typical for human skin or red eyes. To set a typicalness level, we analysed the distribution of 5000 manually labelled skin pixels and 5000 red eye pixels. In addition, we took into account the theoretically possible values of skin and red eye tints. In a certain sense, the typicalness level is similar to posterior probabilities but we have made dozens of changes manually in the initial stage and automatically in the final stage by maximizing the detection quality criterion for photos from our training set.

For each pixel of the image, the special colour mark is set if the typicalness level for skin is equal to 0 and the typicalness level for red eye is greater than 1. White points in Fig. 7.5 correspond to *colour marks* for the photo from Fig. 7.1.

In contrast to other red eye detection methods, the proposed method does not use symmetrical matching or correlation filters but employs four directional edge-detection filters. It allows the detection of regions of red eyes of different sizes

Fig. 7.5 *Colour marks* for photo from Fig. 7.1

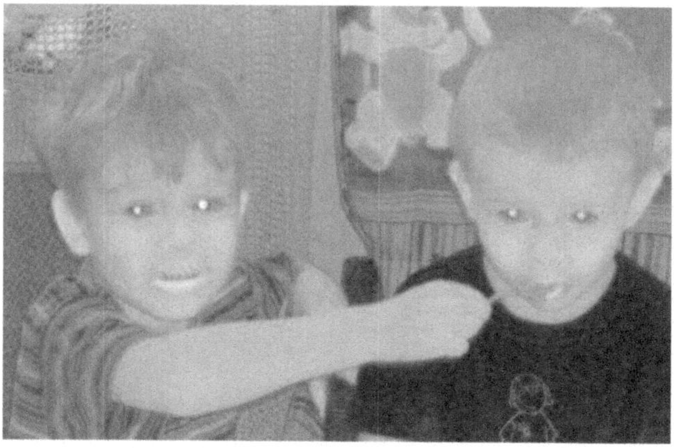

Fig. 7.6 Redness of photo from Fig. 7.1

without the application of several filters with kernels of different sizes and makes it possible to reveal regions of red eyes of any form, not just round.

Course gradient filters, such as those described by Pratt (2001), can be applied as directional edge-detection filters. However, we discovered that filters based on a ratio instead of a difference are more stable to noise and compression artefacts. We carry out filtration of a redness channel. Many of our experiments showed that a good estimation of redness is a linear combination of brightness and intensity of the red channel. For YCbCr colour space, a reasonable evaluation of redness is Y + 5Cr. Figure 7.6 demonstrates the redness for the photo from Fig. 7.1. We propose the following four-directional edge-detection filters (see Fig. 7.7):

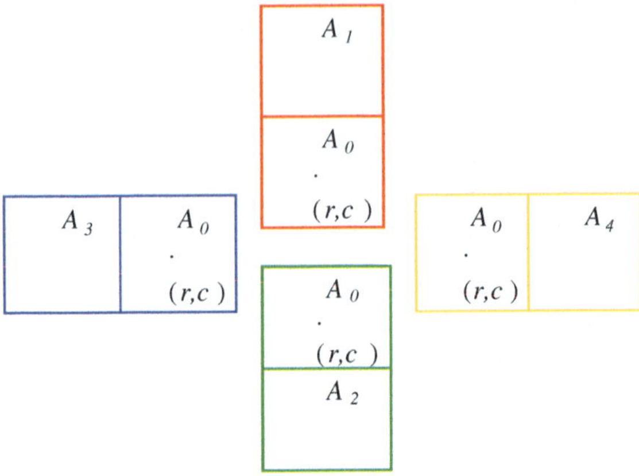

Fig. 7.7 Four directional edge detection filters

$$A_0(r,c) = \sum_{i=r-1}^{r+1} \sum_{j=c-1}^{c+1} (\mathrm{Y}(i,j) + 5 \times \mathrm{Cr}(i,j)),$$

$$A_1(r,c) = \sum_{i=r-4}^{r-2} \sum_{j=c-1}^{c+1} (\mathrm{Y}(i,j) + 5 \times \mathrm{Cr}(i,j)),$$

$$A_2(r,c) = \sum_{i=r+2}^{r+4} \sum_{j=c-1}^{c+1} (\mathrm{Y}(i,j) + 5 \times \mathrm{Cr}(i,j)),$$

$$A_3(r,c) = \sum_{i=r-1}^{r+1} \sum_{j=c-4}^{c-2} (\mathrm{Y}(i,j) + 5 \times \mathrm{Cr}(i,j)),$$

$$A_4(r,c) = \sum_{i=r-1}^{r+1} \sum_{j=c+2}^{c+4} (\mathrm{Y}(i,j) + 5 \times \mathrm{Cr}(i,j)),$$

$$E_1(r,c) = \frac{A_0(r,c)}{1 + A_1(r,c)},$$

$$E_2(r,c) = \frac{A_0(r,c)}{1 + A_2(r,c)},$$

$$E_3(r,c) = \frac{A_0(r,c)}{1 + A_3(r,c)},$$

$$E_4(r,c) = \frac{A_0(r,c)}{1 + A_4(r,c)},$$

where filtration operates for all pixels of an image, r is the row, c is the column, and Y and Cr are colour channels of an image in YCbCr colour space. We add one to the denominator to avoid singularity.

The computational complexity of the given approach is close to five convolutions with a 3 by 3 kernel. We assign a special *edge mark* for an image pixel if the maximal result of filtrations by $E_1(r, c)$, $E_2(r, c)$, $E_3(r, c)$, or $E_4(r, c)$ at the pixel is greater than a threshold value T. In addition, we keep the index 1, 2, 3, or 4 of the filter, which produces the maximal response. Threshold T is not constant for all pixels of the image; it increases when the colour of the image pixel is typical for human skin colour and decreases when the colour of the pixel is typical for red eye. We selected filter coefficients and thresholds by means of maximization of the detection quality criterion for photos from our training set.

The method allows detection of pixels of red eye whose colour is close to the colour of skin tones and pixels of so-called "gold eyes". The described approach also makes it possible to calculate informative features for the subsequent classification of regions. Figure 7.8 demonstrates *edge marks* for photos from Fig. 7.1. The pixels marked by E_1 are coloured in red, those marked by E_2 are coloured in green, those marked by E_3 are coloured in blue, and those marked by E_4 are coloured in yellow. Notice that regions of red eyes generate the specific template, which differs from the other marked zones.

We combine the outcomes of these two marking stages, that is, *colour* and *edge marks*, by an OR logical operation. After that, we carry out labelling of connected regions. The proposed segmentation approach provides excellent detection results. For our training set, more than 99.9% of HQ and about 97.5% of LQ red eyes are detected. The average number of FPs per image is 1897. This huge number of FPs is reduced drastically in the classification stage.

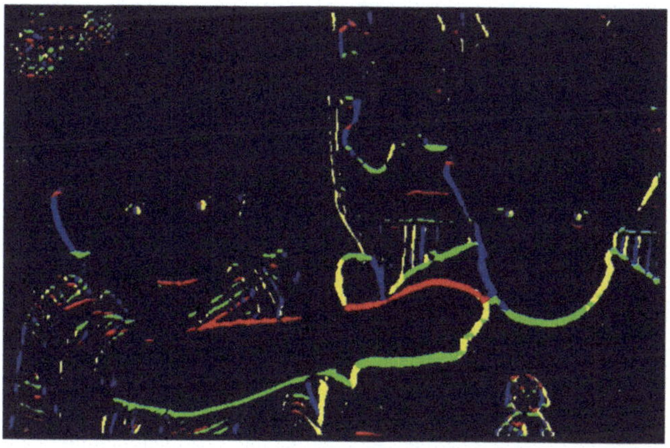

Fig. 7.8 *Edge marks* for photo from Fig. 7.1

7.4.2 *Classification*

The aim of the classification stage is to eliminate FP. The application of adaptive boosting for both feature selection and classification is valuable. AdaBoost is a family of learning algorithms for producing a classifier committee that is a weighted combination of weak classifiers. The weak classifier is any classifier that is able to achieve a slightly better error rate than random guessing. AdaBoost has three main advantages: good generalization capability, low computational complexity of the final ensemble, and simple implementation in software. Nevertheless, sometimes the method can lead to overfitting if the training procedure runs too many iterations, especially in the case of a noisy training set. Figure 7.9 illustrates the general scheme operation of the AdaBoost classifier committee.

There are several versions of AdaBoost algorithms which differ in terms of the approach used for the optimization of weights w_i. Some realizations of these algorithms offer the possibility of adjusting parameters of simple weak learners and, in particular, optimizing thresholds. We used the GML AdaBoost Matlab toolbox (Vezhnevets and Vezhnevets 2005) for feature selection, building the classifier ensemble, and adjusting the parameters of weak learners. The toolbox contains an implementation of the Real and Gentle AdaBoost algorithms. Real AdaBoost is a generalization of the basic AdaBoost approach introduced by Fruend and Schapire (1996). Gentle AdaBoost (Friedman et al. 2000) is a more robust and stable version of real AdaBoost. As a rule, Gentle performs identically to Real AdaBoost on regular data but is considerably better on noisy data and much more resistant to outliers.

GML AdaBoost toolbox supports weak learners as Classification and Regression Trees (CART). CART is a tree with leaves representing the classification result and nodes representing some predicate. Branches of the tree are marked *true* or *false*. The classification process in the case of a decision tree is a process of tree traversal.

For effective learning, it is essential to use balanced sets of positive and negative samples, that is, sets of approximately equal sizes. However, we have a set of

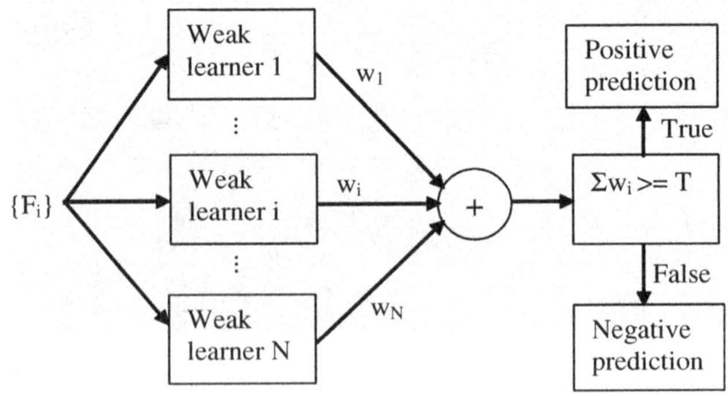

Fig. 7.9 The scheme of the AdaBoost ensemble

negative samples, that is, false regions, that is several times larger than the set of positive samples, that is, red eye regions. A convenient way of reducing the number of negative samples is to apply a cascade of classifiers where the first classifiers of the cascade have to eliminate the majority of false regions. The basic idea is to construct a cascade that consists of trained classifiers in a cascade layer. Each layer produces a binary decision, that is, to accept or reject a given region as a candidate for a red eye region. If the current layer rejects a sample, then the entire cascade rejects it and the classification procedure stops. If the sample is accepted then it is passed to the next layer of the cascade. Cascades are built hierarchically; each subsequent layer is more accurate than its predecessor.

Figure 7.10 demonstrates our classification cascade, where the first three classifiers have to eliminate a huge number of FPs. Let us consider the classifiers of the cascade. We calculate a set of features for every connected region: $(r1, c1)$ is the coordinate of the left-top corner of the bounding box, $(r2, c2)$ is the coordinate of the right-bottom corner; and $p(r, c)$ is an indicator function that is equal to 1 if the pixel with coordinate (r, c) belongs to the connected region and equal to 0 otherwise. The following features are calculated (Safonov 2007).

- N, the total number or area of pixels of a connected region.
- N_r, the number of pixels that have colour marks.
- N_1, N_2, N_3, and N_4, the numbers of pixels of a connected region marked by the corresponding directional edge-detection filters.

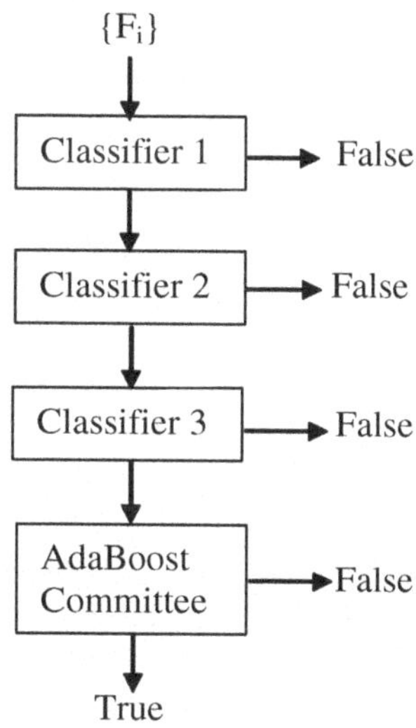

Fig. 7.10 A cascade of classifiers

- The dimensions of the bounding box:

$$L = c2 - c1,$$

$$W = r2 - r1.$$

- Compactness:

$$Kc = \frac{N}{L \times W}.$$

- Elongation 1:

$$Ke = \frac{\min(L, W)}{\max(L, W)}.$$

- The number of directional filters, whose indexes are stored in arrays of marks for a given connected region:

$$M = \operatorname{sign}(N_1) + \operatorname{sign}(N_2) + \operatorname{sign}(N_3) + \operatorname{sign}(N_4).$$

- The ratio between the maxima among *N1*, *N2*, *N3*, and *N4* and the total number of pixels in a connected region:

$$Ka = \frac{\max(N_1, N_2, N_3, N_4)}{N}.$$

The first elementary classifier of the cascade is the following condition: if $N_r = 0$ OR $M < 2$, then the region is rejected. That rule decreases the number of FPs significantly; after application of the classifier, the average number of FPs per image drops to 60 (Fig. 7.11). The second classifier is a decision tree with thresholds for *M*, *Ke*, *Kc*, and *Ka*. After application of the classifier, the average number of FPs per image drops to 39. The third classifier employs rules based on the anthropomorphic ratio between the face and pupil sizes of a human as well as

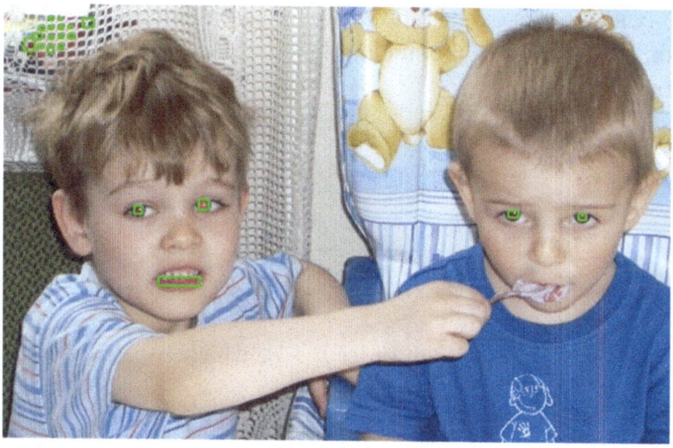

Fig. 7.11 Regions that survived after the first classifier of the cascade was applied to the photo from Fig. 7.1

the distance between eyes. The number of FNs increases insignificantly after application of these three classifiers .

Initially we calculated 80 features for each surviving region. The features describe the size and shape of a region, the distribution of colour of a region and its vicinity, local contrast, the percentage of skin tones in the surroundings, the percentage of tones that are non-typical for human faces, the similarity of a region to its vicinity, the presence of whitish and highlighted pixels, the percentage of edges in the surroundings, and the relationships between sub-regions marked by directional filters and marked via 3D colour tables. We selected 48 features that are applicable for weak learners by means of AdaBoost's ability to select effective and informative features. The computational complexity and memory consumption for the calculation of various features differ significantly. Depending on the hardware platform, we utilize different sets from those features for building the final ensemble in the cascade. The most important features besides to those enumerated above are the following.

- The number of pixels in the vicinity of a connected region with a colour like a human skin tone, Ks.
- The number of pixels with colours typical for glint inside a bounding box of the given connected region, Kg.
- The ratio between the number of pixels marked with a colour mark and the total number of pixels in a connected region, Kr:

$$Kr = \frac{Nr}{N}.$$

- The local contrast, that is, the ratio between the differences and sum of maximal and minimal intensities in a bounding box.
- Elongation 2:

$$Ke2 = \sqrt{\frac{I_{20} + I_{02} + \sqrt{(I_{20} - I_{02})^2 + 4I_{11}^2}}{I_{20} + I_{02} - \sqrt{(I_{20} - I_{02})^2 + 4I_{11}^2}}},$$

where I_{pq} is a central moment with order $p + q$, which is calculated as:

$$I_{pq} = \sum_{i=r1}^{r2} \sum_{j=c1}^{c2} (i - r_m)^p (j - c_m)^q p(i,j).$$

- The average intensities of the colour channels *Y*, *Cb*, and *Cr* in the region.

We carried out many experiments with different ensembles and its parameters. To avoid overfitting, for model selection and parameter optimization, we used a special *Dual Leave-Group-of-Sources-Out* (LGOSO) *Cross-Validation* (CV) procedure (Safonov et al. 2015). Cavley and Nalbot (2010) demonstrate the drawbacks of parameter optimization in traditional k-fold cross-validation and claim that robust unbiased performance evaluation is likely to require more rigorous and computationally intensive protocols such as nested or double cross-validation. Regions of red eyes in photos taken by the same camera have similar characteristics. The use of photos originated by identical cameras in training and testing sets leads to overfitting. We combine images into groups corresponding to the camera name extracted from EXIF-information.

Figure 7.13 demonstrates the scheme of LGOSO CV. This procedure comprises at least three nested cycles. The innermost cycle is the common LGOSO CV for N – 1 groups, leaving out the *j*th group for testing. This cycle estimates the quality metrics *M* for the *i*th set of parameters. A middle-level cycle or cycles identify the parameters that maximize *M* over *i*. The optimal parameter set is then used for training on data of all groups without the *j*th group and tested on the *j*th data. So, the top-level cycle, as in common LGOSO CV, puts one group's data in the test set and uses the other data for learning and model optimization in the inner cycles. The quality metrics Qc (see Sect. 7.6) calculated in the top-level CV characterizes all stages of the recognition algorithm.

Several implementations taking into account the trade-off between quality of detection and correction, processing time, and memory consumption are possible. In general, Gentle AdaBoost, which utilizes CART with a tree depth of three, provides a better result: the accuracy is about 1%. Figure 7.12 demonstrates the Receiver Operator Characteristic (ROC) curve as the dependence of the number of

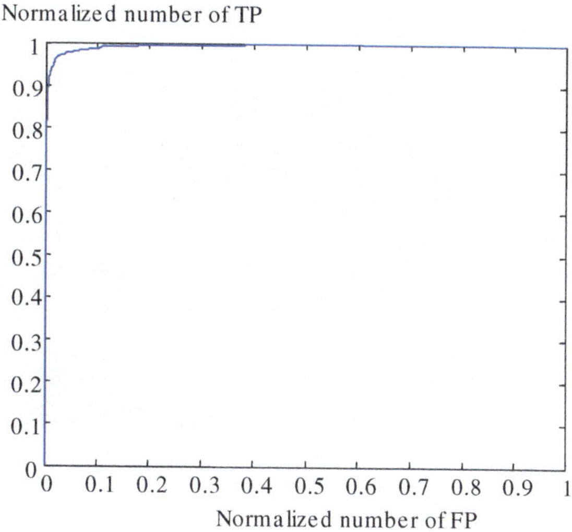

Fig. 7.12 ROC curve for AdaBoost ensemble

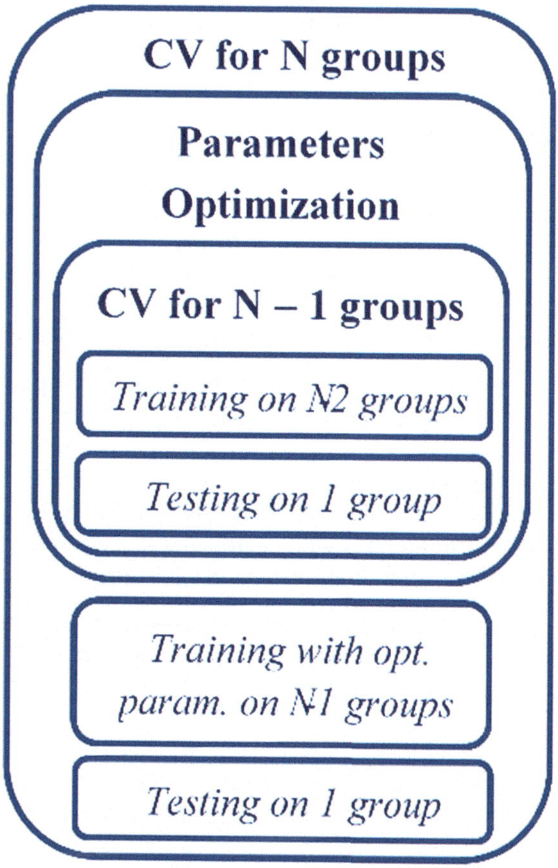

Fig. 7.13 Scheme of dual leave-group-of-sources-out cross-validation

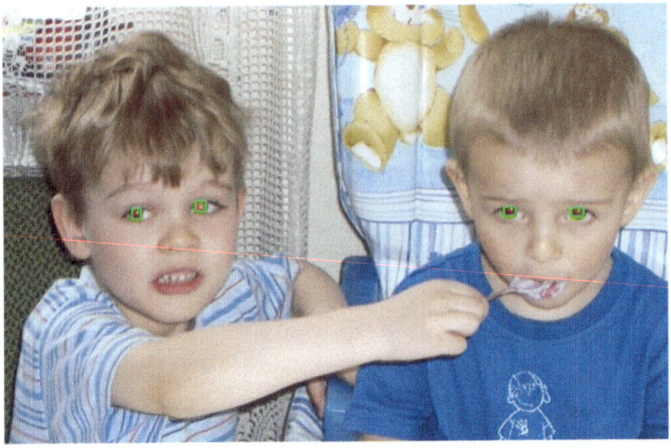

Fig. 7.14 Detection outcome for photo from Fig. 7.1

False Positives (FP) from the number of True Positives (TP) for one of the best versions of the committee that contains 180 decision trees for 32 features. Figure 7.14 shows the detection outcome for the photo from Fig. 7.1.

7.5 Correction

Firstly, we concatenate adjacent pixels having similar colours and the segmented regions. This makes it possible to add pink and violet tones, which frequently occur on the border of the red area because of the mixture of the red colour of the pupil and iris and/or glint inside a pupil. We adapt a general approach of growing a seeded region (Adams and Bischof 1994). It comprises the following iterative steps:

- calculation of the mean brightness and mean value of the *Cr* component of pixels of given region, Y_m and Cr_m;
- if a pixel with coordinates (r, c) does not belong to the region and at least two adjacent pixels belong to the region, then the Euclidian distance between $(Y(r, c), Cr(r, c))$ and (Y_m, Cr_m) is calculated; if the distance is less than the threshold T_d, then the pixel with coordinates (r, c) is marked as an attachable;
- concatenation of all marked pixels at the end of iteration.

Iterations continue unless at least one pixel is marked as an attachable. The next optional stage is the filling of holes: we append pixels lying inside the inner contour

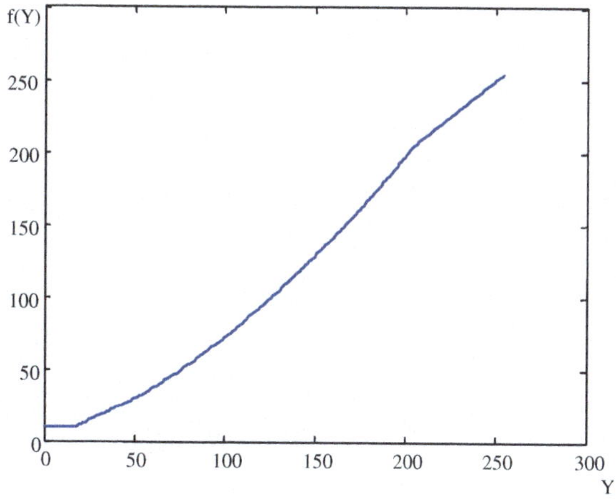

Fig. 7.15 Plot of a darkening function

to the region. This allows the glint and inner pixels deformed by noise or JPEG compression artefacts to be taken into account.

Finally, we carry out modification of the colours of regions. The modification consists in conversion of pixels to grey colour, darkening, and blending with the initial image according to the formulas:

$$Y(r,c) = f(Y(r,c) \times \alpha(r,c) + Y(r,c) \times (1 - \alpha(r,c)),$$

$$Cb(r,c) = 128 \times \alpha(r,c) + Cb(r,c) \times (1 - \alpha(r,c)),$$

$$Cr(r,c) = 128 \times \alpha(r,c) + Cr(r,c) \times (1 - \alpha(r,c)),$$

$$\alpha(r,c) = N/8,$$

where modification is performed for all pixels of the connected region; function *f(Y (r, c))* decreases the brightness of pixels with coordinates *(r, c)*, Fig. 7.15 shows a plot of that function; and N_n is the number of pixels belonging to a given connected region and adjacent to the pixel with coordinates *(r, c)*.

The correction outcomes look attractive; observers estimate the results positively. Figure 7.16 demonstrates the correction outcome for the photo from Fig. 7.1.

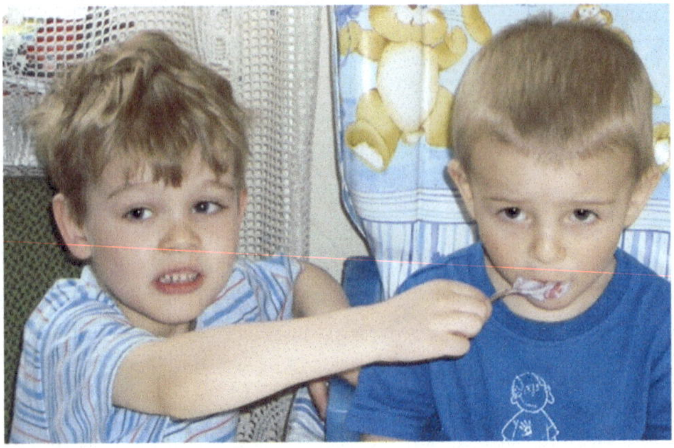

Fig. 7.16 Corrected photo from Fig. 7.1

7.6 Quality Metrics

Formulation of a quality metrics makes it possible to choose the best solution and is necessary for the adjustment of the parameters of an algorithm in the best way. A mature red eye correction method produces the least number of unwanted cases. Initially we enumerated such cases; then, we set priorities by means of the Analytic Hierarchy Process (AHP) (Saaty 2001). Obviously, good correction solutions have low values of FN and FP; ideally, FN and FP are equal to zero. However, the severity of FPs differs significantly. Almost invisible small FPs in the foreground are undesirable but sometimes allowable. Visible FPs in the foreground, especially on human faces and bodies, are absolutely unallowable; such artefacts irritate more than red eyes. Therefore, we divided all FPs into two groups: FP_c is the number of critical FPs and FP_n is the number non-critical FPs.

An identical situation occurs with regard to FNs: the visibility and need for correction for various types of red eyes are different. Several red eye regions are relatively large and easily distinguishable; other regions are small and have a dilute tint of red. Detection of the first type of red eye region is mandatory whereas detection of the second type of region is desirable. Accordingly, we divided all FNs into two groups: FN_m is the number of regions whose detection is mandatory and FN_d is the number of regions whose detection is desirable. One more unwanted situation is the correction of only one eye from a pair. For software, this is not so crucial, because the user has the option of correcting the second eye manually, but for embedded implementations, this situation is quite unpleasant for users. N_P is the number of faces where only one eye from a pair has been corrected.

The retouching quality is important too. We distinguish two factors (Safonov et al. 2008).

- The corrected eye looks worse than the original red eye; for example, when only part of the red region is corrected; these cases are irritating;

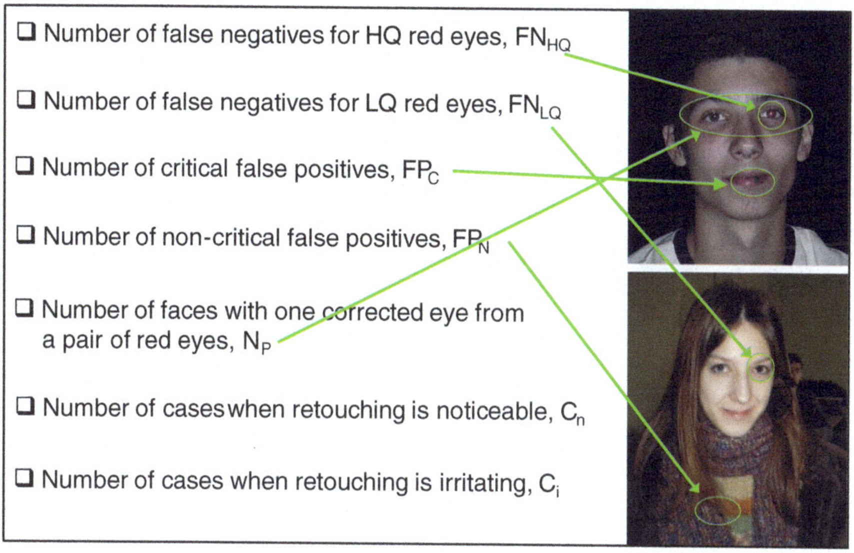

Fig. 7.17 Illustration of various types of errors

Table 7.3 AHP table

Required quality	FNm	FNd	FPc	FPn	Np	Ci	Cn	Geometric mean	Weight
FNm	1.00	5.00	0.20	5.00	1.00	0.20	5.00	1.26	0.13
FNd	0.20	1.00	0.33	5.00	0.20	0.20	5.00	0.68	0.07
FPc	5.00	3.00	1.00	3.00	5.00	5.00	5.00	3.43	0.36
FPn	0.20	0.20	0.33	1.00	0.20	0.20	1.00	0.34	0.04
Np	1.00	5.00	0.20	5.00	1.00	1.00	5.00	1.58	0.16
Ci	5.00	5.00	0.20	5.00	1.00	1.00	5.00	1.99	0.21
Cn	0.20	0.20	0.20	1.00	0.20	0.20	1.00	0.32	0.03

- The retouching is noticeable on close examination but does not irritate.

Accordingly, C_i is the number of irritating cases and C_n is the number of situations when retouching is only slightly noticeable. Figure 7.17 illustrates various types of unwanted factors in red eye correction techniques.

For prioritization of undesired factors, we constructed an AHP table (see Table 7.3) according to the observer's opinions. We used the simplest method of filling the table: if the item on the left is more important than the one at the top, then the cell value is 5; if the severity of two items is the same, then cell value is 1; if the item at the top is more important than the one on the left then the cell value is 1/5. Taking into account weights from the AHP table, we propose the following correction quality criterion Qc:

Table 7.4 Quality metrics for tested algorithms

	FN_m	FN_d	FP_c	FP_n	N_p	C_i	C_n	Qc
Proposed	2	21	0	2	13	5	13	**0.60**
Fotonation	7	21	0	3	12	4	13	0.56
Arcsoft	10	22	0	2	13	4	7	0.55
Canon	12	31	2	3	9	3	7	0.48
Kodak	24	39	1	1	2	5	4	0.47
HP RedBot	18	17	3	4	5	4	10	0.46
HP PS 475	11	27	4	9	14	3	6	0.25
Microsoft	23	31	6	6	15	15	25	0
Cyberlink	30	27	21	18	6	9	6	0

It emphasize the highest quality metrics.

$$Q_c = \frac{N_t - 1.3FN_m - 0.7FN_d - 3.6FP_c - 0.4FP_n - 1.6N_p - 2.1C_i - 0.3C_n}{N_t},$$

where N_t is the total number of red eyes.

7.7 Results

We experimented with the following solutions with automatic red eye correction: Nikon View with Fotonation red eye detector (Corcoran et al. 2005); HP Photosmart 475 photoprinter, HP RedBot Web service; Kodak Easy Share software; Canon Easy-PhotoPrint; Microsoft Digital Image Starter Edition; ArcSoft Photoprinter software utility; and Cyberlink PhotoNow, which uses Viola and Jones's (2001) face detector.

For the assessment of detection quality, a testing set containing 50 photos of various sizes was used. The number of red eyes N_t is 152. For the estimation of correction quality, the processing images were printed by a dye-sublimation photo-printer with a resolution of 300 dpi. Table 7.4 shows a comparison of the correction quality of existing automatic red eye correction solutions and the proposed method. The table shows that the proposed method provides the best outcomes. Table 7.5 demonstrates the correction examples for the proposed method and the Arcsoft and Fotonation algorithms.

Table 7.5 Examples of automatic red eye correction by the proposed method and the Arcsoft and Fotonation algorithms

Original	Proposed	ArcSoft	Fotonation
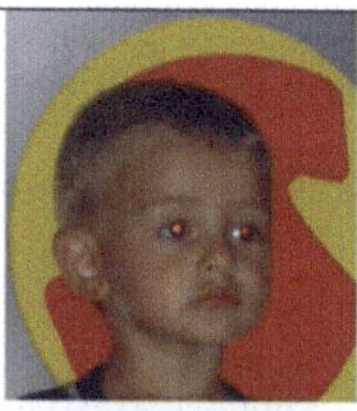	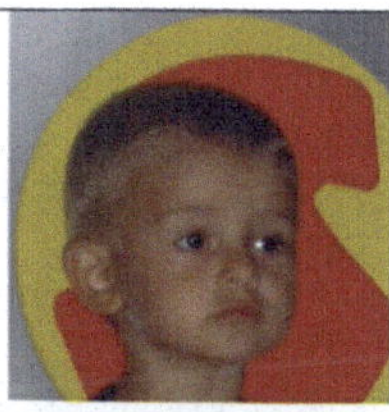	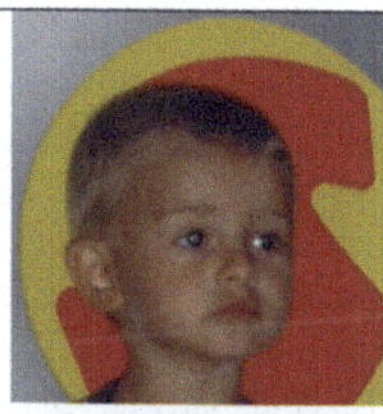	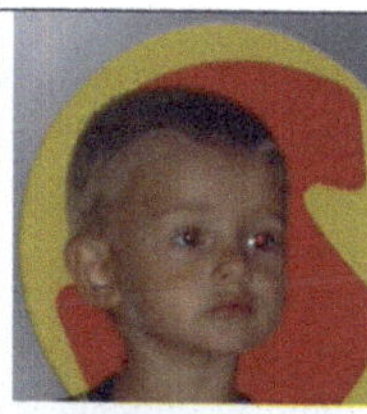
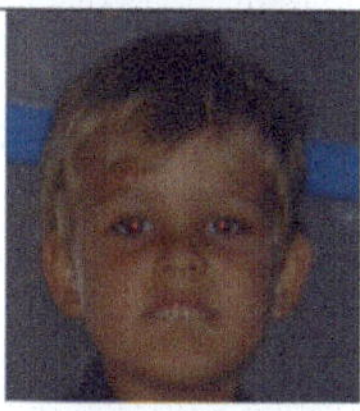	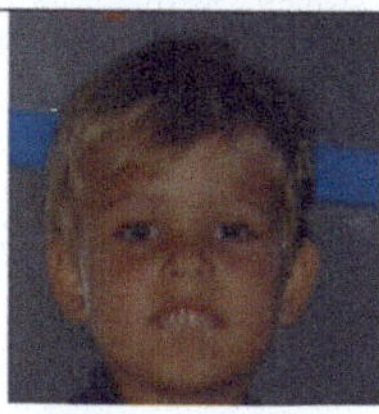	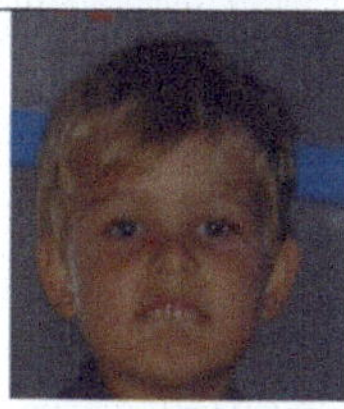	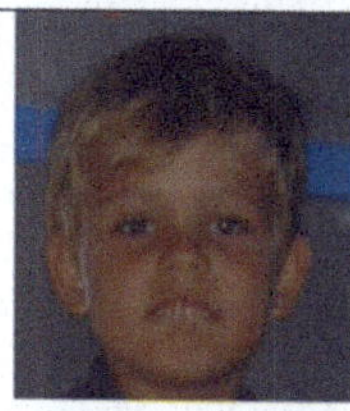
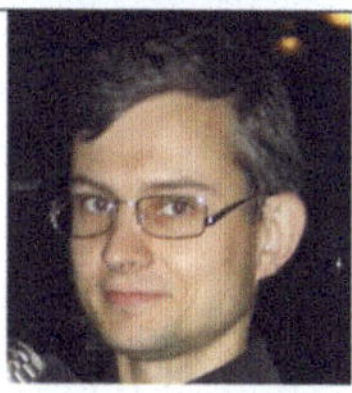	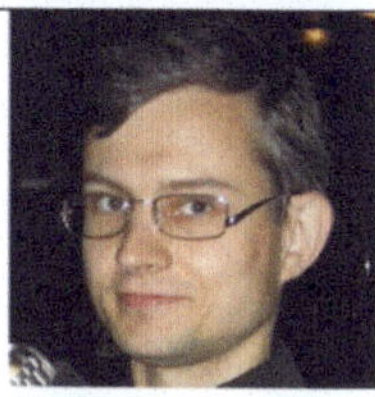	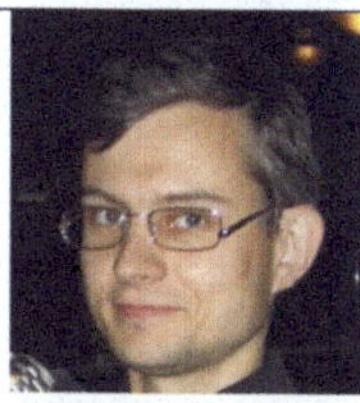	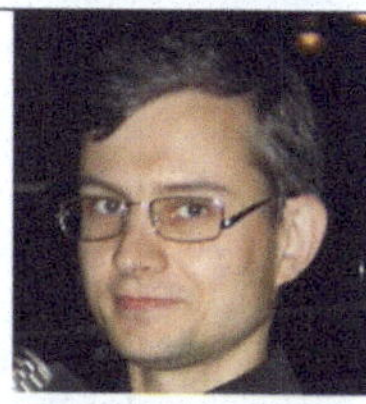
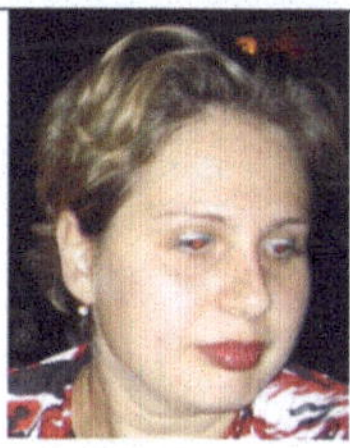	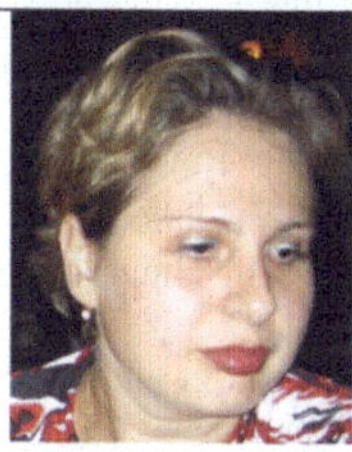	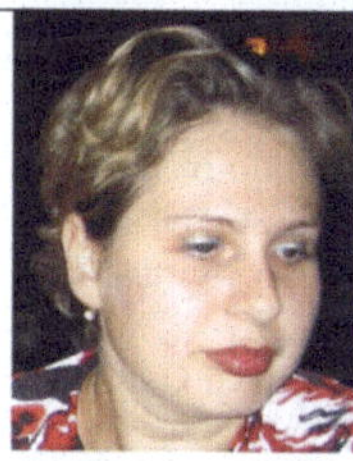	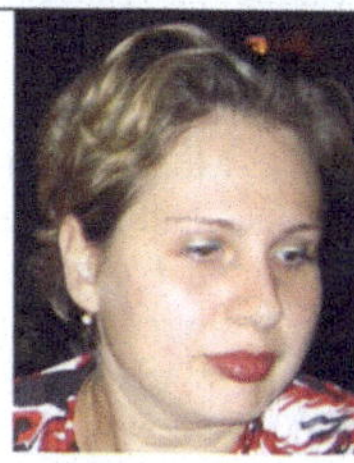
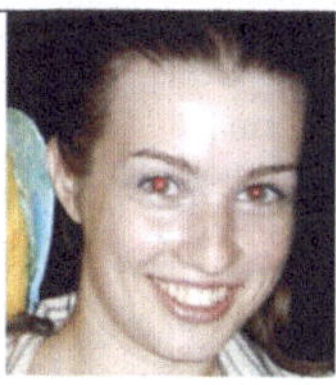	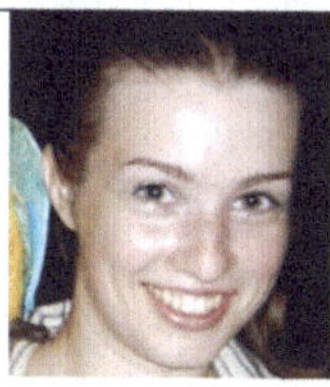	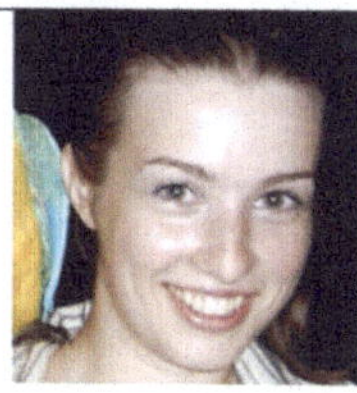	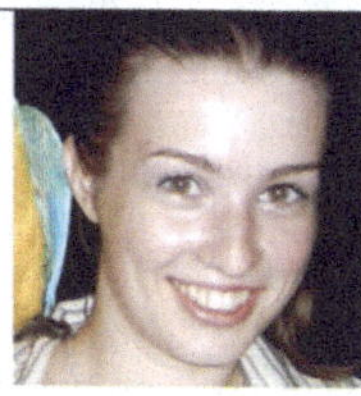

References

Adams, R., Bischof, L.: Seeded region growing. IEEE Trans. Pattern Anal. Mach. Intell. **16**(6), 641–647 (1994)

Cavley, G.C., Nalbot, N.L.C.: On over-fitting in model selection and subsequent selection bias in performance evaluation. J. Mach. Learn. Res. **11**, 2079–2107 (2010)

Chen, T.-X., Chen, X., Platt, J.C., Yan, J., Zhang, H.-J.: Red-eye detection based on red region detection with eye confirmation. US Patent 6,895,112 (2005)

Corcoran, P., Bigioi, P., Steinberg, E., Pososin, A.: Automated in-camera detection of flash-eye defects. IEEE Trans. Consum Electron. **51**, 11–17 (2005)

Czubin, K., Smolka, B., Szczepanski, M., Hardeberg, J.Y., Plataniotis, K.N.: On the redeye effect removal algorithm. In: The First European Conference on Colour Graphics, Imaging and Vision, France, pp. 292–297 (2002)

Egorova, M.A., Murynin, A.B., Safonov, I.V.: An Improvement of face detection algorithm for color photos. Pattern Recogn. Image Anal. **19**(4), 634–640 (2009)

Freund, Y., Schapire, R.: Experiments with a new boosting algorithm. In: International Conference on Machine Learning, pp. 148–156 (1996)

Friedman, J., Hastie, T., Tibshirani, T.: Additive logistic regression: a statistical view of boosting. Ann. Stat. **38**(2), 337–374 (2000)

Gasparini, F., Schettini, R.: Automatic redeye removal for smart enhancement of photos of unknown origin. In: 8th International Conference on Visual Information Systems, vol. 3736, pp. 226–233 (2005)

Gaubatz, M., Ulichney, R.: Automatic red-eye detection and correction. In: International Conference on Image Processing, vol. 1, pp. 804–807 (2002)

Gomez, G., Morales, E.: Automatic feature construction and a simple rule induction algorithm for skin detection. In: Proceedings of the ICML Workshop on Machine Learning in Computer Vision, pp. 31–38 (2002)

Hardeberg, J.Y.: Red eye removal using digital color image processing, pics 2001: image processing, image quality. In: Image Capture Systems Conference, pp. 283–287 (2001)

Held, A.: Model-based correction of red eye defects. In: 10th Color Imaging Conference: Color Science and Engineering Systems, Technologies, Applications, pp. 223–228 (2002)

Ioffe, S.: Red eye detection with machine learning. In: International Conference on Image Processing, vol. 2, pp. 871–874 (2003)

Keelan, B.W.: Handbook of Image Quality: Characterization and Prediction. Marcel Dekker, Inc (2002)

Luo H., Yen, J., Tretter, D.: An efficient automatic redeye detection and correction algorithm. In: IEEE International Conference on Pattern Recognition, vol. 2, pp. 883–886 (2004)

Luo, H., Yen, J., Tretter, D.: Detecting and correcting red-eye in a digital image. US Patent 7,116,820 (2006)

Luo, H., Yen, J., Tretter, D.: Detecting and correcting redeye in an image. US Patent 7,333,653 (2008)

Oberhardt, K.,Taresch, G., Jacob, F., Damm, T., Schindler, H.-G.: Method for the automatic detection of red-eye defects in photographic image data. US Patent Application 20,030,044,178 (2003)

Patti, A., Konstantinidies, R., Tretter, D., Lin, Q.: Automatic digital redeye reduction. In: International Conference on Image Processing, pp. 55–59 (1998)

Pratt, W.: Digital Image Processing, Wiley, Inc (2001)

Saaty, T.L.: Decision Making for Leaders: The Analytic Hierarchy Process for Decisions in a Complex World, New Edition, Analytic Hierarchy Process Series (2001)

Safonov, I.V., Rychagov, M.N., Kang, K.M., Kim, S.H.: Automatic red eye correction and its quality metrics. In: Proceedings of IS&T/SPIE Electronic Imaging 0277-786x, v. 680 (2008)

Safonov, I.V.: Automatic red eye detection. In: Proceedings of 17th International Conference on Computer Graphics, GraphiCon, pp. 112–119 (2007)

Safonov, I.V., Gartseev, I., Pikhletsky, M., Tishutin, O., Bailey, M.J.A.: An approach for model assessment for activity recognition. Pattern Recogn. Image Anal. **25**(2), 263–269 (2015)

Schapire, R., Singer, Y.: Improved boosting algorithms using confidence-rated predictions. Mach. Learn. **37**(3), 297–336 (1999)

Schettini, R., Gasparini, F., Chazli, F.:A modular procedure for automatic redeye correction in digital photos. In: Proceedings of IS&T/SPIE Electronic Imaging (2004)

Schildkraut, J.S., Gray, R.T.: A fully automatic redeye detection and correction algorithm. In: International Conference on Image Processing, pp. 801–803 (2002)

Smolka, B., Czubin, K., Hardeberg, J.Y., Plataniotis, K.N., Szczepanski, M., Wojciechowski, K.: Towards automatic redeye effect removal. Pattern Recogn. Lett. **24**, 1767–1785 (2003)

Ulichney, R., Gaubatz, M., van Thong, J.M.: RedBot—A Tool for Improving Red-Eye Correction. IIP Lab (2003)

Vezhnevets, A., Vezhnevets, V.: Modest AdaBoost—teaching AdaBoost to generalize better. In: Proceedings of 15th International Conference on Computer Graphics. GraphiCon, pp. 322–325 (2005)

Viola, P., Jones, M.: Robust real-time object detection. Technical Report CRL 2001/01, Compaq Cambridge Research Laboratory (2001)

Volken, F., Terrier, J., Vandewalle, P.: Automatic red-eye removal based on sclera and skin tone detection. In: Proceedings of Third European Conference on Color in Graphics, Imaging and Vision, pp. 359–364 (2006)

Willamowski, J., Csurka, G.: Probabilistic Automatic red eye detection and correction. In: Proceedings of 18th International Conference on Pattern Recognition, vol. 3, pp. 762–765 (2006)

Wu, D.: Automatic red eye removal. US Patent 7,852,377 (2010)

Zhang, L., Sun, Y, Li, M., Zhang, H.: Automated red-eye detection and correction in digital photographs. In: International Conference on Image Processing, ICIP 04, vol. 4, pp. 2363–2366 (2004)

Chapter 8
Image Upscaling

8.1 Introduction

Interpolation is model-based recovery of continuous data from discrete data within a known range of abscissa. The quality of the interpolated image generated by any of the single input image interpolation algorithms is inherently limited by the amount of data available in the image. Image zooming cannot produce the high frequency components lost during the low-resolution sampling process unless a suitable model for zooming can be established. There is no rigorous way to check whether an interpolation model corresponds to the physical reality without introducing at least some assumptions. The three most important assumptions for interpolation are as follows.

1. The underlying data are continuously defined.
2. Given data samples, it is possible to compute a data value of the underlying continuous function at any abscissa.
3. The evaluation of the underlying continuous function at the sampling points yields the same value as the data themselves.

In digital photography, the quality of the processed photo is measured by how pleasant the image looks after modifications. This means that the image interpolation must save or enhance perceptual image characteristics, such as perceptual sharpness, and avoid introducing any unnatural features such as artificial patterns, ringing, and so on.

A band-limited function can be exactly reconstructed by convolution with the sinc kernel. Band-limited also means infinite, but real images are finite. Generally, the interpolated function can be found using the following formula:

$$f(x) = \sum_{k \in Z^q} f_k \varphi(x - k), \quad x \in R,$$

I.V. Safonov et al., *Adaptive Image Processing Algorithms for Printing*,
Signals and Communication Technology,
https://doi.org/10.1007/978-981-10-6931-4_8

where f_k is the value of the function at known points and φ is the convolution kernel. There are a wide variety of interpolating kernels, such as sinc, Lanzcos, linear, B-spline, Gauss, Hermite, and so on. In addition, a large palette of apodization windows has been proposed for the practical realization of the sinc function, such as Hanning, Kaiser-Bessel, Cauchy, Dirichlet, Riemann, and many others. Usually $\varphi(x)$ is required to be a finite support function, and to simplify numerical implementation $\varphi(x)$, it is chosen to be separable as well. This means that data can be processed in a separable fashion, line-by-line, column-by-column, and so forth. A comprehensive description of generalized interpolation and choice of convolution kernel functions is presented in Thévenaz et al. (2000).

Apart from choosing the appropriate interpolating functions, there exist several approaches for processing 2D data (greyscale images), for example, two-step interpolation grid filling (Li and Orchard 2000; Giachetti and Asuni 2011), which allows upscaling by a factor of two; image triangulation, where the triangles are filled with new intensity values (Yu et al. 2000; Su and Willis 2004); and so on.

Regularization-based image interpolation algorithms pose the image interpolation as a functional minimization problem (Liu 2012; Farsiu et al. 2004). The functional contains the data-fitting term and the stabilizer term. The data-fitting term restricts the high-resolution image to match the low-resolution image. The stabilizer term makes the high-resolution image fit a priori information. Similar approaches are MAP (maximum a posteriori estimation), POCS (projection-onto-convex sets), and PDE (partial differential equations)-based algorithms (Belekos et al. 2010; Chan and Shen 2005).

Modern methods include learning-based techniques using local self-examples (Freedman and Fattal 2011), more sophisticated learning techniques such as neural networks (Dong et al. 2014) or random forest (Schulter et al. 2015), or techniques based on constructing a pre-built dictionary containing pairs of corresponding high- and low-resolution patches (Peleg and Elad 2014). But such methods are highly computationally complex, requiring much effort during learning or significant storage space for learnt coefficients.

In the next chapter we describe interpolation method, intended for implementation in digital signal processor (DSP), i.e. in the system with strict limitations on memory and number of required operations.

8.2 DSP-Oriented Edge-Directional Interpolation Method

8.2.1 Upsampling

This chapter describes a fast and computationally efficient interpolation method aimed at implementation on a DSP system. This places strict limitations on memory and the number of operations. At the same time, the method should provide the user with high-quality perceptually pleasing images.

Image magnification is an ill-posed problem, which allows numerous solutions due to the lack of information presented in the low-resolution image. That is why the magnification of an image is generally a trade-off between the appearance of artefacts (as a sequence of wrong assumptions) and blur (since the new image will be band-limited like the original image). For example, the nearest neighbour solution preserves edge sharpness but is unacceptable because it causes a checkerboard effect and aliasing.

The idea of following edges during image upscaling has many applications. Famous methods of edge-directed interpolation were presented by Kimmel (1999) and Li and Orchard (2000). Kimmel's method was initially designed for colour filter arrays (CFA) interpolation and was adapted by Muresan and Parks (2000) for greyscale image interpolation. Each pixel is assigned a weight according to the likelihood of it occurring at the edge. Calculation of weights is based on directional derivatives. Then interpolation is carried out such that more weight is given to the pixels lined up along the edge than to the pixels across the edge. Allebach and Wong (1996) proposed another edge-directed method. The aim is to avoid interpolation across edges, which will cause the result to look blurred. The authors first estimate sub-pixel edge positions and construct an edge map and then, during the rendering phase, use a modified bilinear interpolation method. A very interesting approach was proposed by Li and Orchard (2000): their so-called NEDI solution performs magnification by the power of 2 and is based on the assumption that there is geometric duality between the low-resolution covariance and the high-resolution one. The edge-directed property of covariance-based allows tuning the interpolation coefficients to match an arbitrarily oriented step edge. They propose to solve the small optimization problem for each unknown pixel value. Another non-linear interpolation technique is presented by Jensen and Anastassiou (1990). They estimate the edge direction for every pixel in a low-resolution image using an assumption that, within a small enough neighbourhood around a pixel lying on an edge between two objects, these objects can be separated by a line into two regions, each with a nearly constant intensity value. The idea of finding pixels that are not separated by an edge was explored by Quak and Schumaker (1990). This idea was extended by Yu et al. (2000) for data interpolation; they propose a more sophisticated approach involving a cost function and repetition of the optimization process. The authors construct a data adaptive triangulation mesh (similar to a triangle strip) and interpolate colour values inside the triangle linearly, depending on the distance between triangles' vertices and the point being interpolated.

Suppose the low-resolution image is X and the high resolution image to be generated is Y. For each unknown pixel in image Y, the following procedure is carried out.

All pixels of the source image are processed in a window-based manner. To interpolate a pixel of Y, all neighbouring pixels contained in a window centred at this pixel's coordinates are considered. First, the prevailing edge direction is determined (Su and Willis 2004). For this purpose, the absolute values of the difference between neighbouring pixels are used. If the diagonal direction prevails and an edge passes through the square formed by four nearest neighbours of a pixel

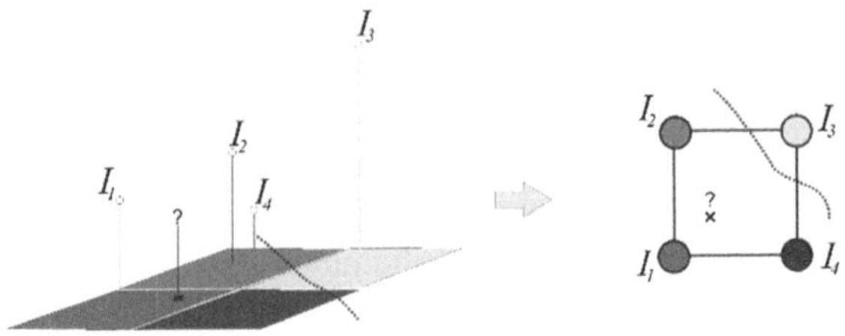

Fig. 8.1 If the edge outlines vertex I_3, then for the intensities of vertices the following holds: $|I_2 - I_4| < |I_1 - I_3|$, and vertex I_3 will not be used for interpolation of the point marked by "?"

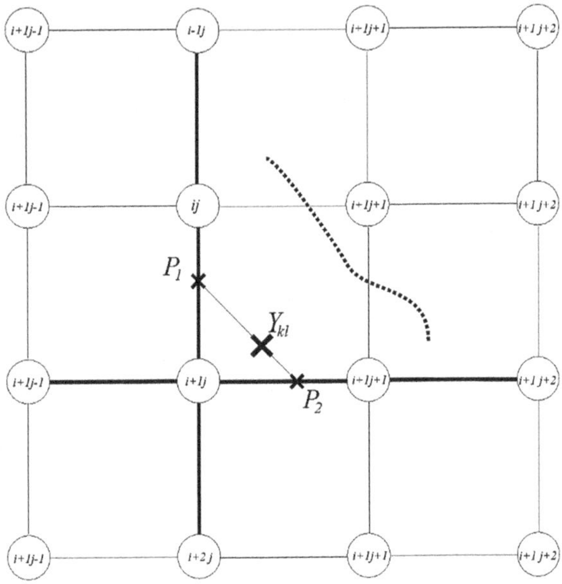

Fig. 8.2 Layout for finding the interpolated value Y_{kl}. First a diagonal is found which outlines the vertex with the greatest difference from the other three vertices. Then the intermediate values P_1 and P_2 are found using cubic polynomials

being interpolated, cutting one corner, the absolute difference between the diagonal pixels located on different sides from the edge will be greater than the difference between the other two diagonal pixels (see Fig. 8.1). In this case, to preserve the edge sharpness, an *outlier* (a pixel that is cut by an edge) is not used in the interpolation; that is, the two diagonal vertices with a smaller difference are connected.

Next, after one or another diagonal is connected, it is time to find the interpolated value Y_{kl}. Here, cubic polynomials are applied, in contrast to the work of Quak and Schumaker (1990), who use linear interpolation inside a triangle. Let us interpolate the value of Y_{kl} of a high-resolution image Y (see Fig. 8.2). The first points P_1 and P_2 are found using neighbouring pixels. The simplest way is to use the linear method, but for enhanced quality interpolation, points P_1 and P_2 are found using cubic polynomials. Then, having the values of P_1 and P_2, the value of Y_{kl} is found using linear fitting.

Such a technique suppresses the appearance of the checkerboard artefact, which occurs when interpolation is performed along the image coordinate axes, revealing the underlying sampling grid.

In cases in which a vertical or horizontal edge direction prevails, first, values across the edge are interpolated using a cubic polynomial and then the unknown value along the edge is found using linear fitting.

This technique is further extended for more stable detection of the edge direction. In this case, more of the surrounding pixels are involved, not just the four closest ones. For determination of the diagonals, we use the absolute values of differences between pixels located in the 6 × 6 neighbourhood of the pixel being interpolated.

Let the four closest neighbours of the interpolated value Y_{kl} be $X(i, j)$, $X(i + 1, j)$, $X(i, j + 1)$, and $X(i + 1, j + 1)$. Then,

$$X_x = \frac{1}{12} \sum_{\substack{k=-1...1 \\ l=-1...2}} |X(i+k+1, j+l) - X(i+k, j+l)|,$$

$$X_y = \frac{1}{12} \sum_{\substack{k=-1...2 \\ l=-1...1}} |X(i+k, j+l+1) - X(i+k, j+l)|,$$

$$X_{d1} = \frac{1}{9} \sum_{\substack{k=-1...1 \\ l=-1...1}} |X(i+k+1, j+l+1) - X(i+k, j+l)|,$$

$$X_{d2} = \frac{1}{9} \sum_{\substack{k=-1...1 \\ l=-1...1}} |X(i+k+1, j+l) - X(i+k, j+l+1)|.$$

This technique reduces the number of undesired artefacts sufficiently.

Cubic polynomial interpolation is executed very fast under the assumption that the distance between pixels in the input image I is equal to 1. The points to be interpolated are $\vartheta_1, \vartheta_2, \vartheta_3$, and ϑ_4 at the locations [−1, 0, 1, 2]. The aim is to find value ϑ_Δ at point Δ, where $0 < \Delta < 1$. Then ϑ_Δ is found according to the following formula:

$$\begin{aligned} \vartheta_\Delta = \vartheta_2 + \Delta\left\{\vartheta_3 - \vartheta_2 - \tfrac{\Delta-1}{15}[5(\Delta(C_1 - C_2) + C_2) - 7(C_1 + C_2)]\right\}, \\ C_1 = \vartheta_3 - 2\vartheta_2 + \vartheta_1, \\ C_2 = \vartheta_4 - 2\vartheta_3 + \vartheta_2. \end{aligned}$$

Post-processing sharpening is performed using tone mapping. The image is divided into overlapping windows whose size depends on the image magnification factor. Then minimal (min) and maximal (max) values are found inside each window, and the centre pixel of the window is mapped using some tone-mapping curve, as follows:

$$Y' = f\left(\frac{I - min}{max - min}\right) \times \frac{max - min}{255} + min.$$

The tone-mapping curve is a monotonously growing function that maps values from segment [0, 1] in itself, with the derivative at the middle point being greater than one. The resulting edge sharpness depends on the curve's shape, especially its slope angle at $x = 0.5$: the greater the angle, the sharper the edges. The function for the curve is selected as follows:

$$f(x) = \begin{cases} \frac{(2x)^n}{2}, & 0 \le x \le 0.5 \\ 1 - \frac{[2(1-x)]^n}{2}, & 0.5 < x \le 1 \end{cases},$$

where the curve's slope angle at $x = 0.5$ is controlled by the parameter n. This curve can be implemented as a Look-Up Table (LUT) with a constant slope for all edge intensities. This method is direction independent, so in some sense it helps to suppress interpolation artefacts.

In cases where the colour image is being interpolated, first the channel L_{in} with the maximal value is found and processed. The new value is L_{out}. Values of the rest of the channels are found as follows:

$$C_{out} = \frac{C_{in}}{L_{in}} L_{out}.$$

8.2.2 Implementation Issues

Let us compare the computational complexity of bilinear, bi-cubic, and edge-directional interpolation (without a sharpening step). Let an image of size M by N be interpolated to have a size of M' by N' pixels.

8.2.2.1 Bi-cubic Interpolation

Bi-cubic interpolation takes $MN' + M'N'$ 1D cubic interpolation evaluations in a separable manner. This means that the image is interpolated, for example, first in the horizontal direction and then in the vertical direction. If $N' = kN$ and $M' = kM$, the total number of 1D cubic evaluations will be:

$$C_1 = kMN + k^2MN = (k + k^2)MN.$$

However, this approach requires auxiliary data storage of the $4M'$ for storing the intermediate results.

8.2.2.2 Bi-linear Interpolation

By analogy to the bi-cubic interpolation, the total number of 1D linear interpolation evaluations is

$$C2 = (k + k2)MN.$$

Here, the auxiliary data storage has a size of $4M'$.

8.2.2.3 Edge-Directional Method (EDM)

This method is not readily separable. But at the points with vertical or horizontal edges it is possible to use the same principles as above. In this case, vertically or horizontally elongated interpolation will be performed in a similar way to bi-cubic (or bi-linear) interpolation, and the intermediate results can be stored in a separate additional buffer in floating point format (to avoid extra reductions to integers). The size of this additional buffer is $2M'$. In the case of vertically or horizontally elongated features, the most optimal method is to interpolate cubically across the edge and linearly along the edge in order to better preserve the sharpness and save some time.

Let us estimate the computational complexity of the method, supposing that edges in the image are distributed in the following way:

- D pixels with a diagonal edge direction (the processing complexity of both diagonals is the same).
- S pixels with a vertical or horizontal edge (the processing complexity of the horizontal edge is the same as that of the vertical one).

All of these edge direction estimations are performed using an input image, and for every pixel, a 4 × 4 neighbourhood is estimated, as described in the previous sections. Note that the computational complexity is different for pixels whose edge directions are different. For pixels lying within an area with diagonal edge direction D, the number of linear evaluations is $3M'N'$ (this accounts for two evaluations in the horizontal and vertical directions plus one in the diagonal direction), and for S pixels the total number of cubic evaluations is $2MN'$ plus the number of linear evaluations, which is equal to $M'N'$. The overall complexity is

$$3M'N'D/MN + M'N'S/MN$$

linear evaluations, and

$$2MN'S/MN$$

cubic evaluations.

Usually linear evaluation requires one floating point multiplication and cubic evaluation requires eight integer and floating point multiplications. Let us assume that $M' = kM$ and $N' = kN$.

$$C3 = 3k2D + k2S + 16kS = 3k2MN + S(16k - 2k2)$$

If $D \sim S$, then

$$C3 \sim 3k2MN + MN(8k - k2) = (2k2 + 8k)MN$$

Bi-linear and bi-cubic interpolations require, correspondingly,

$$C1 = (k + k2)MN,$$
$$C2 = 8(k + k2)MN.$$

We can see that generally, the new EDM requires fewer multiplications and divisions than bi-cubic interpolation but more than the bilinear method.

8.2.2.4 Sharpening

The most time consuming operation during the edge post-processing procedure is going through the local window to find the minimal and maximal values. But since the edge direction is known and the minimal and maximal values are most probably separated by the edge, it is possible to make this window narrower so that it is elongated *across* an edge. Typically, the window size is $[k + 1]$ by 3, where k is the magnifying ratio. Therefore, processing of one pixel requires $2 * 3 * k$ comparisons (for minimal and maximal values), and then the application of tone mapping, which additionally requires one division and one multiplication. Tone mapping is conducted using an LUT for the 1D tone mapping curve. However, Eq. (7) may be implemented using 2D LUT, saving one division and one multiplication per pixel:

$$Y' = L_{ij} + min,$$

where

$$L_{ij} = f\left(\frac{i}{j}\right) \times \frac{j}{255},$$

for $0 \leq i \leq 255$, $0 \leq j \leq 255$.

Processing of colour images adds one division and one multiplication for each additional colour channel for the preservation of ratios between colour channels.

The real computational time depends on the architecture of the system, the time of memory access, the size of the cache memory, and the execution of conditional expressions.

8.2.3 Results

For comparison of results, the well-known image "Barbara" was chosen (Fig. 8.3). This image was enlarged four times and then reduced four times. Then the Mean Squared Error (MSE) and Peak Signal-to-Noise Ratio (PSNR) were computed (see Sect. 5.3.3 for details).

For comparison, well-known methods such as the bilinear and bi-cubic ones were employed as well as the results produced by several commercial products. The latter include Genuine Fractals and PhotoZoom. The results of the comparison are presented in Fig. 8.4.

It is well known that the perceptual quality of the interpolated image does not readily correspond to the MSE and PSNR values. A high-quality method implemented in Genuine Fractals gave the worst results in terms of MSE and PSNR.

The following figures demonstrate cross-sections of the resulting image. Figure 8.5 shows the location of the cross-section, shown with a yellow line, and Fig. 8.6 shows the cross-section itself.

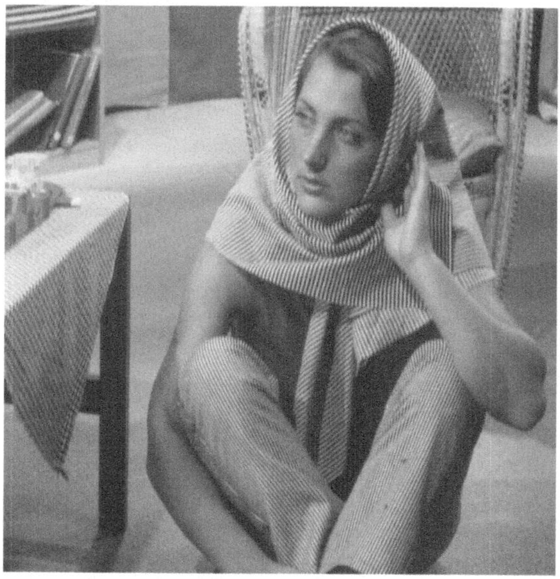

Fig. 8.3 "Barbara" image, greyscale, 512 by 512 pixels

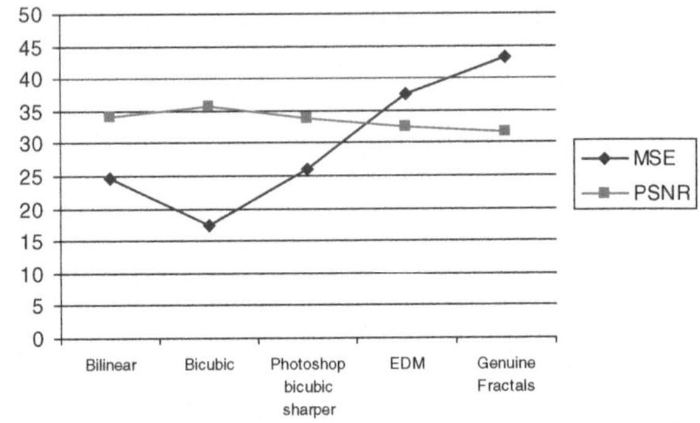

Fig. 8.4 MSE for different interpolation methods for enlargement/reduction of the image “Barbara” (Fig. 8.3)

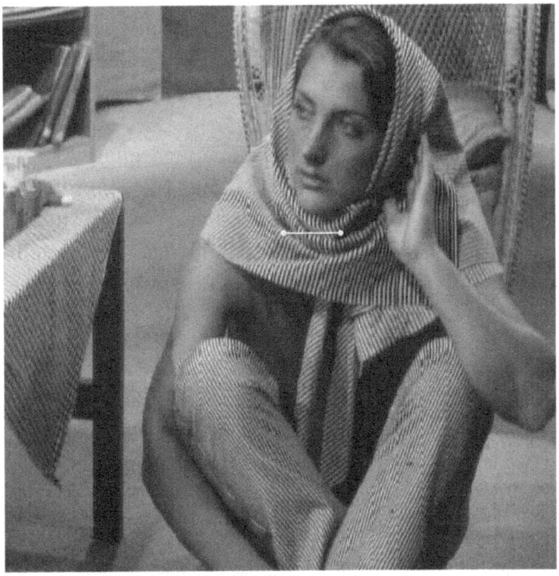

Fig. 8.5 Location of the cross-section shown in Fig. 8.6

Execution times for the cited methods vary significantly. The following chart in Fig. 8.7 presents the time (in seconds) spent on enlarging the image “Barbara” by a factor of four. Here, the EDM post-processing step was included in the timing. Interpolation was performed on a regular PC with a Pentium IV CPU at 3.2 GHz. As can be seen, EDM requires more time than the usual linear methods but its results are comparable with complicated high-quality methods, which in turn are much slower than EDM.

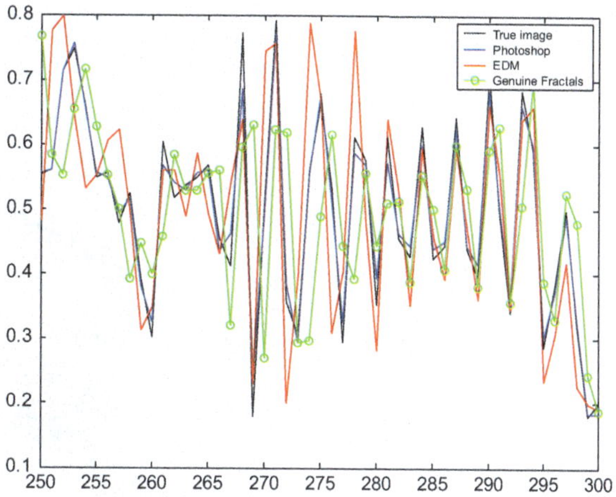

Fig. 8.6 Cross-section of the image along the line shown in Fig. 8.5

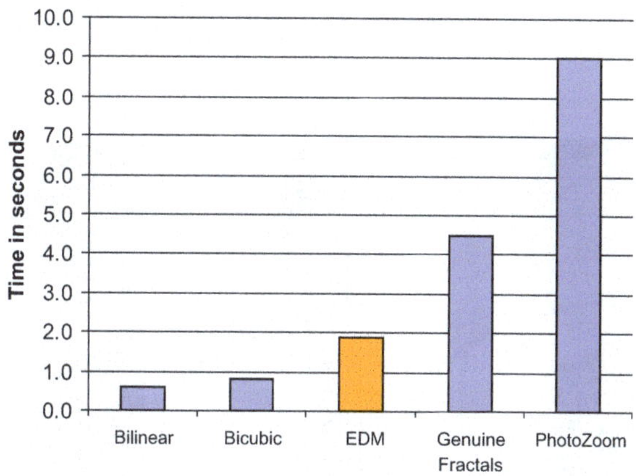

Fig. 8.7 Time (in seconds) spent enlarging the image “Barbara” (Fig. 8.5) by a factor of four

8.2.4 Examples

In this section, several examples will demonstrate the advantages of the proposed technique.

We can see from Figs. 8.8–8.13 that the results of the proposed method are quite pleasing, without artefacts of bi-cubic interpolation, and comparable to high-quality commercial products, but the new method is faster and requires less resources.

The method is quite fast in comparison with regular linear techniques and makes it possible to interpolate the image in one pass. Among the advantages of

Fig. 8.8 Lighthouse. A famous coloured image with fine textured features and edges at various orientations

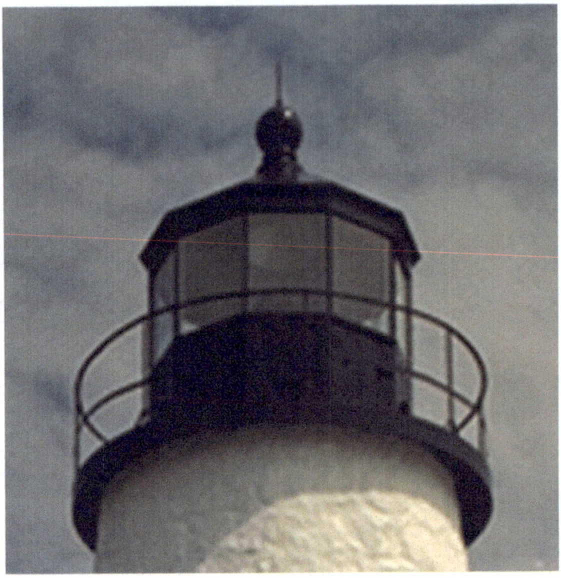

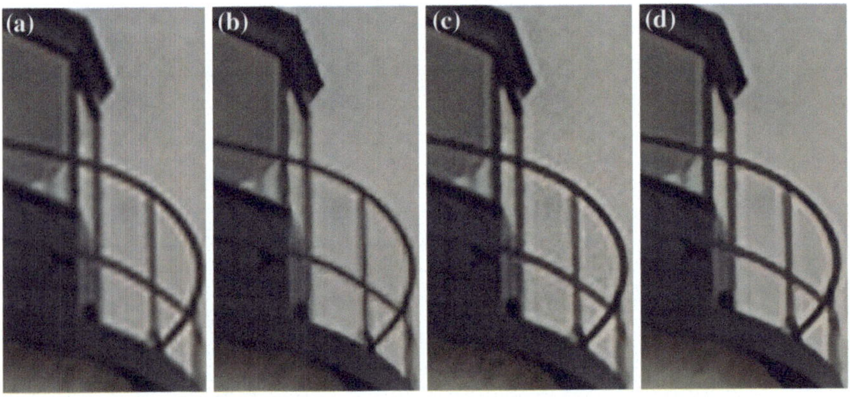

Fig. 8.9 Results of four-times (in each direction) enlargement for Lighthouse image **a** Bi-cubic. **b** Genuine fractals. **c** PhotoZoom. **d** Edge-directed method

the new method is its independence of the magnifying ratio. This means that the image can be enlarged or reduced in one pass, providing arbitrarily sized interpolated images. This also means that this method can be applied for operations other than size changing ones, such as rotation, affine transforms, and various warpings (Fig. 8.14).

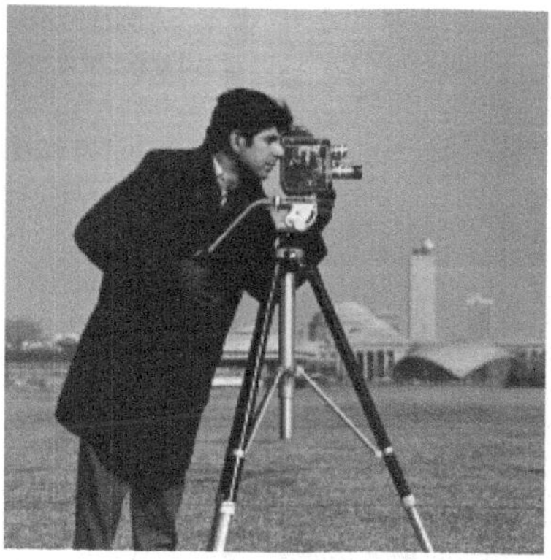

Fig. 8.10 "Cameraman", a very popular greyscale image in image processing

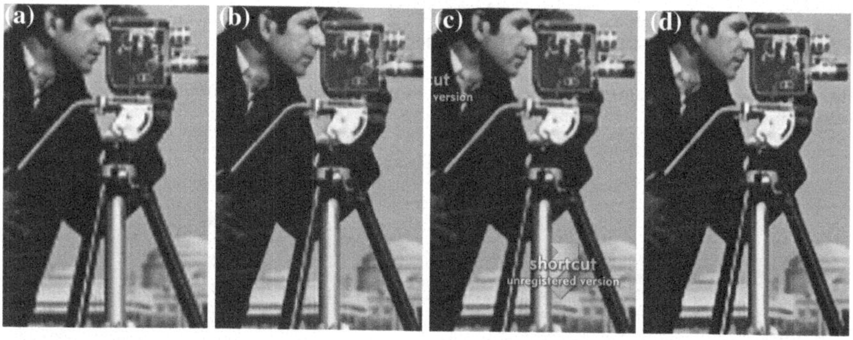

Fig. 8.11 Results of four-times (in each direction) enlargement for Cameramen image **a** Bi-cubic. **b** Genuine fractals. **c** PhotoZoom. **d** Edge-directed method

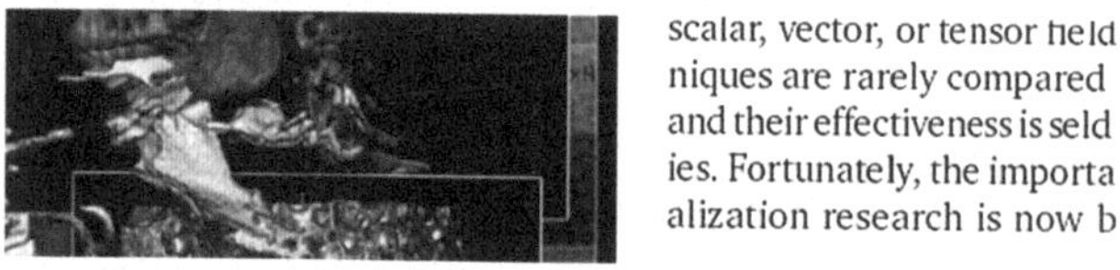

Fig. 8.12 Text with figures

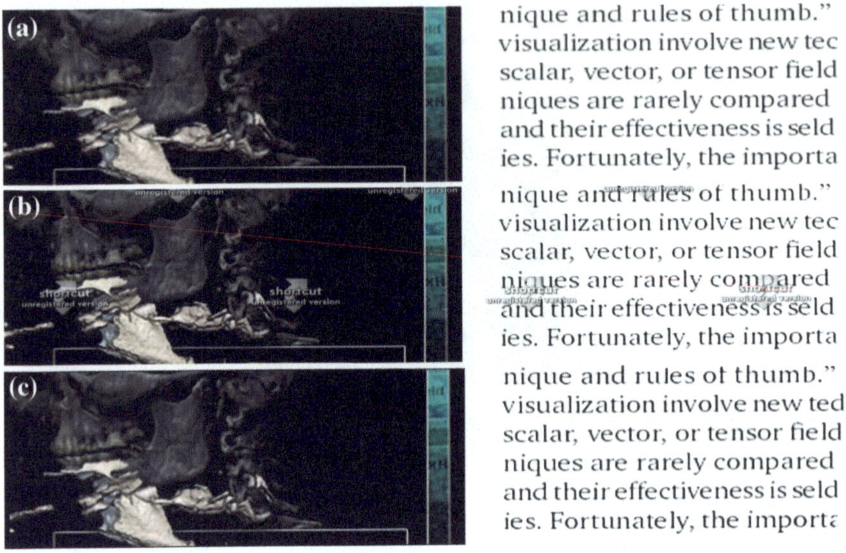

Fig. 8.13 Results of four-times (in each direction) enlargement **a** Bi-cubic. **b** PhotoZoom. **c** Edge-directed method

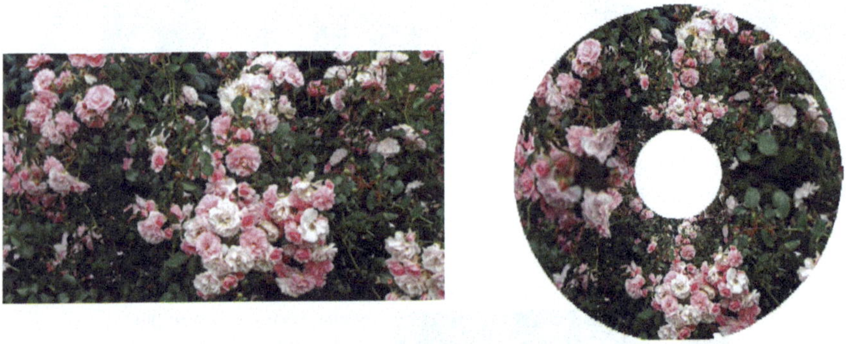

Fig. 8.14 Example of image warping

8.3 Edge-Directional Interpolation Algorithm Using Structure Tensor

The second upsampling method that we describe is aimed at fast and effective edge-directional interpolation of power-of-two interpolation ratios. The method is based on structure tensor and individual interpolation kernels for each direction. The main difference between the proposed algorithm and state-of-the-art algorithms is quantization of the direction vector into six directions and using optimal 4 × 4 kernels for each direction for faster implementations.

The algorithm consists of the following steps.

1. Initial approximation of the high-resolution image.
2. Finding the direction for every pixel of a high-resolution image.
3. First interpolation step: the interpolation of a central pixel inside every 4 × 4 block of low-resolution pixels.
4. Second interpolation step: the interpolation for the rest of pixels using the approach from the first interpolation step.

8.3.1 Initial Approximation

The initial approximation of the high-resolution image is used to find directions for every interpolated pixel. Experiments have shown that, in practice, the directions obtained in the next step do not depend on the interpolation method used. For simplicity, we use the standard bi-cubic interpolation method to construct the initial approximation of the high-resolution image. Let u be the input low-resolution image and v the interpolated image. The approximation is expressed as follows:

$$\begin{aligned} v_{2i,2j} &= u_{i,j}, \\ v_{2i+1,2j} &= \frac{9}{16}\left(u_{i,j} + u_{i+1,j}\right) - \frac{1}{16}\left(u_{i-1,j} + u_{i+2,j}\right), \\ v_{2i,2j+1} &= \frac{9}{16}\left(v_{i,2j} + v_{i,2j+2}\right) - \frac{1}{16}\left(v_{i,2j-2} + v_{i,2j+4}\right). \end{aligned}$$

8.3.2 Finding Directions

8.3.2.1 Construction of the Structure Tensor

The structure tensor has the following matrix:

$$T_{i,j} = \begin{bmatrix} \langle v_x \rangle_{i,j}^2 & \langle v_x v_y \rangle_{i,j} \\ \langle v_x v_y \rangle_{i,j} & \langle v_y \rangle_{i,j}^2 \end{bmatrix},$$

where v_x and v_y are partial derivatives of v, and $\langle \ldots \rangle_{i,j}$ is averaging over the small neighbourhood of the pixel (i, j).

We calculate the partial derivatives as follows:

$$v_x = v * h_1(x) * h_2(y),$$
$$v_y = v * h_2(x) * h_1(y),$$

where h_1 and h_2 are the Gaussian filter and shifted derivatives of the Gaussian filter respectively:

$$h_1(t) = \exp\left(-\frac{(t+0.5)^2}{2\sigma_1^2}\right),$$
$$h_2(t) = -(t+0.5)\exp\left(-\frac{(t+0.5)^2}{2\sigma_2^2}\right),$$

where $\sigma_1 = \sigma_2 = 0.5$.

The averaging $\langle ... \rangle_{i,j}$ is the convolution with the shifted Gaussian filter with the kernel

$$\exp\left(-\frac{(t-0.5)^2}{2\sigma_3^2}\right),$$

where $\sigma_3 = 1.5$.

Normalization of kernels h_1, h_2, and h_3 is not necessary here. A half-pixel shift is used to improve the accuracy of the method when applied to discrete images.

8.3.2.2 Structure Tensor Analysis

Local structure directions are obtained using the analysis of eigenvectors and eigenvalues of the matrix (1).

Let λ_1 and λ_2 be the eigenvalues of T such that $|\lambda_1| \geq |\lambda_2|$ and let **p** be the eigenvector corresponding to λ_1. The ratio between λ_1 and λ_2 defines the type of structural element in the analysed pixel. If $|\lambda_1|$ is significantly greater than $|\lambda_2|$, then

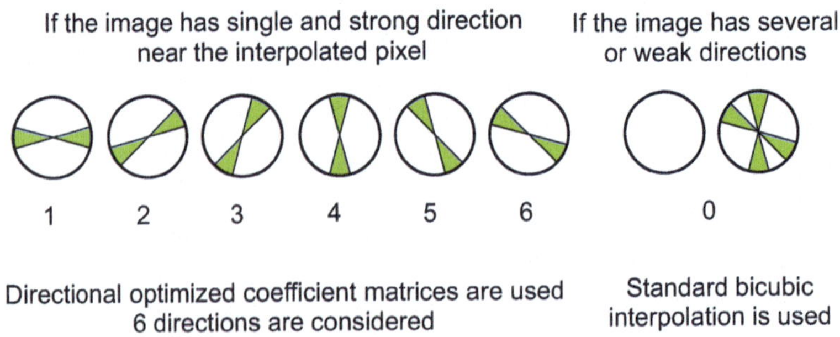

Fig. 8.15 Finding edge directions using the structure tensor

the pixel is a part of a linear structure like an edge or ridge with the direction **p** and edge-directional interpolation will be effective. Otherwise, if $|\lambda_1| \sim |\lambda_2|$ or $\lambda_1 \sim 0$, then there is no dominant direction in the analysed pixel.

The direction of **p** is quantized into one of the six directions (0°, 30°, 60°, 90°, 120°, and 150°). The output of the direction-finding step is the following (see Fig. 8.15): (i) If $|\lambda_1| \leq 2|\lambda_2|$ or $|\mathbf{p}| \sim 0$, there is no distinct direction in the analysed pixel, and the output is zero; (ii) Otherwise, the output is the direction index of **p** (range of 1–6).

8.3.3 First Interpolation Step

At the first interpolation step, pixels with coordinates of (2*i*,2*j*) are copied directly from the low-resolution image:

$$v_{2i,2j} = u_{i,j},$$

while pixels with coordinates (2*i* + 1, 2*j* + 1) are updated using a 4 × 4 block of surrounding pixels from the low-resolution image (see Fig. 8.16).

Let a be the vector constructed from 16 pixels from the 4 × 4 block, while $q^{(d)}$ is the interpolation kernel corresponding to the direction d. The value of the interpolated pixel is computed as

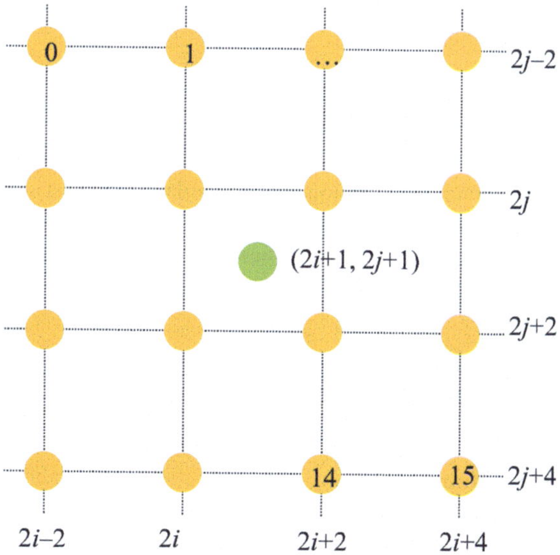

Fig. 8.16 The first interpolation step. The interpolated pixel is in the centre of a 4 × 4 grid of low-resolution pixels, shown in orange, of the low-resolution image; the interpolated pixel is green

$$v_{2i+1,2j+1} = \sum_{k=0}^{15} a_{n,k} q_k^{(d)}.$$

The interpolation kernels $q^{(d)}$ were calculated experimentally using the reference images from the LIVE database (Sheikh et al. 2006). The reference images were downsampled by a factor of two and then a set of correspondences between vectors a and values v was constructed for each direction d for all pixels. The interpolation kernels $q^{(d)}$ were obtained by minimizing the sum of squared errors:

$$\sum_n \left(v_n - \sum_{k=0}^{15} a_{n,k} q_k^{(d)} \right)^2.$$

Restrictions based on kernel symmetry and rotation are added to minimize the number of coefficients and to reduce the condition number of the least squares minimization problem. The restrictions include:

1. Central symmetry of all the kernels: $q_k^{(d)} = q_{15-k}^{(d)}$, horizontal and vertical symmetry for $q^{(1)}$ and $q^{(4)}$, and four-directional symmetry for $q^{(0)}$.
2. The kernel $q^{(4)}$ is equal to the kernel $q^{(1)}$ with 90° rotation.
3. The kernels $q^{(3)}$, $q^{(5)}$, and $q^{(6)}$ are equal to the kernel $q^{(2)}$ with 90° rotation and transposition.

The coefficient values of the kernels calculated for the images from the LIVE database are available on the website: http://imaging.cs.msu.ru/en/publication?id=319.

8.3.4 Second Interpolation Step

The second interpolation step is similar to the first step but instead of a 4 × 4 block of pixels of the low-resolution image, rotated by 45°, a 4 × 4 pixel block is used. It contains both pixels from the low-resolution image and pixels interpolated at the previous step (see Fig. 8.17).

8.3.5 Results

The performance of the proposed algorithm is shown in Fig. 8.18. The objective quality comparison using PSNR and SSIM (structured similarity) metrics is presented in Table 8.1. The algorithm DCCI (Zhou et al. 2012) is used in the comparison as it has shown excellent performance among low-complexity state-of-the-art algorithms (Yu et al. 2013). It can be seen that the proposed

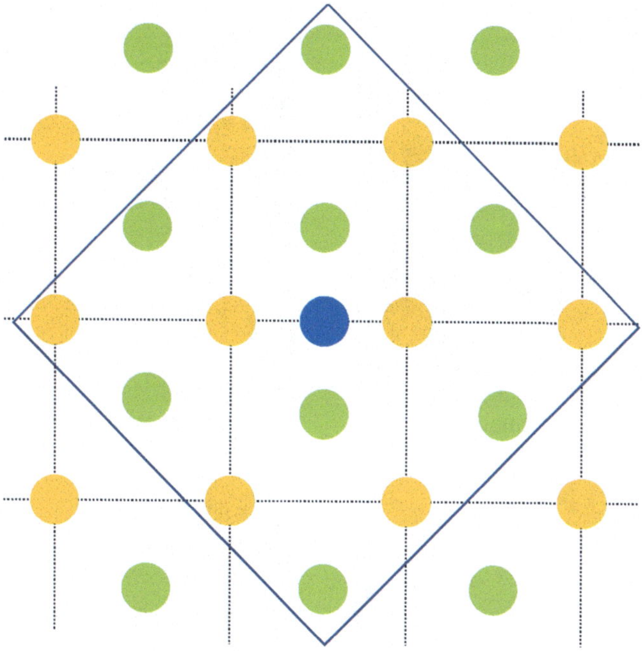

Fig. 8.17 The second interpolation step. Pixels of the low-resolution image are orange; pixels interpolated in the previous step are green

algorithm shows a quality improvement even for highly textured images like "Goldhill" and "Baboon". For low-textured images like "Cameraman", the proposed algorithm shows a worse PSNR but a better SSIM than DCCI.

Although the modern image interpolation algorithms based on LR-to-HR (low resolution to high resolution) mapping (Dong et al. 2014; Dai et al. 2015) show better quality than the proposed algorithm, they have significantly higher computational complexity. Also they may produce flicking and vibration of fine details when applied to videos containing noise. The proposed algorithm does not produce such effects. It is suitable for video resampling and can be used as part of multi-frame super-resolution algorithms.

The algorithm showed excellent performance and quality in comparison to state-of-the-art image interpolation algorithms.

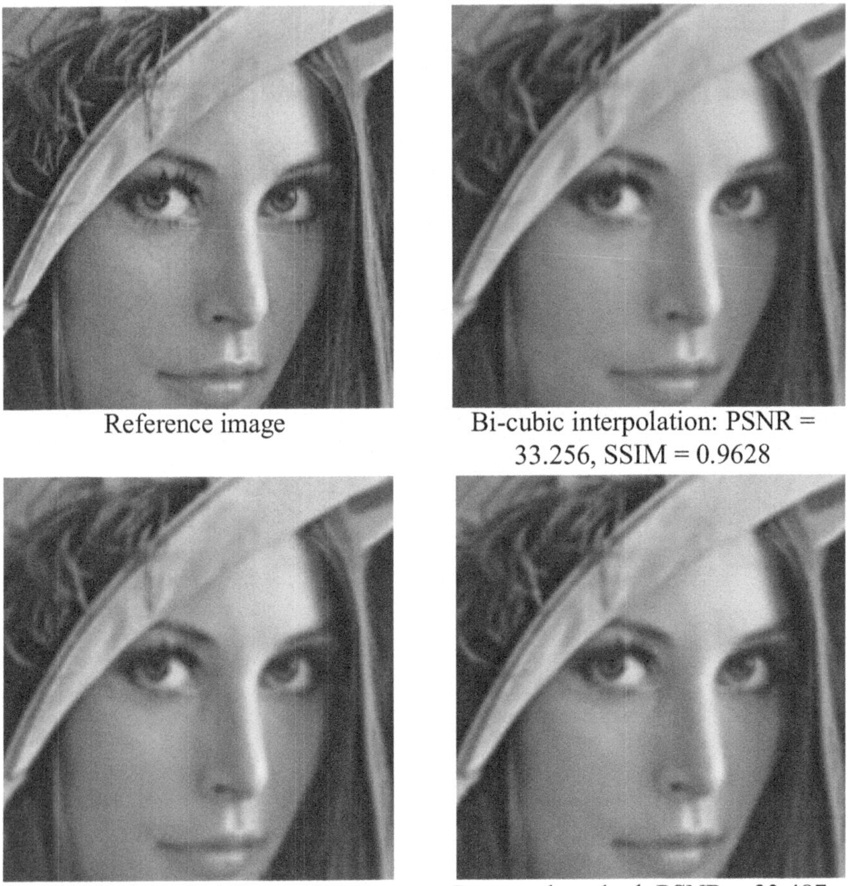

Reference image

Bi-cubic interpolation: PSNR = 33.256, SSIM = 0.9628

DCCI (Zhou et al., 2012): PSNR = 33.460, SSIM = 0.9631

Proposed method: PSNR = 33.487, SSIM = 0.9638

Fig. 8.18 The results of the proposed image interpolation algorithm

Table 8.1 Objective quality comparison

Image	Bi-cubic		DCCI Zhou et al. (2012)		Proposed	
	PSNR	SSIM	PSNR	SSIM	PSNR	SSIM
Lena	33.256	0.9629	33.46	0.9632	33.487	0.9638
Peppers	32.101	0.952	32.226	0.9518	32.232	0.9523
Goldhill	30.949	0.9243	30.87	0.9224	30.95	0.9244
Baboon	23.581	0.7972	23.585	0.7965	23.612	0.7994
Cameraman	25.499	0.9152	25.678	0.9163	25.619	0.9169

References

Allebach, J., Wong, P.: Edge-directed interpolation. In: Proceedings of International Conference on Image Processing, vol. 3, pp. 707–710 (1996)

Belekos, S., Galatsanos, N., Katsaggelos, A.: Maximum a posteriori video super-resolution using a new multichannel image prior. IEEE Trans. Image Process. **19**(6), 1451–1464 (2010)

Blu, T., Thvenaz, T., Unser, M.: Generalized interpolation: higher quality at no additional cost. In: Proceedings of International Conference on Image Processing, vol. III, pp. 667–671 (1999)

Chan, T., Shen, J.: Image Processing And Analysis: Variational, PDE, Wavelet, And Stochastic Methods. Society for Industrial and Applied Mathematics (2005)

Dai, D., Timofte, R., Gool, L.V.: Jointly optimized regressors for image super-resolution. Eurographics **34** (2015)

Dong, C., Loy, C.C., He, K., Tang, X.: September. Learning a deep convolutional network for image super-resolution. In: European Conference on Computer Vision, pp. 184–199. Springer International Publishing (2014)

Farsiu, S., Robinson, M., Elad, M., Milanfar, P.: Fast and robust multiframe super resolution. IEEE Trans. Image Process. **13**(10), 1327–1344 (2004)

Freedman, G., Fattal, R.: Image and video upscaling from local self-examples. ACM Trans. Graph. (TOG) **30**(2), 12 (2011)

Giachetti, A., Asuni, N.: Real-time artefact-free image upscaling. IEEE Trans. Image Process. **20** (10), 2760–2768 (2011)

Jensen, K., Anastassiou, D.: Spatial resolution enhancement of images using nonlinear interpolation. In: Proceedings of IEEE International Conference on Acoustics, Speech, and Signal Processing, ICASSP-90, vol. 4, pp. 2045–2048 (1990)

Kimmel, R.: Demosaicing: image reconstruction from color CCD samples. IEEE Trans. Image Process. **8**(9), 1221–1228 (1999)

Li, X., Orchard, M.: New edge directed interpolation. In: IEEE International Conference on Image Processing (2000)

Liu, Z.: Adaptive regularized image interpolation using a probabilistic gradient measure. Opt. Commun. **285**(3), 245–248 (2012)

Muresan, D., Parks, T.W.: New image interpolation techniques. IEEE 2000 Western New York Image Processing Workshop (2000)

Peleg, T., Elad, M.: A statistical prediction model based on sparse representations for single image super resolution. IEEE Trans. Image Process. **23**(6), 2569–2582 (2014)

Quak, E., Schumaker, L.: Cubic spline interpolation using data dependent triangulations. Comput. Aided Geom. Design **7**, 293–301 (1990)

Schulter, S., Leistner, C., Bischof, H. Fast and accurate image upscaling with super-resolution forests, In Proceedings of the IEEE Conference on Computer Vision and Pattern Recognition, pp. 3791–3799 (2015)

Sheikh, H., Sabir, M., Bovik, A.: A statistical evaluation of recent full reference image quality assessment algorithms. IEEE Trans. Image Process. **15**(11), 3440–3451 (2006)

Su, D., Willis, P.: Image Interpolation by Pixel-Level Data-Dependent Triangulation. In: Computer Graphics Forum, vol. 23(2), pp. 189–201. Blackwell Publishing Ltd. (2004)

Thévenaz, P., Blu, T., Unser, M.: Interpolation revisited. IEEE Trans. Med. Imaging **19**(7), 739–758 (2000)

Yu, X., Morse, B., Sederberg, T.: Image reconstruction using data-dependent triangulation. IEEE Comput. Graphics Appl. **21**(3), 62–68 (2001)

Yu, S., Zhu, Q., Wu, S., Xie, Y.: Performance evaluation of edge-directed interpolation methods for images. Comput. Vis. Pattern Recognit. (2013)

Zhou, D., Shen, X., Dong, W.: Image zooming using directional cubic convolution interpolation. IET Image Proc. **6**(6), 627–634 (2012)

Chapter 9
Changing the Aspect Ratio for Borderless Printing

9.1 Introduction

The use of digital photography and the photo printing market are growing very quickly, and users usually prefer borderless printing. However, the aspect ratio of digital images and photo paper very often do not match. The aspect ratio of an image describes the proportional relationship between the corresponding width and height of the captured or printed image. This ratio is commonly expressed as two integer numbers separated by a colon, for example, 16:9. The number before the colon represents the width of the image and the number after it represents its height. Both numbers represent a relationship, rather than a specific measurement. This means that for an $x : y$ aspect ratio, it does not make any difference how big or small the image is; if its width is divided into x units of equal length and its height y is measured using this same length unit, the height will be measured as y units.

The most common aspect ratios of digital images produced by consumer Digital Still Cameras (DSCs) or camera phones are 4:3, 3:2, and, more recently, 16:9 in consumer cameras. However, the most popular photo print size is 15 by 10 cm (6 by 4 inches). The paper size also does not match the image size when an image with an aspect ratio of 3:2 (prosumer and professional DSCs) is printed onto 22 by 15 cm paper.

Thus, in the case of borderless printing, some cropping or trimming of the image borders is necessary; that is, it is necessary to discard (not print) parts of the image. In some cases, this can deteriorate the composition and/or lose some important details of the photo; some parts of human bodies or faces can even be lost. For example, Fig. 9.1 demonstrates conventional trimming of a photo prepared for printing where a man's face is damaged.

I.V. Safonov et al., *Adaptive Image Processing Algorithms for Printing*,
Signals and Communication Technology,
https://doi.org/10.1007/978-981-10-6931-4_9

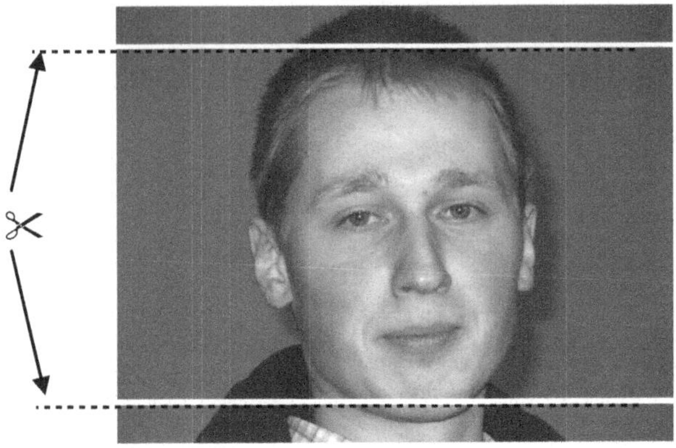

Fig. 9.1 Conventional trimming of photo during printing

To avoid possible artefacts, smart trimming methods are considered as an effective tool for changing the original aspect ratio of an image by trimming strips from the top and bottom (and/or left and right) sides of the image. The word "smart" means that such trimming will not corrupt the main objects in the image.

In practice, there are at least two approaches apart from digital processing methods to change an aspect ratio of a photo.

The *crop to fill* procedure consists in applying an equal amount of trimming to the top and the bottom of the image, or, alternatively, 20% of the needed trimming amount to the top and the remaining 80% to the bottom of the image. While, in general, this procedure produces satisfactory prints because photographers tend to frame the picture such that the main subject is in the middle of the picture, it often leads to catastrophic failures such as cropping off a subject's face, head, or other part of the body when it is located near the top or bottom side of the original image.

Another approach called *crop to fit* involves padding the right and left sides of a 4:3 image with white space(s) to obtain the desired aspect ratio. However, this approach does not make full use of the paper size.

Accordingly, the main problem in high-quality automatic trimming is to determine the main subject and background of the image.

The methods of detecting the main subject can be put into two major categories.

Pixel-based methods are designed to locate interesting pixels or *spots* ("blocks"), which usually do not correspond to entities of objects or subjects in an image.

Region-based methods are designed to locate interesting regions which correspond to semantically meaningful entities of objects or subjects in an image (Ma and Guo 2004). Most pixel-based approaches to region-of-interest detection are essentially edge detectors.

9.2 Related Work

There has been little research on automatic trim and crop due to the apparent difficulty involved in performing such a task. None of the known conventional image manipulation software applications use scene content in determining the automatic crop amount.

For example, a program entitled XV (http://www.trilon.com/xv), a freeware package developed by John Bradley at the University of Pennsylvania, USA (Department of Computer and Information Science), provides an autocrop function for manipulating images and operates in the following way.

1. The program examines a border line of an image, in each of four directions, namely from the top, bottom, left, and right sides.
2. The program checks the variation within the line. In greyscale images, a line must be completely uniform to be cropped. In colour images, both the spatial correlation and the spectral one should be low, except for a small percentage of pixels, for the line to be qualified for cropping. In other words, a line will not be cropped if it contains a significant amount of variation.
3. If a line along one dimension passes the criterion, the next line (row or column) inward is then examined.
4. The final cropped image is determined when the above recursive process stops.

This program essentially tries to remove relatively homogeneous margins around the borders of an image. It does not examine the overall content of the image. In practice, the *XV* program is effective in cropping out the dark border generated due to imprecise alignment during the scanning process. However, disastrous results can often be produced due to the apparent lack of understanding of the scene. In some extreme cases, the entire image can be cropped.

Bollman et al. (1999) provide a method for cropping images based upon the different intensity levels within the image. A typical image contains some regions where the intensity level and colour are uniform and other regions where they vary considerably. For instance, the "background" of an image may be uniform with a distinct "edge" separating the background from the "foreground". For example, a portrait typically comprises a subject set against a uniform backdrop or background such that a sharp edge or boundary exists between the subject and the background. With this system, an image to be cropped is scaled down to a grid and divided into non-overlapping blocks. The mean and variance of the intensity levels are calculated for each block. Based on the distribution of variances in the blocks, a threshold is selected for the variance. All blocks with a variance higher than the threshold variance are selected as regions of interest. The regions of interest are then cropped to a bounding rectangle. However, such a system is only effective when uncropped images contain regions where intensity levels are uniform and other regions where intensity levels vary considerably.

The effectiveness of such a system is expected to be comparable to that of the XV program. The difference is that the XV software examines the image in a line-by-line fashion to identify uniform areas, while the above method examines the image in a block-by-block fashion to identify uniform areas. In summary, neither technique can deal with images that have non-uniform backgrounds.

In the latest version of the package Microsoft Digital Image Suite 2006, an *Intuitive Crop* function is realized. It offers the possibility of intelligently analysing the composition of the photo, such as faces in portraits or family photos, when a consumer clicks on the *Crop* task and suggests a way to crop the photo that will properly frame the subject(s) and create the best possible image.

In the first place, the program suggests a cropping mode that is optimal, from the point of view of the automatic classifier, that is, *Custom* mode. Further, the user has the possibility of choosing the ratio of an image from a large list of standard printing formats. Besides, the user can set the ratio directly in pixels. There is also an opportunity to show guidelines for the rule of thirds.

Di Gesu et al. (1997) used the Discrete Moment Transform (DMT) and Discrete Symmetry Transform (DST) to measure local central moments about each pixel and local radial symmetry. To exclude trivial symmetry cases, nonuniform region selection is needed. The specific DMT operator acts like a detector of prominent edges (occlusion boundaries) and the DST operator acts like a detector of symmetric blobs. The results from the two operators are combined via a logic AND operation. Some morphological operations are needed to dilate the edge-like raw output map generated by the DMT operator.

An original method of detecting *salient regions* in an image which combines knowledge about the human visual system with computer vision techniques has been proposed by Milanese (1993). The suggested computational model of visual attention is structured into three major stages.

- First, multiple feature maps are extracted from the input image (orientation, curvature, colour contrast, and so on).
- Second, a corresponding number of "conspicuity" maps are computed using a derivative of the Gaussian model, which enhances regions of interest in each feature map.
- Finally, a nonlinear relaxation process is used to integrate the conspicuity maps into a single representation by finding a compromise among inter-map and intra-map inconsistencies.

To determine an optimal tonal reproduction, a system that partitions the image into blocks, combines certain blocks into sectors, and then determines the difference between the maximum and minimum average block values for each sector has been invented (Boyack 1999). A sector is labelled as an active sector if the difference exceeds a pre-determined threshold value. All weighted counts of active sectors are plotted versus the average luminance sector values in a histogram, which is then shifted via some predetermined criterion so that the average luminance sector value of interest will fall within a destination window corresponding to the tonal reproduction capability of a destination application.

Automatic identification of perceptually important regions in an image has been performed by Osberger et al. (1998). Several features influencing human visual attention to each region of a segmented image were extracted in order to produce an importance value for each feature in each region. The features mentioned include low-level factors (contrast, size, shape, colour, motion) and higher-level factors (location, foreground/background, people, context), but only contrast, size, shape, location, and foreground/background (determining the background by determining the proportion of the total image border that is contained in each region) were implemented in the resulting solution.

Segmentation of the foreground and background in colour images can also be realized by the integration of multiple cues (Huang et al. 1995). The algorithm represents a multi-level segmentation scheme that includes colour clustering, unsupervised segmentation based on the Minimum Description Length (MDL) principle, edge-based foreground/background separation, and integration of both regions as well as edge-based segmentation. In particular, the MDL-based segmentation algorithm was used to further group the regions from the initial colour clustering, and the four corners of the image were used to adaptively determine an estimate of the background gradient magnitude.

Another data-driven region selection method using colour region segmentation and region-based saliency measurement has been formulated by Syeda-Mahmood (1997). A collection of 220 primary colour categories was predefined in the form of a colour look-up-table. Pixels are mapped to one of the colour categories, grouped together through connected component analysis, and further merged according to compatible colour categories. Two types of saliency measures, namely self-saliency and relative saliency, are linearly combined using heuristic weighting factors to determine the overall saliency. In particular, self-saliency includes colour saturation, brightness, and size, while relative saliency includes colour contrast and size contrast between the region concerned and the surrounding region that is ranked highest among its neighbours according to size, extent, and contrast, in successive order.

In summary, almost all of the reported methods have been developed for targeted types of images: photography where the main subject is a person against a relatively simple static background (Marichal et al. 1996; Osberger et al. 1998); museum images, where there is a prominent main subject centred in the image against a large area of relatively clean background (Huang et al. 1995), and toy-world images, where the main subjects are a few distinctively coloured and shaped objects (Milanese 1993; Syeda-Mahmood 1997). These methods were either not designed for unconstrained photographic images or, if designed with generic principles, were only demonstrated for their effectiveness on rather simple images. The criteria and reasoning processes used were somewhat inadequate for less constrained images, such as photographic images.

The photographer intends to convey to the viewer his or her interest in one or more main subjects. Mechanisms for doing so include controlling the position of the main subject within the frame, arranging for contrast in tone, colour, or texture between the main subject and its surroundings, and highlighting the main subject with the brightest light. To the degree that a photographer is successful in emphasizing the main subject in a particular picture, a group of independent third-party viewers will agree well with one another as to the main subject, and we argue that this agreement will reflect the intention of the photographer. In practice, it is expected that an automatic system will not have the specific knowledge that the photographer may have about the people, setting, event, and so on. Therefore, the goal of our own development is to create automatic algorithms to identify, from the perspective of a third-party viewer, the main subject in a photograph and then to trim it.

9.3 Automatic Photo Image Trimming

At the time of creation of the algorithm, according to a study of a photographic image database of over 2000 images (Li and Jain 2005), over 70% of photographic images contain people and about the same percentage of images contain sizable faces. Indeed, people are the single most important subject in photographs. Therefore, an algorithm that can effectively detect the presence of human faces is useful in identifying the main subject of an image.

Although face recognition research started very early, there was not much attention to the face-detection problem in digital printing until recently. Over the last decades, greater attention has been given to the face-detection problem and there has been a large increase in the number and variety of applications attributed to face detection.

Given an image of arbitrary size, the task is to detect the presence of any human faces appearing in the image and, if there are any, to return their positions. The fact that human faces may appear at different scales and orientations and with different head poses makes face detection a challenging task. In addition, conditions such as lighting, non-rigidity of human faces, variation in facial expressions, and the presence of facial features such as glasses, make-up or beards contribute significantly to the variation of facial appearance in an image.

According to Li and Jain (2005), face detection methods can be mainly divided into two categories: (a) *appearance-based* and (b) *feature-based methods*. While the latter approach first extracts many point features and then matches these features with the face model, the former directly models the pixel intensities.

Obviously, a priori knowledge is needed in the feature-based approaches, such as the shape of the head, eye and mouth locations, colour of the face, texture, and 3D model of the face. On the contrary, the appearance-based approaches do not need any a priori knowledge; most of them treat the whole face as a vector of pixel intensities, which are then modelled using some techniques. Appearance-based approaches are known to be better suited to detecting non-frontal faces and more successful in complex scenes; however in simple scenes, feature-based approaches are more successful.

An automatic photo-image trimming algorithm consists of two interconnected parts: (a) analysis of the uniformity of the image's strips and preliminary trimming (see Fig. 9.2) and (b) face detection and corresponding correction of the trimming results (if needed).

To cut off one-third of an image with a size of $4l \times 3l$ pixels (l is an integer) along the vertical direction to transform its aspect ratio from 4:3 to 3:2, we should somehow position the cutting frame of height 8/3 l (see Fig. 9.3). Red dashed lines A and D in Fig. 9.2 depict the cutting frame centred in the image vertically. The green dashed line B is the top side of the frame having bottom position in the image. Finally, the blue dashed line C is at the bottom side of the frame having the top position in the image.

In the beginning, the given colour image is converted into a monochrome one with 256 grey levels and is divided into $P \times Q$ adjacent rectangular regions (depicted by light grey lines in Fig. 9.3).

For each rectangular region, texture features are calculated. The following six features were selected.

1. Mean brightness value.
2. Brightness value deviation σ^2.
3. Value $R = 1 - \frac{1}{1+\sigma^2}$.
4. Value $U = h_1^2 + h_2^2$, where h_l is a fraction of the region's pixels having values no greater than 128 and h_2 is the fraction of pixels having values greater than 128.
5. Value $e = -h_1 \log_2 h_1 - h_2 \log_2 h_2$.
6. Parameter $m = \begin{cases} 1, & \text{if } h_0 > h_1, \\ 2, & \text{else.} \end{cases}$

Having these empirically selected features, $P \times Q$ rectangular regions of the image are clustered into k groups. For the clustering, a *tree clusterization scheme* is utilized. After that, groups of image regions having similar texture properties are identified.

In the next step, for each row i of the image, the number $g(i)$ of different clusters among the image rectangular regions containing the selected row is calculated. The g values are depicted on the right side of Fig. 9.3. This number estimates the *degree of uniformity* of the image row. Rows with low values of g are supposed to be uniform and contain little important information, whereas rows with high values of g are supposed to be heterogeneous and contain a lot of important information.

Further, as shown in Fig. 9.2, depending on the value of g in a specific image region, preliminary image trimming is performed: (a) above line A and below line D; (b) above line B; or (c) below line C.

The values of P, Q, and k are the adjustable parameters of the method. In our experiments with 1024×768 images, we choose $P = 32$, $Q = 32$, and $k = 50$ as they give the best results.

As mentioned earlier, the second part of the algorithm performs face detection.

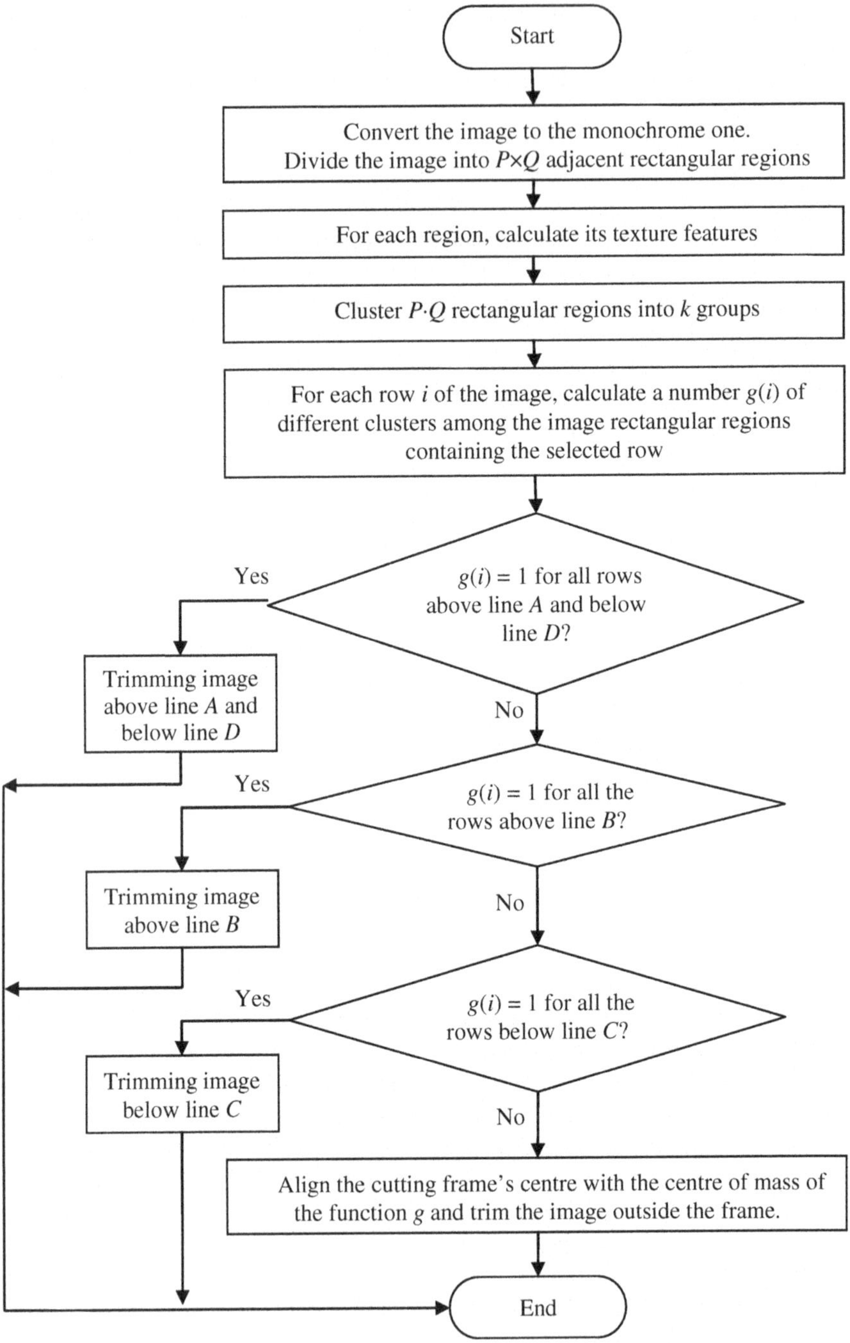

Fig. 9.2 Analysis of uniformity of the image's stripes and preliminary trimming

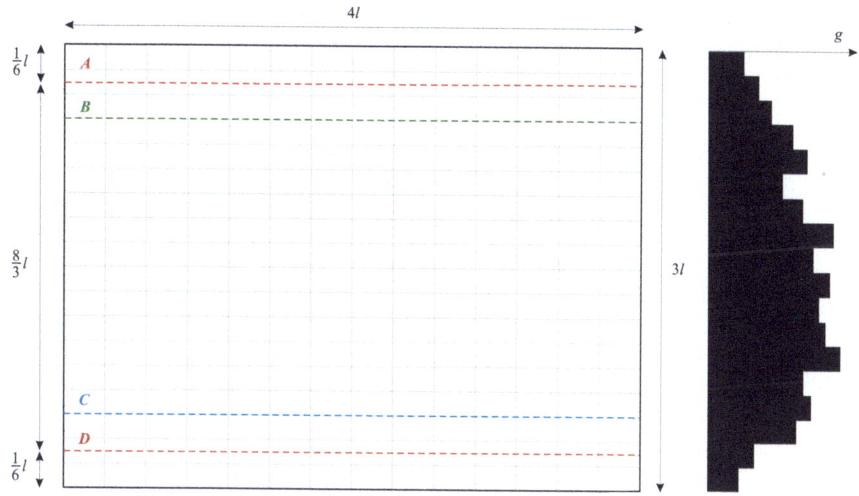

Fig. 9.3 Conventional trimming of photo during printing

There are several known face detection algorithms that can be used to achieve the goal. In the implemented solution, a proprietary face detector developed by Shakenov (2008) has been used. Face detection was carried out using the modified *Haar classifier cascade*. The output of this stage is a set of rectangular regions of different size that are supposed to be human faces in the image. Each region of the set is further analysed to see if it satisfies three conditions relevant to our application: (a) the *size*, (b) *skin*, and (c) *eyes and mouth* conditions.

Verification of the *size* condition is carried out as follows: a region that has a width of w pixels and a height of h pixels and is part of an image that has a width of v pixels and a height of u pixels satisfies this condition if one of the following inequalities holds:

$$\min\left(\frac{u}{h}, \frac{v}{w}\right) < 8 \text{ while } \mathrm{w} < 100,$$

$$\min\left(\frac{u}{h}, \frac{v}{w}\right) < 12 \text{ while } w \geq 100.$$

The *skin* condition is satisfied if the following inequality holds:

$$\frac{N_s}{N} \geq 0,2,$$

where N is the total number of pixels in the sub-region of the form shown in Fig. 9.4, while N_s denotes the number of sub-region pixels such that in RGB colour space their red component value is

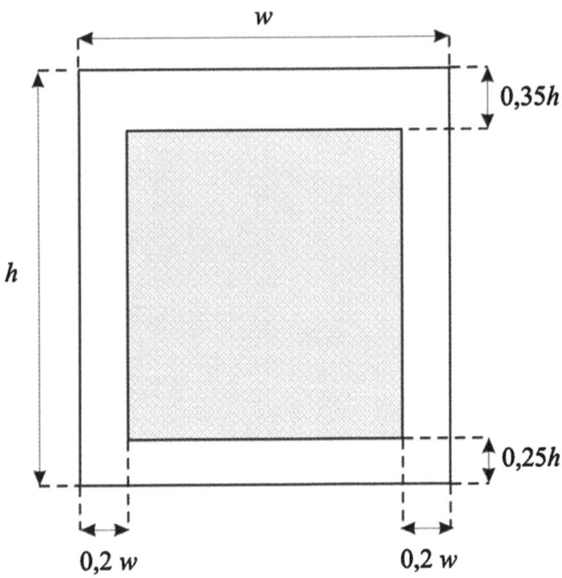

Fig. 9.4 Rectangular region that is supposed to be human faces in an image

1. greater than the green and blue component values, and
2. less than five times the blue component value.

Further, the *eyes and mouth* condition is checked in accordance with the flowchart shown in Fig. 9.5.

The detection of skin pixels is performed using the rules stated above. Around the revealed pixels, an ellipse is plotted by the least squares method. In the further stages, only the region inside this ellipse is considered. Eye and mouth filtering is performed by convolution with the mask having a size of $R_1 \times R_2$, as shown in Fig. 9.6, where $R_1 = 0.09\ h$ and $R_2 = 0.02\ w$. The obtained filtered image is binarized using the threshold $t = m + 2\sigma$, where m is the mean brightness of the filtered image and σ is the mean-square brightness deviation. After the binarization is completed, the connected components are extracted. In each of the circles A, B, and C shown in Fig. 9.7, only the connected component that has the biggest area is extracted.

Finally, the following conditions are checked.

1. Each circle should have one component.
2. The x-coordinate of the component's centre belonging to the circle C should lie in the segment $[x_A + 0,\ 2D,\ x_B - 0,\ 2D]$, where x_A and x_B are the x-coordinates of the selected components in the circles A and B, respectively, and D is the distance between the components.

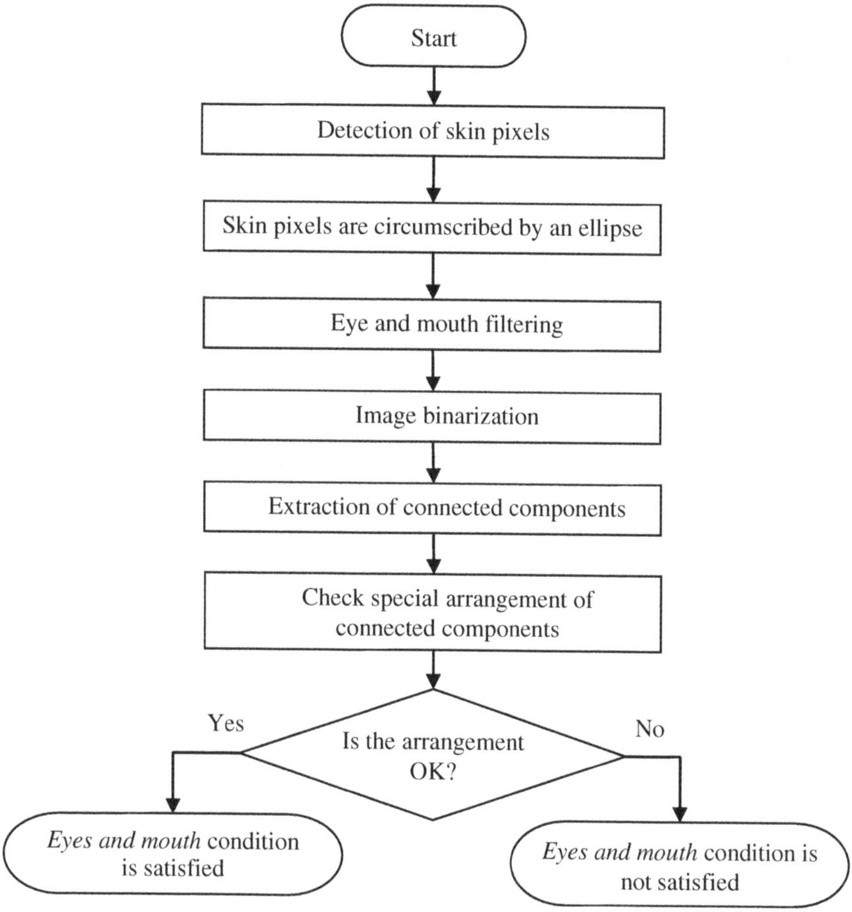

Fig. 9.5 Examination of the *eyes and mouth* condition

The *eyes and mouth* condition is considered fulfilled if both of the above conditions are fulfilled.

In case if some faces are found, a correction of preliminary trimming in the general flowchart of the algorithm is necessary. This correction can be done in the following way (Fig. 9.8).

1. The maximum y_t and minimum y_b coordinates of the rectangles that bound the detected faces are calculated.

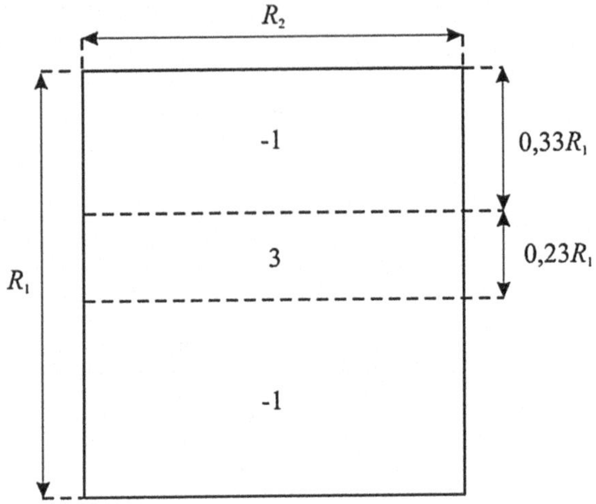

Fig. 9.6 Convolution mask of size $R_1 \times R_2$ for *eye and mouth* filtering

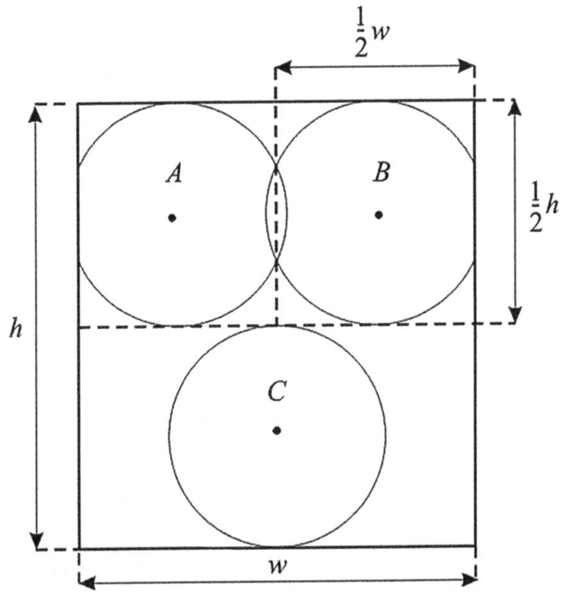

Fig. 9.7 Extraction of connected components

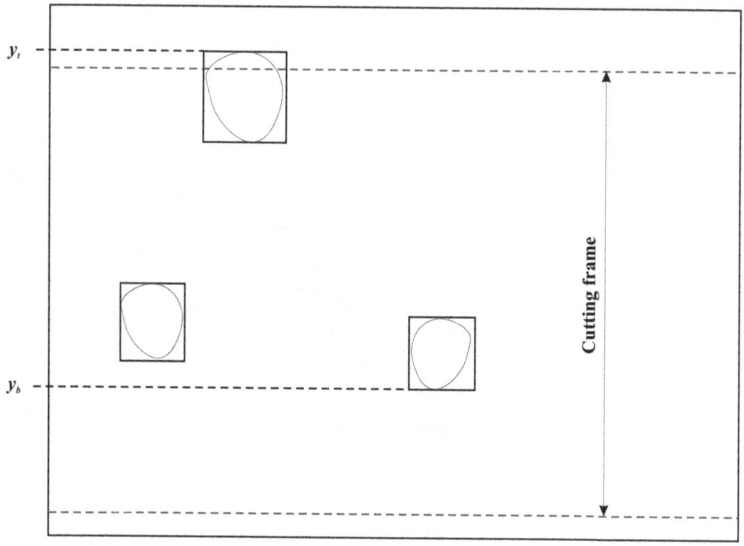

Fig. 9.8 Correction of trimming line position depending on the faces

2. The centre of the cutting frame is aligned with the centre of the vertical segment formed by the points y_t and y_b.

This operation ends the proposed smart trimming method.

Let us note the following.

The face detection procedure may be applied to the image under processing in parallel with the procedure of analysis of uniformity of the image's strips and preliminary trimming. Also, other types of semantic object detectors may be used in the presented algorithm to segment and preserve human bodies, buildings, and so on.

The computational time for the trimming method intended to be implemented in the printer may not be of critical value since typical photo printers may print one colour photo in dozens of seconds. Thus, image trimming in several seconds per image may be convenient.

Figure 9.9 shows examples of the images trimmed by the described method.

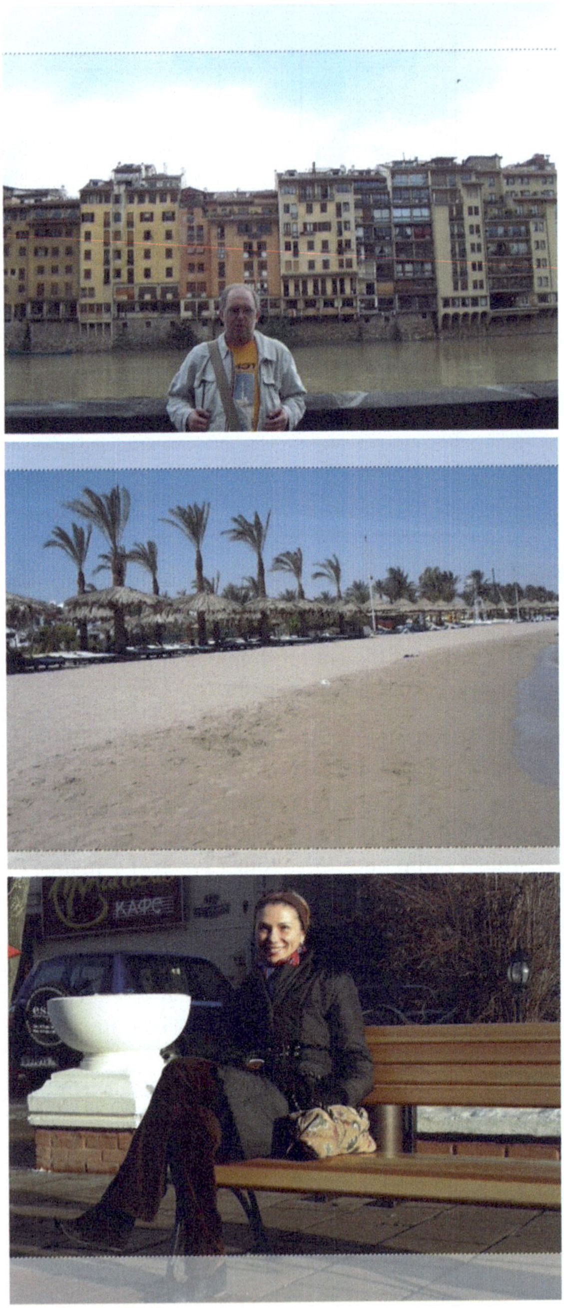

Fig. 9.9 Results of applying automatic photo trimming

9.4 Smart Complementing

9.4.1 Preliminary Remarks

In the preceding part of this chapter, we presented an algorithm for automatic photo image trimming. First, an image was segmented and after that the optimal widths of the strips to be trimmed were defined. However, this method may still result in the loss of some parts of the image. For example, it is not possible to crop the image shown in Fig. 9.1 in order to change its aspect ratio from 4:3 to 3:2 without clipping some parts of the head.

In this section, we describe an alternative way of matching the aspect ratio which is based on complementing an image with fragments extracted from the picture itself. The proposed approach is not applicable for all cases but there are situations in which such a method gives the best results in comparison with alternatives.

For example, Drori et al. (2003) describe the method for completing missing parts of an image resulting from the removal of foreground or background elements from the image. Visible parts of the image serve as a training set to infer the unknown parts and the method iteratively approximates the unknown regions and composites adaptive image fragments into the image. In-painting techniques restore and fix small-scale flaws in an image such as scratches or stains and are mainly oriented toward filling lost parts of the image inside some bounded area (Hirani and Totsuka 1996; Bertalmio et al. 2000). These methods work inside the bounded areas.

However, in our case the area to be filled is not bounded and the image must be complemented seamlessly. Several methods of complementing an image exist. One of the simplest and most visually pleasing methods is by mirroring several columns (or rows) relative to the side of the image. Figure 9.10 illustrates this approach. Without loss of generality, we will consider matching the aspect ratio from 4:3 to 3:2; for other aspect ratios, similar actions can be applied. Mirroring is performed on one or both sides; furthermore, the width of the mirrored strips can vary.

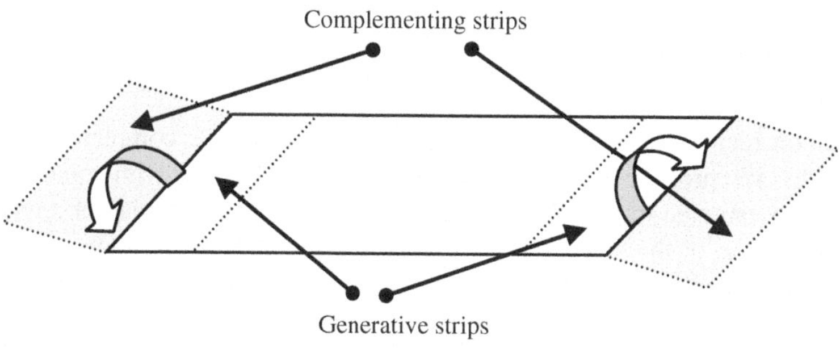

Fig. 9.10 Approach for matching the aspect ratio via mirroring

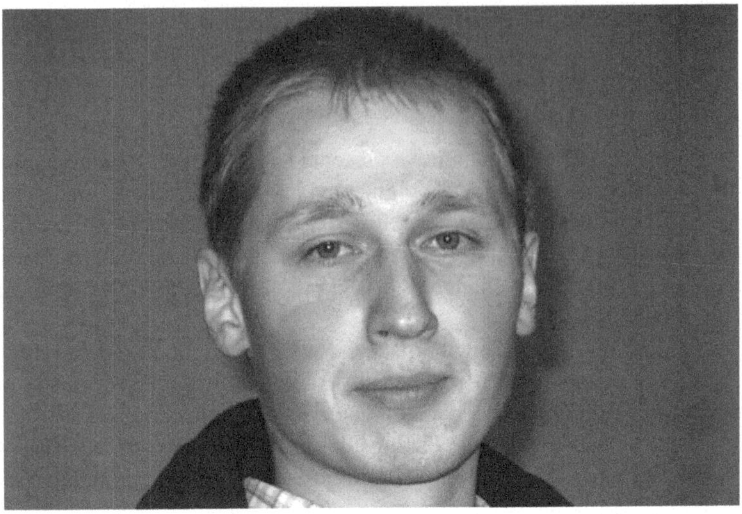

Fig. 9.11 Complementing of photo shown in Fig. 9.1

It is clear that this approach is not universal and that it depends on the content of the photo. We analysed a test set consisting of several hundred photos and found that our approach does not produce visible artefacts in about 30% of photos. It is possible to outline several scene types when mirroring is applicable: scenes with blurred backgrounds, for example, portraits, macro-shots, and some natural scenes (forests, farmlands, and mountains). Figure 9.11 illustrates the complementing of the photo from Fig. 9.1; in this case, artefacts are completely absent.

In general, complementing may be applied in addition to cropping to match the aspect ratio in photo kiosks and photo printers and can provide the user with the ability to choose the most pleasing version. For user convenience, an automatic classifier is developed to estimate the applicability of mirroring properly and exclude clearly unacceptable cases.

9.4.2 Segmentation

Let us consider a case in which an image needs to be complemented with W columns. We propose to analyse the left and right side strips with equal width of $S_d = W/2$ to estimate the applicability of mirroring. If mirroring is not applicable, the width of one of the side strips is decreased up to $W/8$ (at the same time as the opposite strip is enlarged) and analysis is performed for these strips. In the extreme case, only one side strip with width W is considered. In cases in which mirroring of both strips is possible, mirroring is applied; otherwise processing is terminated. Therefore, the key point is how to analyse the side strips.

The algorithm for strip analysis is based on *edge detection* and *edge extent analysis*. Elongation of the edges can be estimated using the extent or area or objects. Mirroring of strips containing some types of textures such as smooth regions will not produce noticeable artefacts. So, mirroring is allowed if there are no sharp edges of big connected regions.

First, the original colour photo is converted to a greyscale image. Then, the image is downsampled by, for example, *nearest neighbour interpolation*, to reduce the computation time and to reveal slightly blurred edges. The downsampled image is filtered with a *Laplacian of Gaussian (LoG) filter* for edge detection according to the following formula:

$$k(x,y) = \frac{(x^2 + y^2 - 2 \times \sigma^2) \times k_g(x,y)}{2\pi\sigma^6 \sum\limits_{x=-N/2}^{N/2} \sum\limits_{y=-N/2}^{N/2} k_g(x,y)},$$

$$k_g(x,y) = e^{-(x^2+y^2)/2\sigma^2},$$

where N is the size of the convolution kernel, σ is the standard deviation, and (x, y) is the coordinates of the Cartesian system with the origin at the centre of the kernel.

Segmentation is performed using *zero-crossings* (Parker 1997). It is assumed that a pixel with coordinates (r, c) belongs to an object if the following holds:

$$\begin{gathered}(|I_e(r,c) - I_e(r,c+1)| >= T \text{ and} \\ I_e(r,c) < 0 \text{ and } I_e(r,c+1) > 0) \text{ or} \\ (|I_e(r,c) - I_e(r,c-1)| >= T \text{ and} \\ I_e(r,c) < 0 \text{ and } I_e(r,c-1) > 0) \text{ or} \\ (|I_e(r,c) - I_e(r-1,c)| >= T \text{ and} \\ I_e(r,c) < 0 \text{ and } I_e(r-1,c-1) > 0) \text{ or} \\ (|I_e(r,c) - I_e(r+1,c)| >= T \text{ and} \\ I_e(r,c) < 0 \text{ and } I_e(r+1,c-1) > 0),\end{gathered}$$

where I_e is the result of *LoG* processing, and the threshold for zero-crossing segmentation I_e is determined by application of the following formula:

$$T = \frac{3}{4H_d W_d} \sum_r^{H_d} \sum_c^{W_d} |I_e(r,c)|,$$

where H_d and W_d are the height and width of the filtered image.

Further, I_e is segmented and labelled with eight-connectivity. Labelled connected regions with an area smaller than some threshold T_a are eliminated from processing. The area of a region is defined as the number of pixels belonging to it. If there are no objects greater than T_a, then mirroring of the strip is performed. If some of these regions are found then region M_l with the biggest area is analysed to find its extent. If the width of M_1 is greater than T_w then this region is rejected and the next biggest region M_2 is considered. In images with natural scenes, the most elongated edge is very often the skyline, so we decided to neglect the region with the largest area. If the area of M_2 is smaller than or equal to T_s, then mirroring is possible for a given strip. Otherwise mirroring is impossible.

Let us consider an example of side strips analysis of the photo from Fig. 9.12. Figure 9.13 demonstrates the absolute value of convolution of the downsampled image with the *Laplacian of Gaussian* filters, which is an image of the edges.

Figure 9.14 demonstrates the result of zero-crossing edge segmentation and labelling of the eight-connected regions with pixels belonging in the right-side strip and with an area greater than T_a. The histogram of areas of connected regions from Fig. 9.14 is shown in Fig. 9.15.

Note here that only one region that is big enough exists. We neglect it and only small regions are left with those areas that are smaller than T_s. The right side strip may be mirrored. A similar analysis is performed for the left side strip. The complementing result is shown in Fig. 9.16. The image is visually pleasant and does not contain visible artefacts.

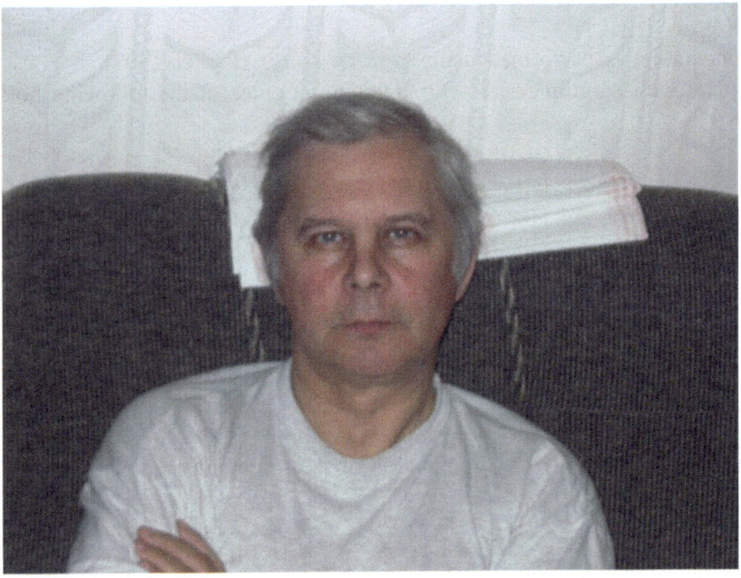

Fig. 9.12 An initial photo

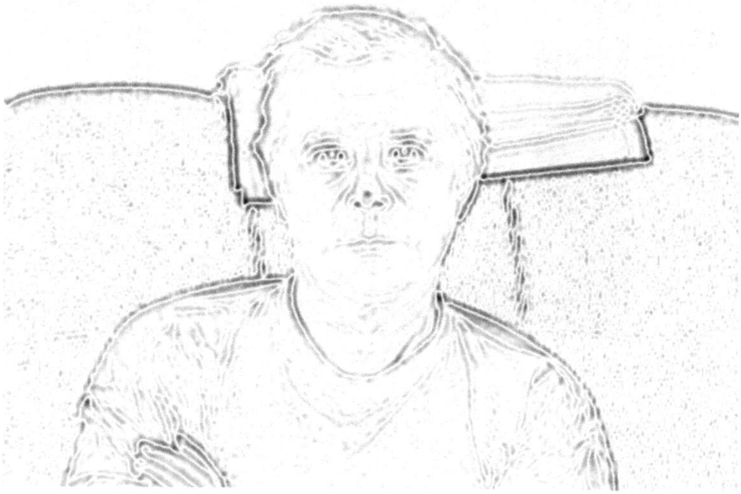

Fig. 9.13 Picture edges for the photo from Fig. 9.12

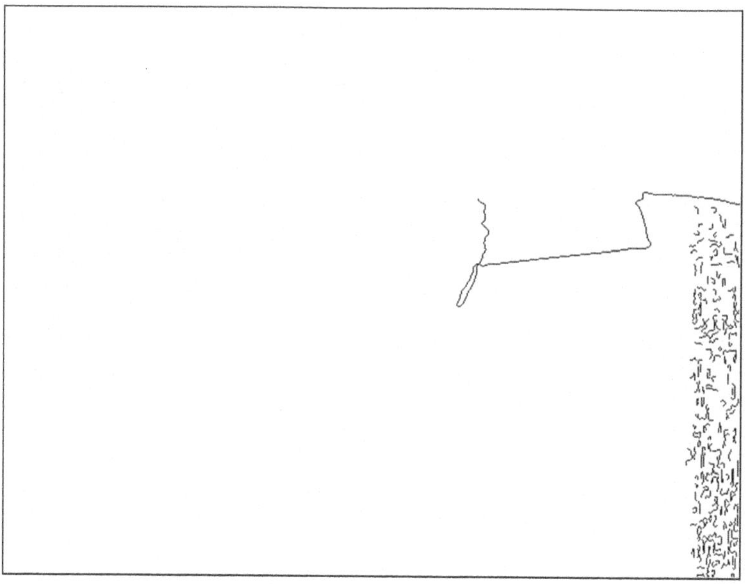

Fig. 9.14 Connected regions with pixels belonging to the right side strip

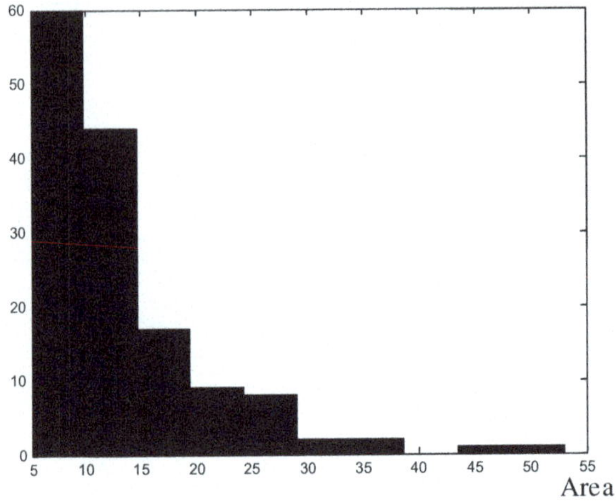

Fig. 9.15 Histogram of areas of connected regions from Fig. 9.14

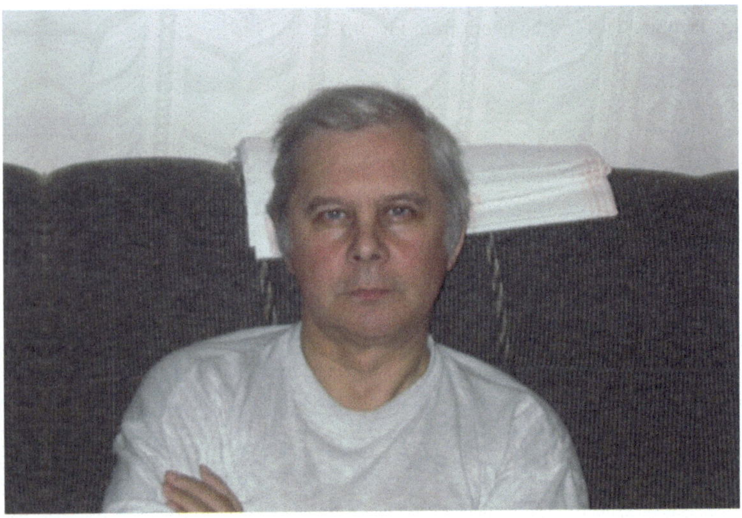

Fig. 9.16 Complementing of photo from Fig. 9.12

9.4.3 Additional Strip Processing

It is worthwhile to apply some modifications to the mirrored strips to suppress slightly noticeable artefacts produced by mirroring. These modifications should make the symmetry of mirrored areas less noticeable, on the one hand, and the boundary less visible, on the other. One of the possible realizations of this transformation is alpha blending of the mirrored strip with its blurred or darkened copy so that the opacity of the alpha channel a varies smoothly from 1 on the bound with the source image to 0.5–0.7 on the edge of the new photo:

$$I = Im \times a + Ip \times (1 - a),$$

where I_m is the mirrored strip, I_p is its darkened and blurred copy.

9.4.4 Results of Complementing

We collected 284 photos with an aspect ratio of 4:3 such that conventional trimming deteriorates their quality. Pictures were taken from private collections and from the Internet. The photos are of various sizes and were taken by different photographers using various models of DSCs and camera phones. Mirroring of 108 photos was applicable as a result of classification. The number of False Negatives (FNs) is 5. An FN occurs in cases in which the source image has a sharp background texture and, after zero-crossing edge segmentation, its edges stitch together in several large regions. The number of False Positives (FPs) is 27. FPs are typical for indoor photos and city/architecture photos. Another serious problem is posed by the imprinted time/date. An example of an FP is shown in Fig. 9.17. Therefore, to improve the quality of photo complementing, detection of the imprinted time/date (Zhu and Lin 2003) as well as detection of skin tones (Wang 2001) in side T strips may be necessary.

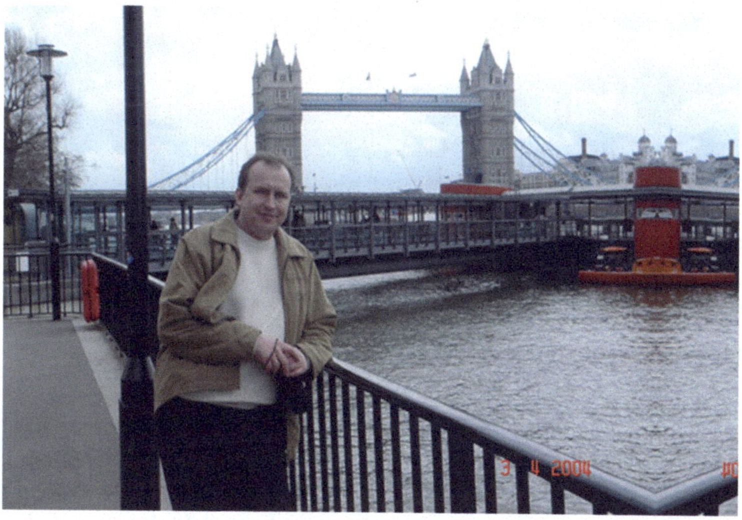

Fig. 9.17 Photo with wrong mirrored strip

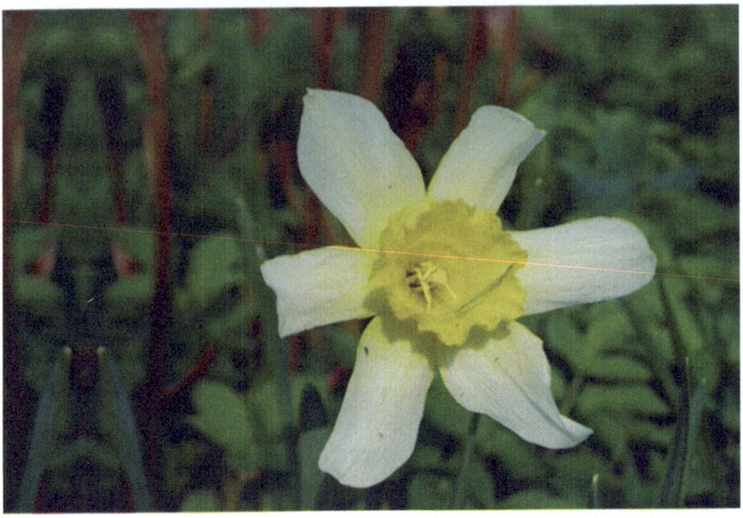

Fig. 9.18 Photo with mirrored left strip

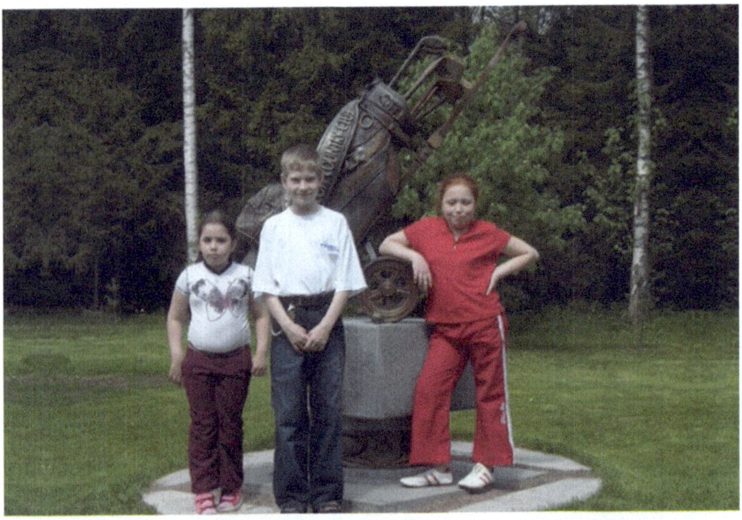

Fig. 9.19 Photo with two mirrored strips

Also, in our opinion, indoor/outdoor classification (Serrano et al. 2002) and city/landscape classification (Vailaya et al. 1998) can significantly improve the results, since in general the applicability of mirroring depends on the type of scene. Examples of proper complementing are shown in Figs. 9.18, 9.19 and 9.20.

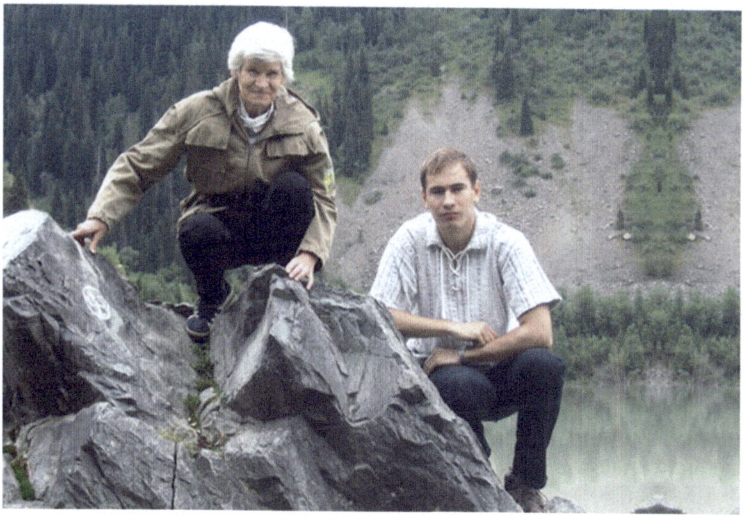

Fig. 9.20 Photo with mirrored right strip

Sometimes the effect of complementing is slightly noticeable under detailed examination. We showed these photos to a group of five observers who were not informed about the processing method. They did not notice the mirrored strips and evaluated the quality of the photos positively.

References

Bertalmio, M., Sapiro, G., Caselles, V., Ballester, C.: Image Inpainting. In: Proceedings of ACM SIGGRAPH, pp. 417–424 (2000)

Bollman, J.E., Rao Ramana, L., Venable, D.L., Eschbach, R.: Automatic image cropping, US patent 5,978,519 (1999)

Boyack, J.R., Juenger, A.K.: Brightness adjustment of images using digital scene analysis. US patent **5**(724), 456 (1999)

Di Gesu, V., Valenti, C., Strinati, L.: Local operators to detect regions of interest. Pattern Recogn. Lett. **18**, 1077–1081 (1997)

Drori, I., Cohen-Or, D., Yeshurun, H.: Fragment-based image completion. ACM Trans. Graph. **22** (3), 303–312 (2003)

Hirani, A.N., Totsuka, T.: Combining frequency and spatial domain information for fast interactive image noise removal. In: Proceedings of ACM SIGGRAPH, pp. 269–276 (1996)

Huang, Q., Dam, B., Steele, D., Ashley, J., Niblack, W.: Foreground/background segmentation of color images by integration of multiple cues. Proc. IEEE Int. Conf. Image Process. **1**, 246–249 (1995)

Li, S., Jain, A.: Handbook of Face Recognition. Springer Verlag, London (2005)

Ma, M., Guo, J.K.: Automatic image cropping for mobile device with built-in camera. In: Proceedings Consumer Communications and Networking Conference, pp. 710–711 (2004)

Marichal, X., Delmot, T., De Vleeschouwer, V., Macq, B.: Automatic detection of interest areas of an image or a sequence of images, pp. 371–374. ICIP, Lausanne, Switzerland (1996)

Milanese, R.: Detecting salient regions in an image: From biology to implementation, PhD thesis, University of Geneva, Switzerland (1993)

Osberger, W., Maeder, A.J.: Automatic identification of perceptually important regions in an image. In: Proceedings 14th International Conference on Pattern Recognition, Brisbane, Australia, August 1998, pp. 701–704 (1998)

Parker, J.R.: Algorithms for Image Processing and Computer Vision. New York, Wiley (1997)

Serrano, N., Savakis, A., Luo, J.: A computationally efficient approach to indoor/outdoor scene classification. In: Proceedings International Conference on Pattern Recognition, pp. 146–149 (2002)

Shakenov, A.K., Efimov, S.V., Nefyodov, A.V.: Method of automatic photograph framing, RU patent 2,329,535, Bull. 20 from July 20, (2008)

Syeda-Mahmood, T.F.: Data and model-driven selection using color regions. Int. J. Comput. Vision **21**(1), 9–36 (1997)

Vailaya, A., Jain, A., Hong, J.Z.: On image classification: city vs. landscape. In: Proceedings IEEE Workshop on Content-Based Access of Image and Video Libraries, pp. 3–8 (1998)

Wang, Y., Yuan, B.: A novel approach for human face detection from color images under complex background. Pattern Recogn. **34**, 1983–1992 (2001)

Zhu, X., Lin, X.: Automatic Date Imprint Extraction from Natural Images. Proc. ICICS-PCM **2003**, 518–522 (2003)

Chapter 10
Content-Aware Image Resizing

10.1 Introduction

Today, many devices for displaying and printing information surround us. They have various resolutions and aspect ratios and are suitable for the reproduction of images of arbitrary sizes. Thus transforming images of different natures for different devices and different targets is an important problem. This transformation to change the aspect ratio and/or resolution is sometimes called *retargeting*. The traditional method of size variation is *resampling* with employment of some interpolation method. In spite of significant progress in interpolation, its use can still worsen images in some cases and it may even be inapplicable. Sometimes some objects become indistinguishable after downsampling, and on the other hand, upsampling blurs them. Variation of the aspect ratio is the most significant modification: object proportions are not preserved and objects in the scene look unnatural.

In the years that preceded the moment of developing the approach submitted in this chapter, several methods aimed at non-proportional image size transformation to preserve the recognizability of the main objects of the scene were proposed. These approaches are problem-specific and work well only for limited types of images. However, the importance of the problem and emerging interest in this field allow it to be asserted that sooner or later content-aware image retargeting algorithms will be significantly improved and their main disadvantages will be surmounted.

In the current chapter, we describe a retargeting algorithm based on the main idea of the *seam carving* approach. Several disadvantages of traditional *seam carving*, especially during upsizing, are conquered by the use of a history map with relaxation. Also, we consider several application scenarios of the retargeting algorithm for demonstration of photos on display devices or in digital photo frames, borderless photo printing, copying documents while changing the size on a digital copier, and during transmission by fax machine.

I.V. Safonov et al., *Adaptive Image Processing Algorithms for Printing*,
Signals and Communication Technology,
https://doi.org/10.1007/978-981-10-6931-4_10

10.2 Related Work

There are many publications related to resizing via scaling or intellectual cropping based on the generation of salience or importance maps as well as other assumptions. However, here we mention only works related to content-dependent non-proportional image retargeting.

The large interest in retargeting arose from the need to display images on rather small displays of mobile devices. Setlur et al. (2005) propose a method of automatic image retargeting with the aim of providing small images effectively by preserving the recognizability and proportions of important image features during downsizing. Given a source image and a specification for the size of the output image, the general form of the suggested retargeting algorithm is represented as follows.

First, the image is separated into regions by means of segmentation. This process involves image segmentation and combination of adjacent regions based on the spatial distributions of colour and intensity. In order to identify important regions, an importance map of the source image based on a computational low-level vision model as well as a face-detection method is generated. If an area of the specified size contains all the important regions, the source image can be cropped. Otherwise, the important regions are removed from the image and the resulting "holes" in the background are filled.

The updated background is then resized to fit the input specification. Regions of importance are then "pasted" back onto the updated background based on importance and relative topology within the scene. If not all of the important regions can fit within the new image, they are resized in inverse proportion to their importance. This technique seems to minimize the loss of detail and distortion that traditional techniques introduce. However, this method does not ensure fully the photorealism of the transformed images; the retargeted pictures often look unnatural.

Another method that uses a nonlinear fisheye view warp that emphasizes parts of an image while shrinking others is described in Liu and Gleicher (2005). The fisheye-view warping employs image information such as low-level salience and high-level object recognition to find the *region of interest* (ROI) of the source image. However, unlike cropping approaches, a non-linear image warping function emphasizes the important aspects of the image while retaining the surrounding context. The fundamental limitation of the method is the assumption that the image has only a single ROI. The method is also not completely photo-realistic.

An impressive retargeting technique was proposed by Avidan and Shamir (2007). A simple image operator called seam carving that supports content-aware image resizing for both reduction and expansion is presented. A seam is an optimal connected path of pixels on a single image from top to bottom or from left to right, where optimality is defined via minimization of an image energy function. The energy function is the outcome of high-pass filtration of the brightness of the initial image.

Dynamic programming is used to solve the minimization problem. By repeatedly carving out or inserting seams in one direction, it is possible to change the aspect ratio of the image. By applying these operators in both directions, the image is thus retargeted to a new size. The technique is simple to understand and relatively fast. In recent times, several software implementations and modifications of seam carving can be found on the Internet. For the purpose of comparison, the seam carving SW implementation with a somewhat unusual name Liquid Resize 0.67 was used.

In spite of some impressive examples, the seam carving approach has serious limitations. In images where salient spatial structures appear, artefacts of significant shape distortion of the objects may appear as a result of both reduction and expansion.

Figure 10.1 demonstrates such artefacts even for modest changes in the number of columns. In addition, duplicate pixels are a common artefact in enlargement seam carving mode (Fig. 10.1c, d).

The seam carving approach was sufficiently improved by Rubinstein et al. (2008). First of all, the technique was extended to retargeting of video because showing widescreen movies on a TV with an aspect ratio 4:3 and vice versa is an important problem. Two main improvements for retargeting of both video and still images were introduced: the dynamic programming method of seam carving was replaced by graph cuts (Boykov and Jolly 2001) that are suitable for 3D volumes (two spatial dimensions and time), and a forward-looking energy function was proposed that measures the effect of seam carving on the retargeted image, not the original one. These improvements are very useful but the majority of limitations remain, especially for enlargement.

Wolf et al. (2007) propose a content-driven video retargeting method. The technique consists of two stages: the video frame is analysed to detect the importance of each region in the frame; then less important regions are shrunk more than important ones. This analysis is done in fully automatic mode and is based on local

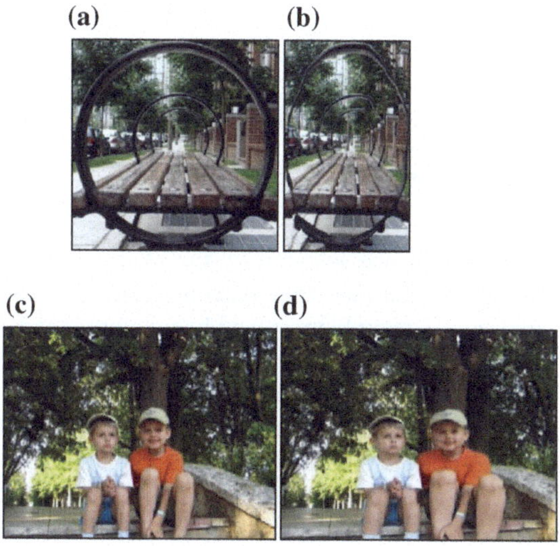

Fig. 10.1 **a** and **c** Original photos; **b** and **d** artefacts introduced by seam carving operator. Image (**a**) is taken from http://www.flickr.com/photos/mag3737/150194930/sizes/m/

saliency, motion, and object detection. Results are shown mainly for reducing video sizes. Chen and sen (2008) describe a method of summarizing a video into a short segment while preserving the important events in the original by application of an idea similar to seam carving.

In Simakov et al. (2008), the problem of summarization or retargeting to smaller sizes is posed as an optimization problem of so-called *bi-directional similarity.* This is a measure of the similarity between two images or videos of different sizes. The authors demonstrate the summarization results, which outperform seam carving ones. However, the summarization approach is significantly slower than seam carving and also has serious limitations.

Intelligent resizing of documents that contain images and scanned images of documents is a crucial task too. In Lin (2006), the content fitting techniques in the context of variable data printing for digital publishing are discussed. The approach is based on cropping images embedded into a document, although in this case the retargeting approach can be beneficial. Several techniques for adaptation of document images to display constraints, including seam carving according to Avidan and Shamir (2007), were compared in Berkner and Erol (2008). The authors noticed that seam carving is unsuitable for retargeting of documents because it is task-independent, whereas task dependency and content-dependent scaling are required for retargeting of documents. In general, for small displays it is true; however there are applications of seam-carving-based retargeting for documents.

As already pointed out in the Introduction, we do not aim to summarize all the related manuscripts devoted to each item under consideration in this book. Our goal is instead to present our own original methods and techniques. Nevertheless, during the preparation of the text, some interesting publications that have recently appeared have been added. One of these is a review paper devoted to a comparative study of image retargeting techniques by Rubinstein et al. (2010); the other two are surveys summarizing and categorizing recent works from the image retargeting literature (Vaquero et al. 2010; Bhujbal and Deshpande 2015).

10.3 Proposed Approach

10.3.1 General Workflow

Our approach was inspired by the seam carving technique, which defines a seam carving operator that supports content-aware image resizing for both reduction and expansion. The problem solved by our approach is image conversion to another size and/or aspect ratio in a manner that preserves the size and aspect ratio of the most important features of the image. To preserve the connectivity of the image features and to avoid the appearance of image artefacts, at least one connected group of the least significant pixels in the vertical or horizontal direction is computed during each resizing step. The nature of the proposed method is iterative and in each resizing step only one connected group of pixels is removed or added to a new resized copy of the image. This group of connected pixels is called a *seam* in Avidan and Shamir (2007); we call it a *path*. Subsequently, the obtained image becomes an input for the next iteration and the process is repeated until the designated retargeting is achieved.

The special function is used to assign a significance cost to every pixel in the processed image. Our cost function has the same purpose as the *energy function* in Avidan and Shamir (2007), but the cost function depends on both the image content and the application method. This function should be selected in one of the following ways:

- use of an additional pre-processing step to determine the image content and cost function which better preserve the content details;
- the cost function is chosen by the operator;
- the cost function is provided with the system that utilizes the present retargeting method.

After the cost function has been chosen, a cost map is computed. Each element of the cost map represents the significance of the image pixel. It is used to assign the weights to the graph arcs that are directed to this element. A regular-mesh graph is constructed to solve the optimization problem of the optimal connected path search in the vertical or horizontal direction. The optimality is defined as the path with the minimum cumulative cost. This path can be found using a dynamic programming algorithm.

The retargeting method defined above has some limitations especially for image enlargement. It is clear that after finding the optimal path there is still a big chance of finding the next path near the same position in the next iteration. This leads to stretching of the artefact by duplicating or reducing the similar paths through all the iterations. The use of forward-energy as in Rubinstein et al. (2008) is useful but does not overcome the drawback completely. We introduce the *history map* to avoid such artefacts and overcome the known limits on resizing. It is used to assign an additional pixel significance that does not depend on the image but does depend on previous iterations of processing. The history map is initialized at the beginning of retargeting and is added to the cost map in each iteration step. During the enlargement operation, the path found is duplicated on the processed image and history map. The elements of the history map that belong to the found path and its duplicated copy are assigned a maximal value. Such an assignment prevents the selection of the same path in the next iteration step since the cumulative cost will be greater than other available costs.

Repetition of this process will most likely create a history map filled with elements with maximal value and the path solver will turn out in the same awkward situation. To avoid this issue, a relaxation procedure is applied. The main task of this procedure is to decrease the value of history elements in time. Thus, oversaturation of the history map is eliminated. A similar approach is applied for image shrinking. The history map is used as an accumulation map of paths to be removed.

The relaxation function is characterized by the following equation:

$$H(i,j) = H(i,j)^{-1} - C,$$

where $H(i,j)$ is the value of the relaxed history map element, $H(i,j)^{-1}$ is the value of the history map element before relaxation, and C is a pre-defined constant which controls the speed of fading of paths in a history map. The higher the value of C, the more frequently the same path will be chosen, which can lead to a stretching artefact. The smaller the value of C, the more uniform the distribution of paths in an image will be. Our modelling shows that good results can be obtained with $C = 2$, which means that a recent path will disappear in a history map in approximately 128 steps.

The procedure of adaptive image retargeting can be represented in the following form (Fig. 10.2).

1. History map initialization.
2. Cost function selection depending on image content.
3. Cost map calculation by means of cost function.
4. History map relaxation.
5. Addition of history map to cost map.
6. Optimal path search.
7. Updating of processed image with at least one extracted or added path.
8. Updating of history map with extracted or added path.

Steps 3–8 are repeated until the designated image size has been achieved.

Figure 10.3 demonstrates the cost map and history map as well as the outcome of the combination of history and cost maps for the initial image from Fig. 10.1c after 100 enlargement steps.

Black colour corresponds to the highest cost; white colour corresponds to the lowest cost. It is possible to see that skin tones and edge regions have high cost

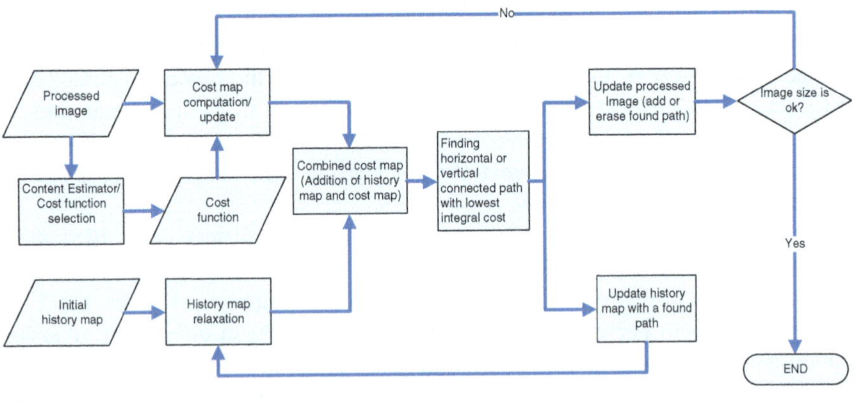

Fig. 10.2 The procedure of adaptive image retargeting

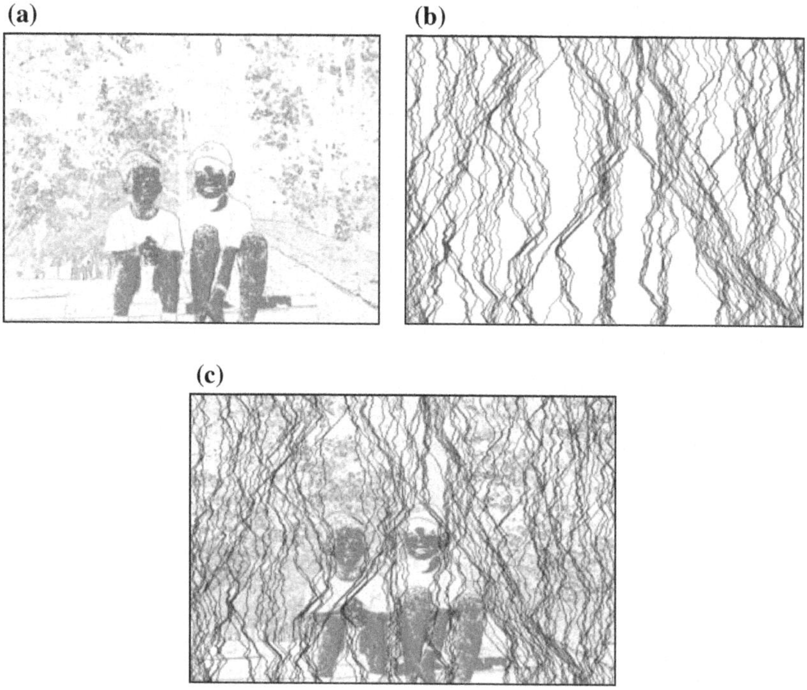

Fig. 10.3 Cost map (**a**), history map (**b**), and cost map combined with history map (**c**), all after 100 enlargement steps

values; the old paths in the history map have lower costs than recent ones. In general, the cost function F can be given by the following equation:

$$F(RGB, x, y) = grad(RGB, x, y) + POI(RGB, x, y),$$

where *RGB* is a processed image in RGB colour space; *x* and *y* are pixel coordinates, *grad(RGB, x, y)* defines the gradient in a pixel with coordinates (*x*, *y*), and *POI(RGB, x, y)* is a pixel of interest function.

The function *grad(RGB, x, y)* defines the general pixel significance in terms of image edges. The Sobel operator, module of colour gradient, Laplacian of Gaussian, and other edge detection techniques can be used for definition of this function. However, it is not only edge pixels that are important in digital photos. The pixel of interest function is introduced to add additional significance to flat but important image features such as human skin, architectural features, and so on.

Figure 10.4 shows a flow chart of the simple photo content estimator presented. At the beginning, this routine performs human detection. An approach described in Gomez and Morales (2002) for automatic detection of human skin defines a *POI* function capable of preventing erasing or stretching of human skin features during the retargeting operation:

$$POI(x,y) = M \times (\frac{b(x,y)}{g(x,y)} < 1.249$$
$$AND \frac{r(x,y)+g(x,y)+b(x,y)}{3r(x,y)} > 0.696$$
$$AND\ (\frac{1}{3} - \frac{b(x,y)}{r(x,y)+g(x,y)+b(x,y)}) > 0.014$$
$$AND \frac{g(x,y)}{r(x,y)+g(x,y)+b(x,y)} < 0.108),$$

where M is a relatively big predefined constant,

$$r(x,y) = \frac{R(x,y)}{R(x,y)+G(x,y)+B(x,y)},$$
$$g(x,y) = \frac{G(x,y)}{R(x,y)+G(x,y)+B(x,y)},$$
$$b(x,y) = \frac{B(x,y)}{R(x,y)+G(x,y)+B(x,y)}.$$

In general, a point of interest, or *POI*, characterizes a specific point location that someone may find useful or interesting.

A stage for additional pixel significance addition is applied to the cost map if a human detector is triggered. City/building and indoor/outdoor detectors are used to detect artificial objects such as building and architecture features. If at least one of these detectors is triggered then an additional cost is assigned to the cost map elements which belong to straight elongated lines. One of the realizations of indoor-outdoor image classification is given in Sect. 10.3.3.

Figure 10.5 illustrates the advantages of our cost function over seam carving for the shrinking procedure. In contrast to conventional seam carving, which distorts the children's bodies and faces, our retargeting mainly alters the background.

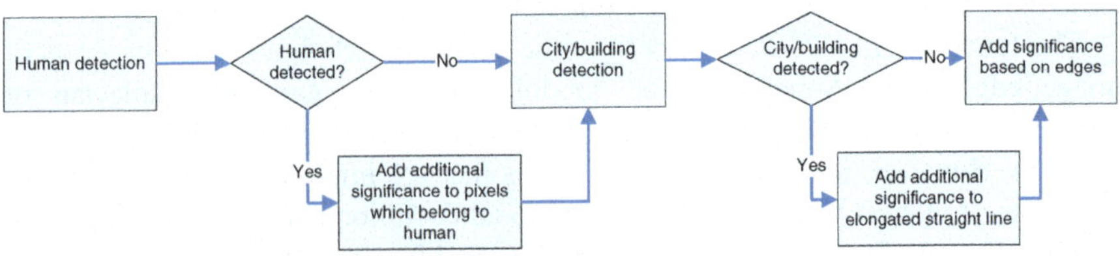

Fig. 10.4 Flow chart of photo content estimator

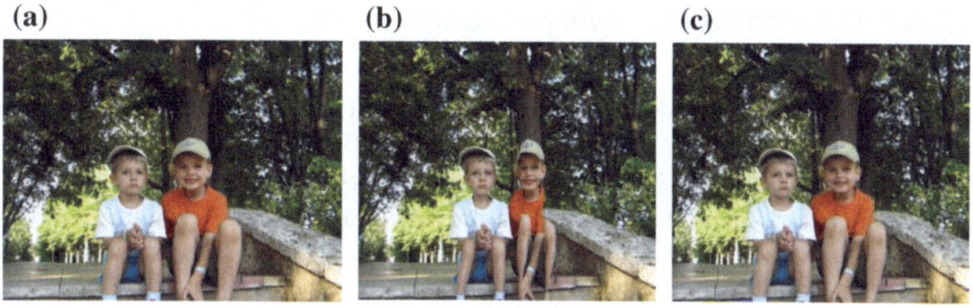

Fig. 10.5 Original image (a), conventional seam carving (b), proposed approach (c)

10.3.2 Additional Processing

The additional processing of pixels along the removed path and pixels on the added path allows for improvements to the image quality. Let us consider the removed path in the shrinking step (Fig. 10.6a); the pixels of the removed path are denoted as black circles. The removal of the path can produce new artificial edges between pixels, which are denoted as grey circles and squares. It is reasonable to modify these pixels using a Gaussian blur. A similar approach using path addition in an enlargement step is also valuable (Fig. 10.6b).

The pixels of the path with minimal cumulative cost are denoted by black circles, while the pixels of the added path are denoted by grey circles. Simple duplication of pixels looks unnatural, and we therefore propose to interpolate the pixels of the added path within the vicinity of the path. These additional processing operations provide better quality photos, but should be applied with care to scanned images of documents.

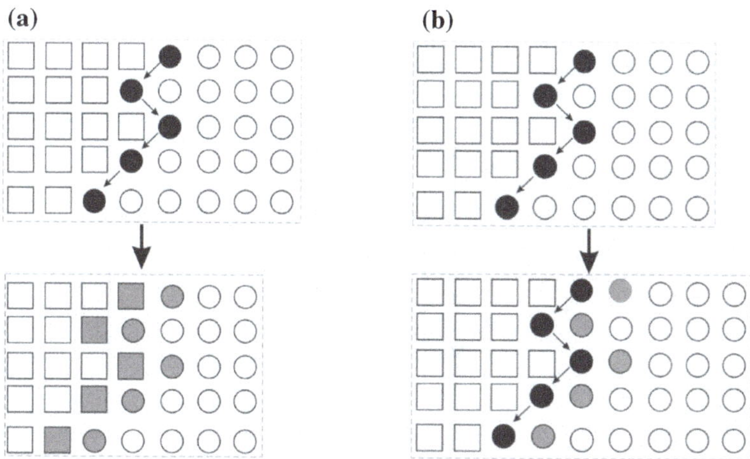

Fig. 10.6 **a** Removal of path on shrinking; **b** addition of a new path for enlargement

10.3.3 Indoor/Outdoor Image Classification

There are number of indoor/outdoor and landscape/city classification approaches. For instance, efficient approaches are described in the following papers (Vailaya et al. 1998; Serrano et al. 2002; Ghomsheh and Talebpour 2012). These detectors are independent of each other and can be applied in an arbitrary sequence within a retargeting procedure. Additional significance based on edges is added to cost map elements at the end of the content estimator routine.

We can convert a given image from RGB into LST colour space using the following relations:

$$L = \frac{k}{\sqrt{3}}(R+G+B),$$
$$S = \frac{k}{\sqrt{2}}(R-B),$$
$$T = \frac{k}{\sqrt{6}}(R-2G+B),$$

where $k = 255/max(R, G, B)$, and $max(R, G, B)$ denotes the maximum value of the red, green and blue channels of the image. This colour space transformation is used to de-correlate colour channels in the original RGB image. Following this, the image is subdivided into fragments (Fig. 10.7), and both colour and texture features are calculated for each fragment (Efimov et al. 2007).

For texture description, a wavelet analysis is used; this provides a texture description in terms of image frequencies at different spatial scales. The particular choice of wavelet filter is not critical for texture analysis. In our case, these features are obtained with the help of a two-level wavelet decomposition using Daubechies' filter (Serrano et al. 2002).

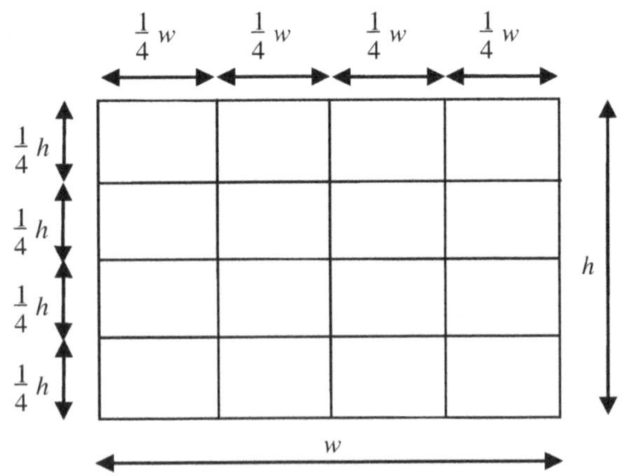

Fig. 10.7 Image subdivision in fragments for indoor/outdoor classification

Texture features $e_1, e_2, \ldots, e_8$ are defined according to the standard equation:

$$e_k = \frac{1}{M_k N_k} \sum_{i=1}^{M_k} \sum_{j=1}^{N_k} c_k^2(i,j),$$

where $k = 1, 2, \ldots, 8$, and $c_k(i, j)$ are sub-band coefficients of the two-level wavelet decomposition of the image fragment at level k.

The low-pass filter $h(*)$ and the high-pass filter $g(*)$ are defined using the optimized coefficients which are found during numerical modelling.

1st level of decomposition

Approximation coefficients	$c_1(i, j)$	$i = 1, 2, \ldots, M_1$
		$j = 1, 2, \ldots, N_1$
Horizontal detail coefficients	$c_2(i, j)$	$i = 1, 2, \ldots, M_2$
		$j = 1, 2, \ldots, N_2$
Vertical detail coefficients	$c_3(i, j)$	$i = 1, 2, \ldots, M_3$
		$j = 1, 2, \ldots, N_3$
Diagonal detail coefficients	$c_1(i, j)$	$i = 1, 2, \ldots, M_4$
		$j = 1, 2, \ldots, N_4$

2nd level of decomposition

Approximation coefficients	$c_5(i, j)$	$i = 1, 2, \ldots, M_5$
		$j = 1, 2, \ldots, N_5$
Horizontal detail coefficients	$c_6(i, j)$	$i = 1, 2, \ldots, M_6$
		$j = 1, 2, \ldots, N_6$
Vertical detail coefficients	$c_7(i, j)$	$i = 1, 2, \ldots, M_7$
		$j = 1, 2, \ldots, N_7$
Diagonal detail coefficients	$c_8(i, j)$	$i = 1, 2, \ldots, M_8$
		$j = 1, 2, \ldots, N_8$

To classify the current fragment into the indoor or outdoor class using these calculated features, a *Support Vector Machine* (SVM) is applied with a radial basis kernel function of

$$K_1(u, v) = \exp(-\|u - v\|^2),$$

where $\|u - v\|$ is a norm of the vector $u - v$. Each of the fragments of the image is classified using an individual SVM.

The final classifier is applied to the outputs of the classifiers associated with the fragments, in order to classify the image as whole. For this classification, an SVM with the polynomial kernel function $K_2(u, v)$ is used:

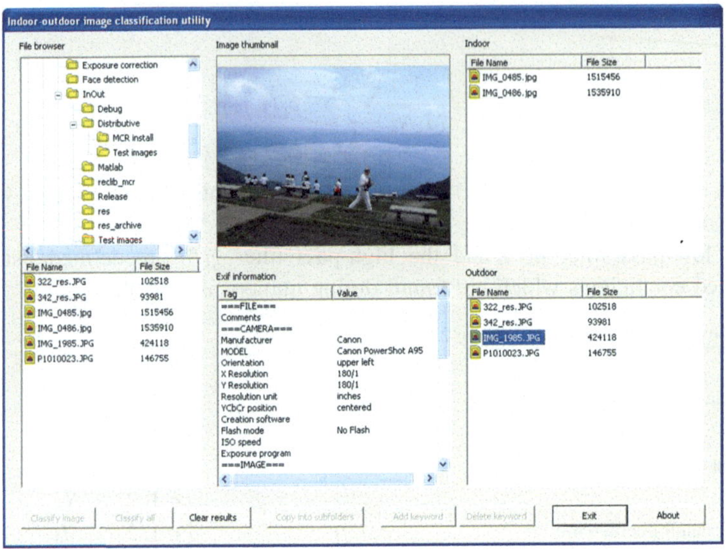

Fig. 10.8 Main window of the indoor/outdoor image classification demo utility

$$K_2(u, v) = \left(\frac{1}{10}(u, v)\right)^3,$$

where (u, v) is the scalar product of the vectors u, v.

The reported accuracy of the methods for image classification into indoor/outdoor classes varies at the moment of algorithm creation between 90 and 94% for different image databases. During testing, the proposed classification method provided a classification accuracy of 92.5% on a test base of more than 1000 images.

The classification algorithm described in this section has been implemented in a software utility. This utility is realized as a stand-alone application (Fig. 10.8) and as a component of the commercial software package *PhotoThru* (Fig. 10.9).

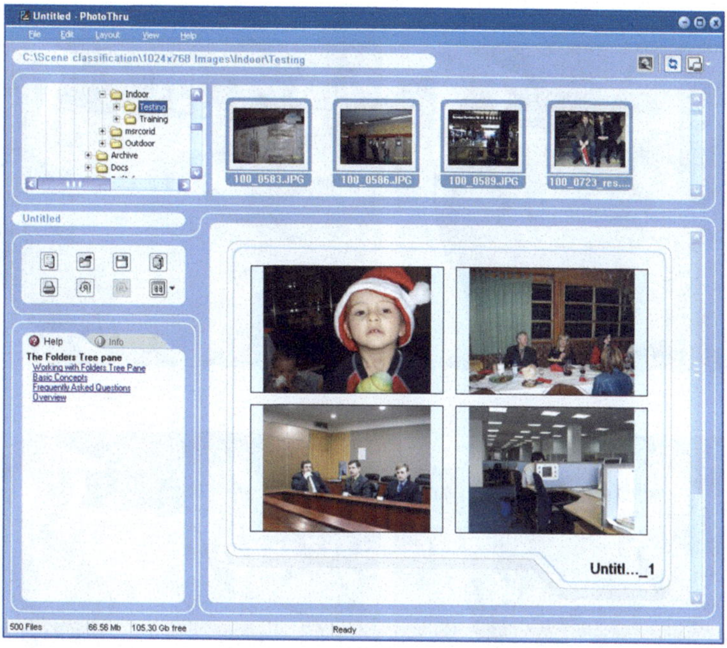

Fig. 10.9 Screenshot of a commercial software package for printing, with the implemented indoor/outdoor classification option

10.4 Results for Various Retargeting Applications

10.4.1 Aspect Ratio Matching in Borderless Photo-Printing

As stated in Chap. 9, in the case of borderless printing, some cropping or trimming of the image borders is necessary; that is, it is necessary to discard (not print) parts of the image. In some cases this can deteriorate the composition and/or lose some important details and some parts of the human body or face can be lost.

For example, Fig. 10.10 demonstrates conventional trimming of a photo with an aspect ratio of 4:3 where during borderless printing on paper of 4 × 6 inches about 6% of the rows are trimmed from the top and bottom. The bird's head is damaged, which is an unwanted situation and will be evaluated negatively by the user. The results of matching the aspect ratio via complementing (Safonov and Tolstaya 2007) are also given in Chap. 9.

In general, the result of complementing (Fig. 10.10c) looks good but close examination of the photo reveals that several columns in the right part of image are mirrored. Enlargement by means of retargeting does not produce any noticeable artefacts.

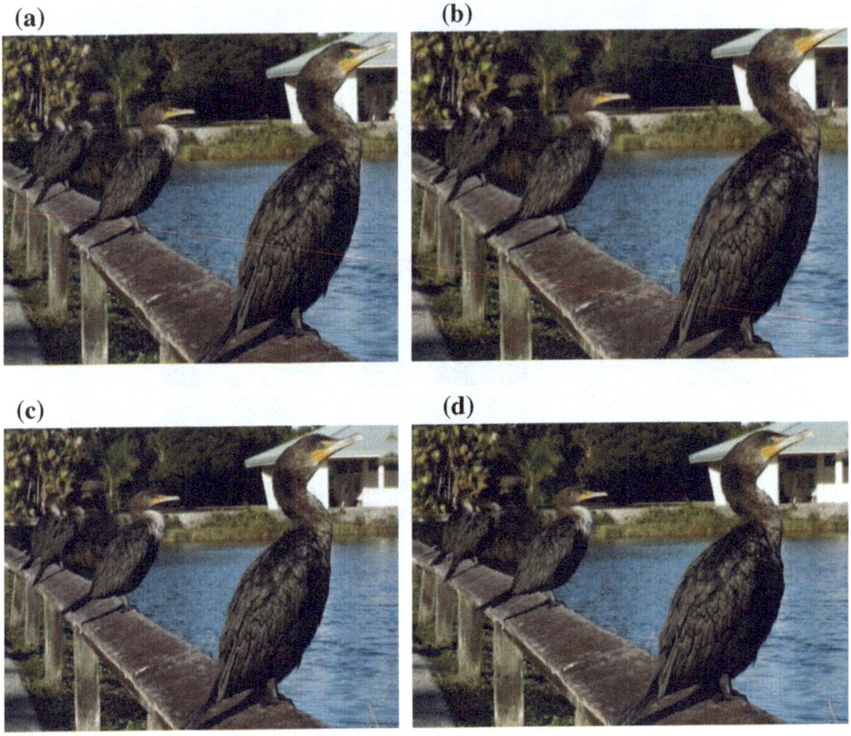

Fig. 10.10 **a** Original image with aspect ratio 4:3; **b** conventional trimming; **c** complementing; **d** retargeting. For **b**, **c**, and **d**, the aspect ratio is 3:2

10.4.2 Fitting of Web Pages

An important application of image retargeting technology is HTML document reformatting. Often, it is cumbersome to resize an HTML page to a size that is smaller than a predefined value due to the page sizes being strictly fixed by developers. Another issue is images on Web pages that cannot be resized without artefacts by traditional approaches. Thus a problem with printing and visualization of HTML pages on devices with small resolution displays such as mobile phones and personal digital assistants (PDA) is arising. PDAs, also known as a handheld PCs, are the mobile devices that function as the personal information managers since the early 2010s after the widespread adoption of highly capable smartphones, in particular those based on iOS and Android.

To overcome such an awkward situation, an HTML document resizing technique is introduced. All images are extracted from the original HTML document. The extracted images are processed by means of retargeting to change the size and/or aspect ratio. Further, a new HTML document is composed with new retargeted images. Text fields are altered too.

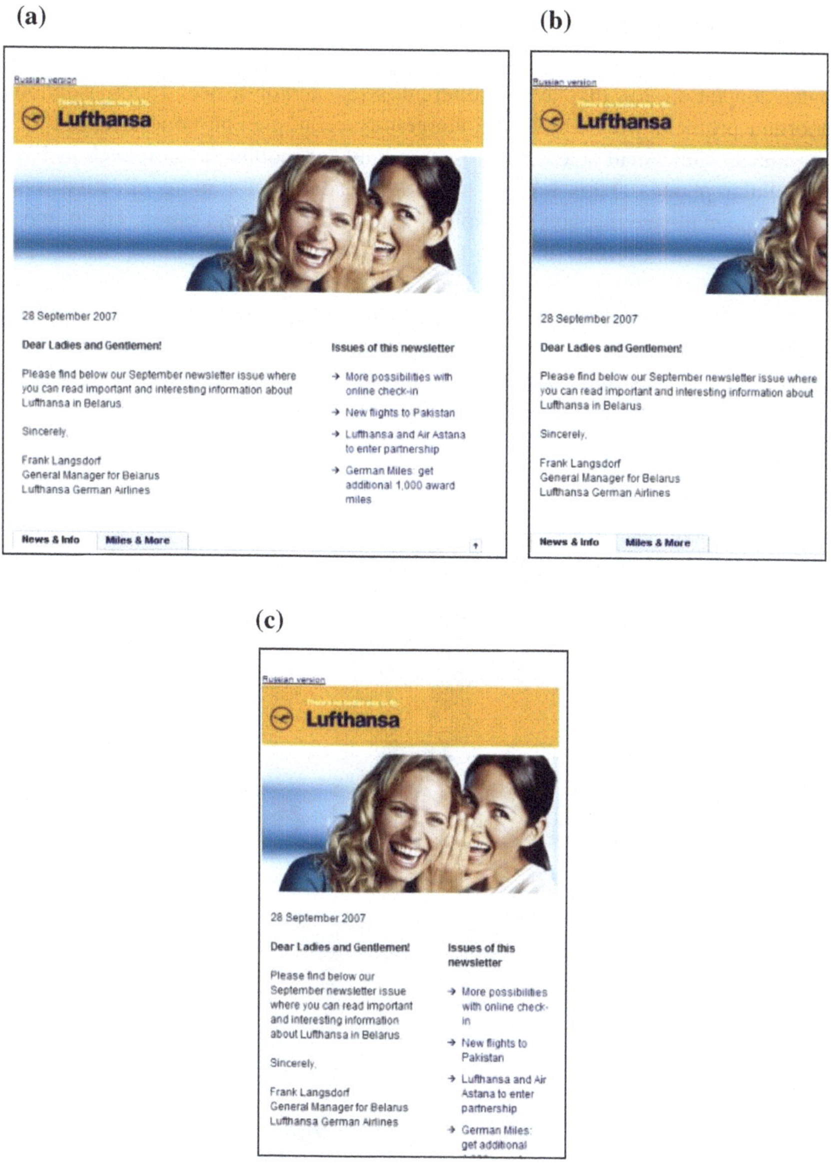

Fig. 10.11 **a** Original HTML document on high-resolution display; **b** the document on low-resolution display, scrolling is required; **c** retargeted document on low-resolution display

Figure 10.11 shows an example of HTML-document retargeting that allows fitting the document to a smaller low-resolution display without the need for scrolling. The conventional method requires scrolling up and down the page.

10.4.3 Matching to Display Size

Image adaptation for display at different sizes and/or aspect ratios is another important problem with which the retargeting technique can cope. The ability to view images on a small-size display is important for mobile devices, such as PDAs and cellular phones. Thus images have to be adapted for more effective reproduction on small-size displays or displays with different aspect ratios. The proposed method preserves objects of importance by minimizing the unimportant space between them.

In Fig. 10.12, a sample of retargeting of a wide image to a rectangular display is shown. The conventional interpolation technique (Fig. 10.12c) causes anisotropic stretching that makes important image objects unrecognizable while the retargeting technique provides an accurate image (Fig. 10.12b).

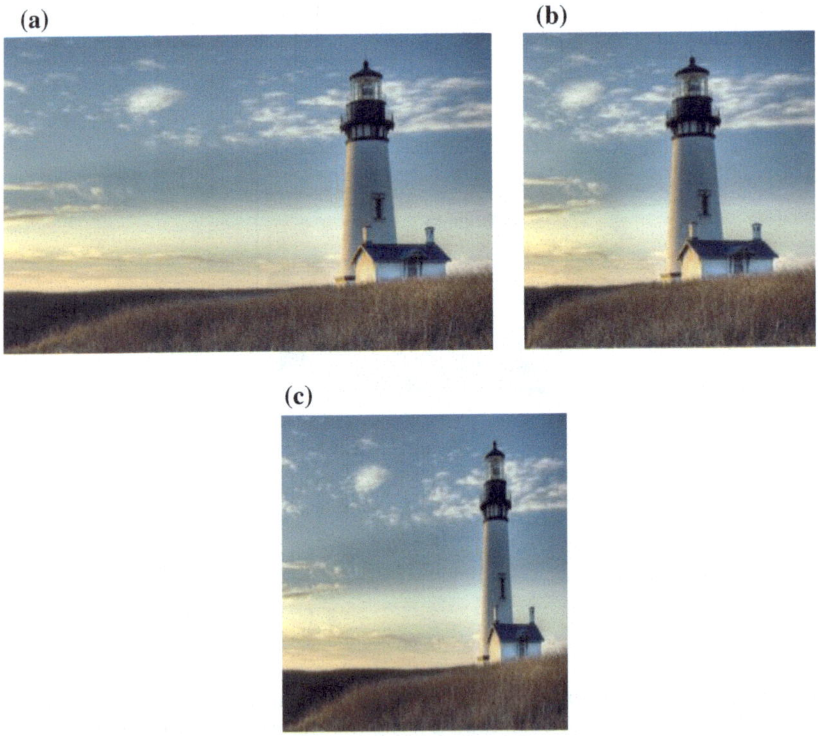

Fig. 10.12 **a** Original image; **b** retargeting 58% width; **c** resampling 58% width

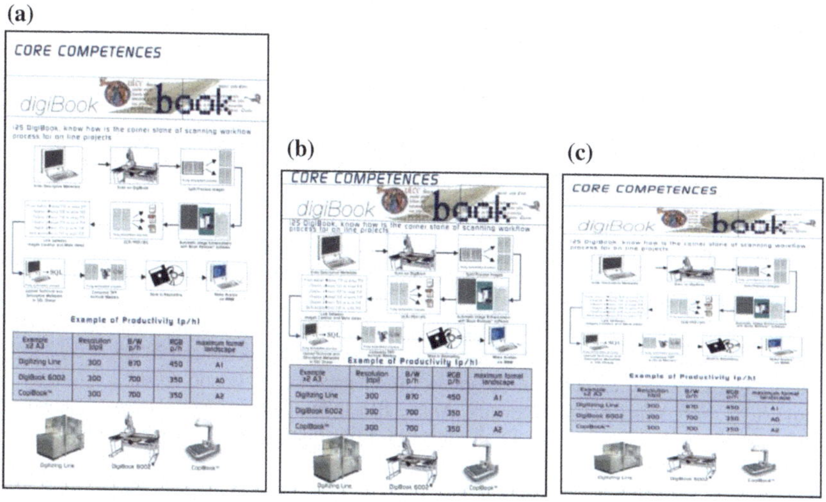

Fig. 10.13 Original document image of *legal* size (**a**); retargeting to *letter* size (**b**); resampling to *letter* size (**c**)

10.4.4 Retargeting in Digital Copiers

Another application of retargeting can be proposed for digital copiers and fax machines for cases when the paper size must be altered. The original paper-based document is scanned and retargeted. The retargeted image is printed. There is one important point about retargeting scanned images: if a paper-based document has a halftone structure then it is necessary to perform de-screening before retargeting. Otherwise the cost in haftoned areas will be too high.

Figure 10.13 illustrates retargeting of the original document from *legal* size to *letter* size during copying. Resampling to *letter* size distorts the proportions of objects in the document, which can sometimes be inadmissible.

An extension of the retargeting method is the change in the aspect ratio of video content based on the seam carving operator. There are many problems in doing so due to the temporal nature of the content, compression of video, and the huge amount of computational resources required for video stream analysis. Fast implementation of the video retargeting can be done with modern GPU APIs such as CUDA (Compute Unified Device Architecture) (Buck 2006), which provides the instruction set for GPU programming as an extension to the C programming language. It is our belief that a retargeting approach could be sped up by a factor of 10 with the GPU implementation.

Acknowledgements We thank the members of the following communities for publicly sharing their images: www.flickr.com mag3737 (A bench in Yaletown).

References

Avidan, S., Shamir, A.: Seam carving for content-aware image resizing. ACM Trans. Graph. **26** (3), 267–276 (2007)

Berkner, K., Erol, B.: Adaptation of document images to display constraints. In: Proceedings of IS&T/SPIE Electronic Imaging, vol. 6806, 68061C (2008)

Bhujbal, M., Deshpande, P.: Survey on content-aware image retargeting methods. Int. J. Eng. Res. Technol. (IJERT) **4**(02), 862–864 (2015)

Boykov, Y., Jolly, M.-P.: Interactive graph cuts for optimal boundary & region segmentation of objects in n-d images. In: International Conference on Computer Vision, (ICCV), vol. I, pp. 105–112 (2001)

Buck, I.: GeForce 8800 & NVIDIA CUDA: a new architecture for computing on the GPU. Workshop General-Purpose GPU Computing: Practice And Experience (2006)

Chen, B., Sen, P.: Video carving, Eurographics (Short papers), pp. 63–66 (2008)

Efimov, S.V., Nefyodov, A.V., Rychagov, M.N.: Block-based image exposure assessment and indoor/outdoor classification. In: Proceedings of 17th International Conference on Computer Graphics. GraphiCon, pp. 84–87 (2007)

Ghomsheh, A.N., Talebpour, A.: A new method for indoor-outdoor image classification using color correlated temperature. Int. J. Image Process (IJIP) **6**(3), 167–181 (2012)

Gomez, G., Morales, E.: Automatic feature construction and a simple rule induction algorithm for skin detection. In: Proceedings of the ICML Workshop on Machine Learning in Computer Vision, pp. 31–38 (2002)

Lin, X.: Intelligent content fitting for digital publishing. In: Proceedings of IS&T/SPIE Electronic Imaging (2006)

Liu, F., Gleicher, M.: Automatic image retargeting with fisheye-view warping. In: ACM UIST, pp. 153–162 (2005)

Rubinstein, M., Gutierrez, D., Sorkine-Hornung, O., Shamir, A.: A Comparative study of image retargeting. In: ACM Transactions on Graphics, Proceedings SIGGRAPH Asia 2010, vol. 29, no. 6, pp. 1–7 (2010)

Rubinstein, M., Shamir, A., Avidan, S.: Improved seam carving for video retargeting. ACM Trans. Graph. **27**(3), 1–9 (2008)

Safonov, I.V., Tolstaya, E.V.: Matching aspect ratio of digital photo via complementing. In: Proceedings of 17th International Conference on Computer Graphics. GraphiCon, pp. 154–157 (2007)

Serrano, N., Savakis, A., Luo, J.: A computationally efficient approach to indoor/outdoor scene classification. In: Proceedings of International Conference on Pattern Recognition, pp. 146–149 (2002)

Setlur, V., Takagi, S., Raskar, R., Gleicher, M., Gooch, B.: Automatic image retargeting. In: ACM International Conference on Mobile and Ubiquitous Multimedia (MUM), pp. 59–68 (2005)

Simakov, D., Caspi, Y., Shechtman, E., Irani, M.: Summarizing visual data using bidirectional similarity. In: IEEE Conference on Computer Vision and Pattern Recognition (CVPR 2008), Anchorage, pp. 1–8 (2008)

Vailaya, A., Jain, A., Hong, J.Z.: On image classification: city vs. landscape. In: Proceedings of IEEE Workshop on Content-Based Access of Image and Video Libraries, pp. 3–8 (1998)

Vaquero, D., Turk, M., Pulli, K., Tico, M., Gelfand, N.: A survey of image retargeting techniques, Proceedings of IS&T/SPIE Electronic Imaging, vol. 7798, p. 779814 (2010)

Wolf, L., Guttmann, M., Cohen-Or, D.: Non-homogeneous content-driven video-retargeting. In: IEEE International Conference on Computer Vision (ICCV '07), pp. 1–6 (2007)

Chapter 11
Sketch for Eco-friendly Printing

11.1 Introduction

Eco-friendly devices attract the attention of people who are responsible for saving the natural environment of our planet. Modern printing devices use various approaches for toner or ink economy at physical and firmware or software levels (Bernal et al. 2008; Safonov et al. 2011). One of them is a special draft or eco-friendly mode in a printing driver. The conventional way in which draft mode works is by decreasing the optical density, that is, the number of dots for the entire page. It significantly deteriorates the quality of the printed hardcopy. The edges of text characters become ragged. Text and halftoned images have low contrast. Human eyes become tired while viewing such hardcopy.

We propose an alternative way of printing pictures extracted from print-jobs. In draft mode, users want to preserve the recognisability of important items in pictures but they can refuse photorealistic quality of illustrations. Sketches or free-hand pencil drawings are fit for the task. Figure 11.1 demonstrates a photo and a colour sketch produced from the photo. Everyone can see the Moscow Kremlin on both pictures. However, the sketch consumes about 80% less toner.

A technique for conversion of pictures to sketches should satisfy the following requirements:

- a sketch created from any picture must remain recognizable and must preserve the main features and colours of the initial image; in particular, colour information is important for plots and diagrams;
- a sketch should contain as many white pixels as possible, because this requirement leads to toner saving;
- an image-to-sketch conversion algorithm must be quite fast in order to perform on-the-fly processing.

Non-photorealistic rendering and drawing of "pencil sketches" from arbitrary images are under continuous discussion in the field of image processing and

I.V. Safonov et al., *Adaptive Image Processing Algorithms for Printing*,
Signals and Communication Technology,
https://doi.org/10.1007/978-981-10-6931-4_11

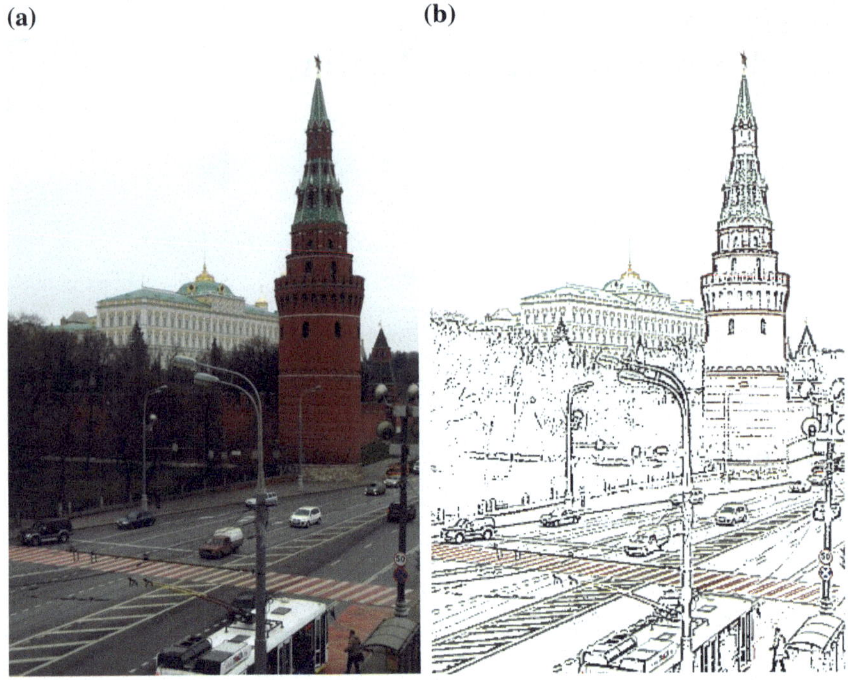

Fig. 11.1 **a** Initial image and **b** colour sketch by our algorithm

computer graphics. In general, there are two approaches to the generation of sketches: adaptive binarization and edge detection.

Mould and Grant (2008) and Xu and Kaplan (2008) describe algorithms for producing stylized black and white images from photographs via binarization. Wen et al. (2006) depict a similar approach to colour sketches. Images produced by those methods are incapable of providing efficient toner saving because they contain many areas of solid colour. The images can be converted into line pencil drawing by posterior morphological filtration, but this requires extra time. An approach that is more practical is adaptive binarization based on the Human Visual System model under conditions of changes in illumination (Vonikakis et al. 2007). The above-mentioned methods describe the main scene objects for images with normal contrast well but lose almost all of the small details as well as colours.

Kang et al. (2007) and Son et al. (2007) describe techniques for the generation of line drawings from photographs by means of sophisticated edge detection techniques. The authors consider various versions of Difference-of-Gaussians (DoG) with thresholding. In general, the methods provide a recognizable attractive picture, but the computational complexity is too high for on-the-fly processing. In addition, they result in many noise-like lines resulting from the contrast textured background and do not preserve tone and hue. The mentioned algorithms provide mediocre results for images with low global contrast. Figure 11.2 demonstrates

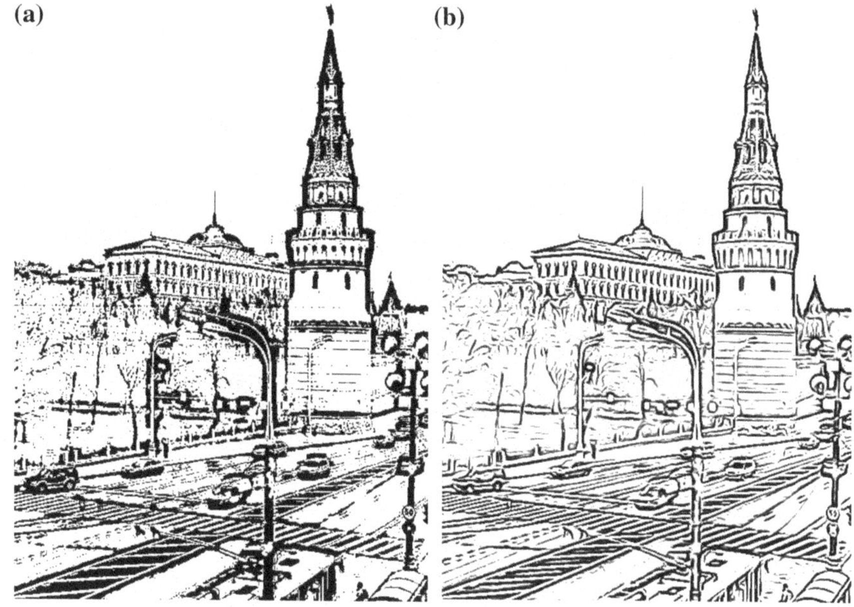

Fig. 11.2 **a** Sketch by adaptive binarization; **b** sketch by coherent line drawing

sketches produced by adaptive binarization by Vonikakis et al. (2007) and the Coherent Line Drawing algorithm by Kang et al. (2007). Both images are recognizable, but the majority of observers prefer the sketch in Fig. 11.1.

Our method is based on an edge-detection filter and the Human Vision model. A universal model of human vision does not exist, but the *Pre-Attentive Human Vision Model* based on feature integration theory is well-known (Itti et al. 1998). Such algorithms build *saliency* or *importance maps*. Usually a saliency map is applied for tasks related to scene analysis in order to determine the most valuable parts of the image, the so-called regions-of-interest (Privitera and Stark 2000).

11.2 Colour Sketch Creation

Our sketch creation algorithm realizes the following three ideas.

1. We carry out a global contrast enhancement of the image.
2. For the sake of preservation of tones and hue, a sketch is constructed by multiplication of the initial image by the mask containing the image edges.
3. To emphasize the main objects of the scene and suppress the textured background, the image is blended with its blurred copy, where the alpha-channel is a saliency map according to the Pre-Attentive Human Vision Model.

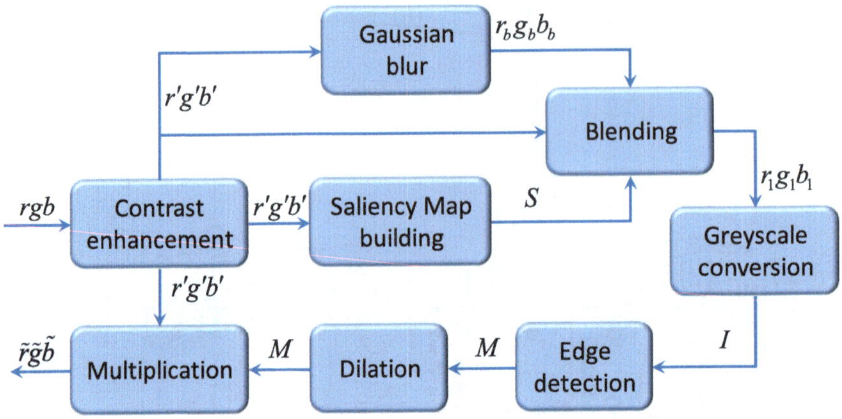

Fig. 11.3 A scheme of the colour sketch generation method

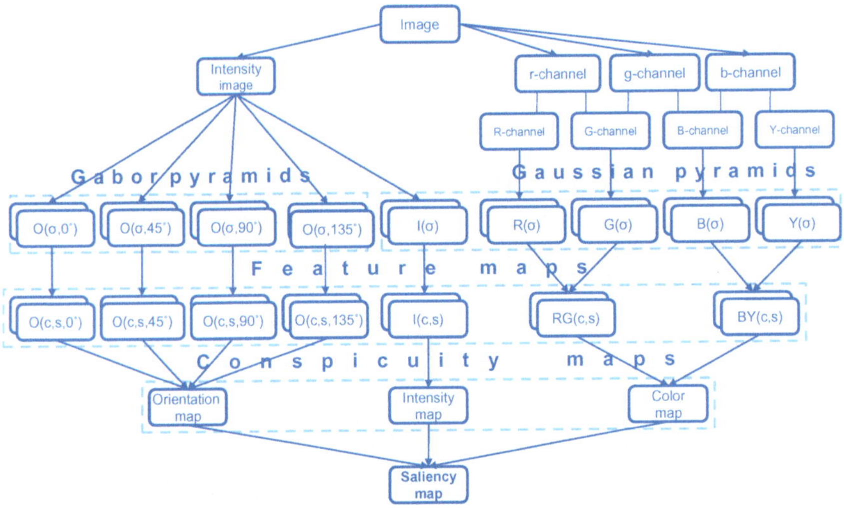

Fig. 11.4 A scheme of saliency map building

Figure 11.3 shows the scheme of the algorithm for colour sketch creation.

In order to make a sketch that looks good and is recognizable even when taken from a low-contrast picture, we carry out global contrast enhancement of the initial image. Chapter 1 presents a description of the algorithm for contrast improvement.

In the next step, we generate the *saliency map* S for enhanced image $r'g'b'$. Figure 11.4 shows a schema for building the saliency map. To achieve time-effective implementation, we use a downsampled copy of image. The intensity map is the greyscale image I:

$$I = \max(r', g', b').$$

We calculate four colour channels, R, G, B, and Y, from r', g', and b':

$$R = r' - \frac{g' + b'}{2},$$

$$G = g\prime - \frac{r' + b'}{2},$$

$$B = b' - \frac{r' + g'}{2},$$

$$Y = \frac{r' + g'}{2} - \frac{|r' - g'|}{2} - b'.$$

We construct eight-level Gaussian pyramids for I, R, G, and Y. From the intensity map I, eight-level Gabor pyramids for different orientations $\theta \in \{0, 45, 90, 135\}$ are created to obtain local orientation information. We compute 42 feature maps by using centre-surround differences:

$$I(c, s) = |I(c) - I(s)|,$$

$$RG(c, s) = |(R(c) - G(c)) - (G(s) - R(s))|,$$

$$BY(c, s) = |(B(c) - Y(c)) - (Y(s) - B(s))|,$$

$$O(c, s, \theta) = |O(c, \theta) - O(s, \theta)|,$$

where $c \in \{2, 3, 4\}$ and $s = c + \delta,\ \delta \in \{2, 3\}$.

We normalize all feature maps by using a local maximum technique and combining them into conspicuity maps by means of cross-scale addition:

$$\bar{I} = \sum_{c=2}^{4} \sum_{s=c+3}^{c+4} N(I(c, s)),$$

$$\overline{C} = \sum_{c=2}^{4} \sum_{s=c+3}^{c+4} [N(RG(c, s)) + N(BY(c, s))],$$

$$\overline{O} = \sum_{\theta \in \{0,45,90,135\}} N(\sum_{c=2}^{4} \sum_{s=c+3}^{c+4} N(O(c, s, \theta))),$$

where $N()$ is a normalization operator. The normalization operator comprises two steps. In the first stage, a Gaussian filter is applied to the image in order to decrease noise. The second stage is the computation of the average local maxima and multiplication of image pixels by the difference between the maximum value of the whole image and the local maximum. The operation helps to prevent the formation of isolated peaks and to avoid having a background that is too bright.

The saliency map S is a weighted sum of conspicuity maps $\overline{I}$ for intensity, $\overline{C}$ for colour, and $\overline{O}$ for orientation:

$$S = \frac{wI \times \overline{I} + wC \times \overline{C} + wO \times \overline{O}}{wI + wC + wO}.$$

The main issue is finding appropriate weights wI, wC, and wO. Due to normalization, various conspicuity maps make different contributions to the final result. Most the existing publications consider summing conspicuity maps in equal proportions. In our case, information about intensity and edge orientation is more important than colour information, so $wI = wO = 2$ and $wC = 1$. Finally, we scale the saliency map S to [0, 1]. Figure 11.5 shows a photo and its conspicuity and saliency maps.

The next step is blending the corresponding channels of the enhanced image and its blurred copy to emphasize regions of interest and to suppress the background slightly. The saliency map is the alpha-channel in the sense of transparency:

$$r_1 = r'S + r_b(1 - S),$$

$$g_1 = g'S + g_b(1 - S),$$

$$b_1 = b'S + b_b(1 - S),$$

where r_b, g_b, and b_b are the Gaussian blur of the r', g', and b' channels respectively.

Blended channels r_1, g_1, and b_1 are converted to a greyscale image I_1 as:

$$I_1 = (r_1 + g_1 + b_1)/3.$$

We create a binary mask M by means of thresholding of the Difference-of-Gaussian filter output:

$$M = \begin{cases} 1, & (I_1 * G_{\sigma_1} - I_1 * G_{\sigma 2}) < D \\ 0, & \text{otherwise} \end{cases},$$

where D is the predefined threshold, $I * G_\sigma$ is the Gaussian blur, $\sigma_1 = 1.4$, and $\sigma_2 = 1$.

Morphological dilation or smoothing of mask M is an optional step. The size of the structure element of the morphological operator or convolution kernel depends on the printing resolution.

The colour channels of our sketch are the outcomes of multiplication of the channel of the enhanced bitmap by the mask M:

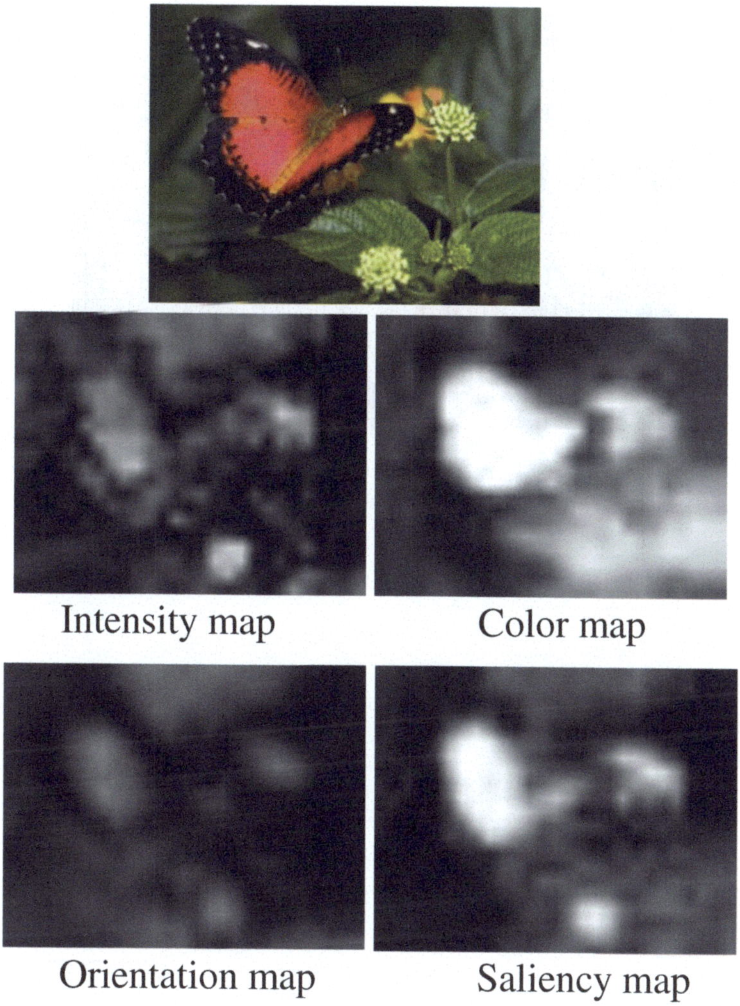

Fig. 11.5 Photo and its conspicuity and saliency maps

$$\tilde{r} = r_1 M, \quad \tilde{g} = g_1 M, \quad \tilde{b} = b_1 M.$$

11.3 Results

To estimate the picture-to-sketch conversion efficiency, we propose to evaluate the percentage of toner saved, which agrees well with the experimental results.

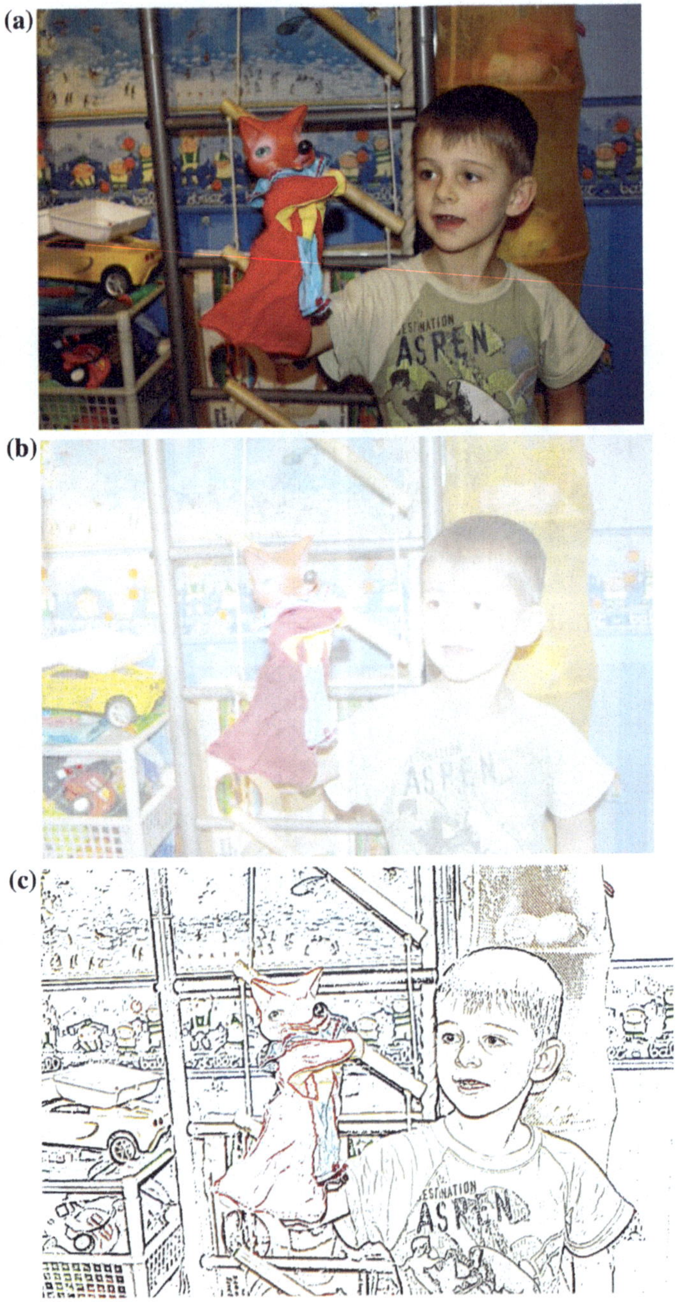

Fig. 11.6 Comparison of colour sketch for toner saving with a decrease in optical density

Each image pixel corresponds to several print dots. In black-and-white printing devices, the dots are black; in colour printing, they are colour. It is possible to construct a function of the percentage of printed dots depending on pixel value by means of prior calibration for every particular printing device. Therefore, it is possible to estimate the number of printed dots in an image.

The simplest way to estimate the percentage of consumable agent saved by using a converted image is:

$$Eb = \frac{Nb - Ns}{Nb} 100\%\,,$$

where Nb is the number of printed dots in the initial image and Ns is the number of printed dots in the sketch.

For colour printing, the percentage of toner or ink saved should be computed for each printed colour separately. Usually the percentage of consumables saved lies in the range of 75–90%.

Figure 11.6 shows an example of conversion of an initial colour image (a) to a colour sketch (c). The percentage of saved toner for a given bitmap is 83%. Image (b) is obtained by conventional approach by increasing the brightness via gamma-correction. Both image (b) and image (c) provide the same percentage of saved ink and allow the image content to be recognized, but (b) looks, according to observer evaluations, dull and ugly, whereas (c), on the contrary, looks nice.

In addition, we conducted a survey among 23 subjects. We showed them 10 sketches and asked them to describe or recognize the picture. All participants gave the correct answers. Therefore, the sketches are good enough for draft printing. Figures 11.7 and 11.8 demonstrate two examples of initial images and

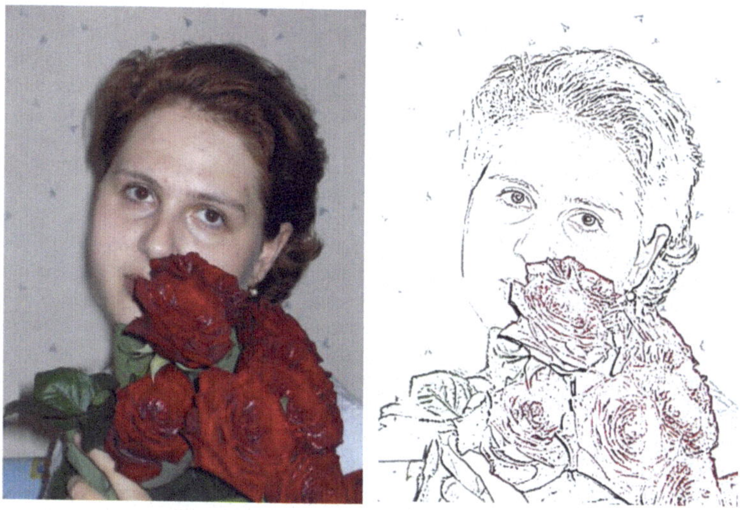

Fig. 11.7 Example 1 of initial photo and corresponding sketch

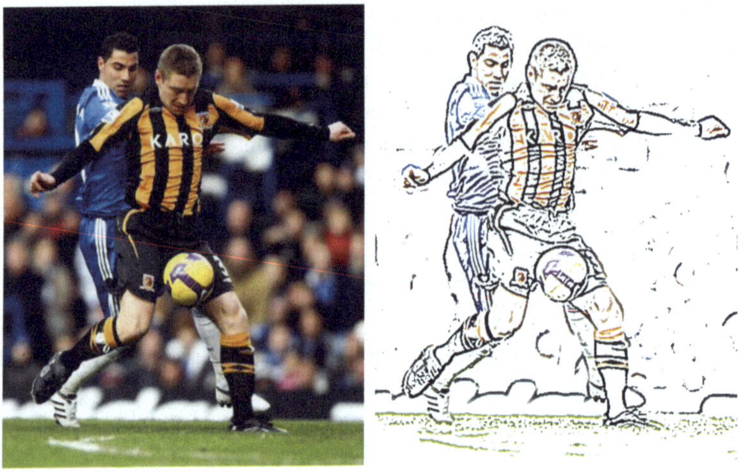

Fig. 11.8 Example 2 of initial photo and corresponding sketch

corresponding sketches from our survey. It is worth noting that the described algorithm was implemented in Samsung Easy Eco Printer Driver.

References

Bernal, E., Allebach, J.P., Trask, J.: Model-based memory-efficient algorithm for compensation of toner overdevelopment in electrophotographic printers. NIP 24 and Digital Fabrication (2008)

Itti, L., Koch, C., Niebur, E.: A model of saliency-based visual attention for rapid scene analysis. IEEE Trans. Pattern Anal. Mach. Intell. **20**(11), 1254–1259 (1998)

Kang, H., Lee, S., Chui, C.K.: Coherent line drawing. In: Proceedings of the 5th ACM International Symposium on Non-photorealistic Animation and Rendering, pp. 43–50 (2007)

Mould, D., Grant, K.: Stylized black and white images from photographs. In: Proceedings of the 6th ACM International symposium on Non-photorealistic animation and rendering, pp. 49–58 (2008)

Privitera, C.M., Stark, L.W.: Algorithms for defining visual regions-of-interest: comparison with eye fixations. IEEE Trans. Pattern Anal. Mach. Intell. **22**(9), 970–982 (2000)

Safonov, I.V., Kurilin, I.V., Rychagov, M.N., Lee, H.K., Kim, S.H., Choi, D.C.: Fast approach for toner saving. In: Proceedings of IS&T/SPIE Electronic Imaging, vol. 7866 (2011)

Son, M., Kang, H., Lee, Y., Lee, S.: Abstract line drawings from 2D images. In: Proceedings of the IEEE Pacific Conference on Computer Graphics and Applications, pp. 333–342 (2007)

Vonikakis, V., Andreadis, I., Papamarkos, N., Gasteratos, A.: Adaptive document binarization: a human vision approach. In: Proceedings of VISAPP, pp. 104–110 (2007)

Wen, F., Luan, Q., Liang, L., Xu, Y.Q., Shum, H.Y.: Color sketch generation. In: Proceeding of the 4th ACM International symposium on Non-photorealistic animation and rendering, pp. 47–54 (2006)

Xu, J., Kaplan, C.S.: Artistic Thresholding. In: Proceedings of the 6th ACM International symposium on Non-photorealistic animation and rendering, pp. 39–47 (2008)

Chapter 12
Content-Based Image Orientation Recognition

12.1 Introduction

Digital cameras have gained great popularity and they produce a huge volume of digital images. However, when the images are viewed they may not always be displayed in their preferred viewing orientation. The preferred viewing orientation is the orientation in which the image was captured.

Displaying images in their correct orientations is necessary in various image processing applications. For example, it is required during index printing, when an aligned set of image miniatures is printed onto photo paper, in image albuming (Loui and Wood 1999; Egorova and Safonov 2010) and display, and in auto-collage applications (Rother et al. 2006); it is also an important step in face detection methods, since many of them require a predefined face orientation. While manually adjusting orientations for several images is trivial, it is desirable to have an automated tool when having several hundred digital photographs taken from a field trip or vacation. One solution is to have the digital camera record, at the time of capture, the orientation information in the image file EXIF tag. However, this information is very often absent or incorrect. A more practical alternative is thus to design systems that are able to determine image orientations using image analysis.

12.2 Content-Based Orientation Recognition Techniques

Technically, the goal of automatic image orientation recognition is to classify an image into one of the four possible orientations corresponding to rotation angles of 0°, 90°, 180°, and 270°. Nevertheless, in practice, it is usually sufficient to determine whether an image is not rotated (0° orientation) or is rotated counter-clock-wise (CCW) or clock-wise (CW) (90° or 270° orientations, respectively), as it is rare that a picture is taken upside down.

I.V. Safonov et al., *Adaptive Image Processing Algorithms for Printing*,
Signals and Communication Technology,
https://doi.org/10.1007/978-981-10-6931-4_12

Automatic image orientation recognition is a relatively new research area in computer vision. Most of the early works focused on documents, and their success was largely due to the constrained nature of the problem (text cues). For natural images, the problem is considerably more challenging. Until the 2000s, there was little work on automatic image orientation recognition for natural images. Humans appear to use scene context and semantic object recognition to identify the correct image orientation. Early works are based on low-level features (such as colour, texture, and edges) that are processed with a learning engine to directly infer high-level information about the image (Wang and Zhang 2001; Vailaya et al. 2002; Szummer and Picard 1998). Modern computer vision algorithms for image recognition and classification are based on the application of neural networks (Fischer et al. 2015) and can solve recognition tasks very effectively; however, such systems require a huge volume of training data and the network itself can be quite complicated and represent a large volume of data, which is why such applications are very often organized as server-based ones. Here we describe a simpler approach with hand-crafted features and the AdaBoost classifier based on weighted voting of an elementary classifier committee (Friedman et al. 2000). The AdaBoost classifier can be stored and executed very effectively, as opposed to the majority of other classifiers.

Existing automatic image orientation recognition methods fall into two main categories. Top-down methods are based on high-level perception cues (i.e., the detection of faces, sky, and walls (Luo and Boutell 2005) or semantic relations in image content (i.e., a textured area in the lower part; Wang et al. 2003). However, top-down methods suffer from instabilities of current object detection and recognition algorithms and are more likely to be biased toward a particular set of training images. On the other hand, bottom-up methods determine image orientations with low-level features; examples include colour moments (Vailaya et al. 2002) and edge direction histograms (Wang and Zhang 2001; Zhang et al. 2002). Compared to high-level cues, low-level features are more robust and reliable.

A comprehensive study of the psychological aspects of image orientation recognition was presented by Luo et al. (2003). The experiment investigates the perception of the orientation of colour photographic images. A collection of 1000 images (a mixture of professional photos and consumer snapshots) was used in this study. Each image was examined by at least five observers and shown at varying resolutions. For each resolution, observers were asked to indicate the image orientation, the level of confidence, and the cues they used to make the decision. The results show that for typical images, the accuracy is close to 98% when using all available semantic cues from high-resolution images or 84% when using only low-level vision features and coarse semantics from thumbnails. The study also revealed that most useful and reliable cues used by humans at various image resolutions are sky, people, colour, texture, trees, and water.

However, while humans recognize thousands of objects and use them to make complex inferences about orientation, robust detection algorithms exist for only a handful of objects. Close-up images, low-contrast images, or images of uniform or homogeneous texture (e.g., sunset/sunrise and indoor images) pose serious problems for robust orientation estimation.

Psychophysical studies by Luo et al. (2003) also confirmed that low-level features are critical for human performance in determining image orientations.

12.3 Image Orientation Recognition

It seems natural to first classify images into portrait and landscape orientations, since the greater part of photos are taken in landscape orientation; among them, photos rotated by 180° are found very rarely. Further discriminating between images rotated by 90° and 270° is easier than discriminating between four possible rotations.

There are many potential features that can be used to represent an image. Features used for orientation recognition typically include colour moments (CMs), such as the component-wise mean and variance within each block, along with edge features, namely edge directions, edge direction histograms, the quantity of edge pixels, and so on, computed from the lightness component. Different features have different abilities to detect whether an image is portrait or landscape oriented. Since global image features are not invariant to image rotation, we prefer to rely on local regional features for classification.

12.3.1 Luminance and Chrominance Features

Usually an image is divided into N × N blocks and features are extracted from those fragments, where N is usually set between 7 and 10. Empirical knowledge suggests that the essential colour information for orientation recognition is usually embedded in the periphery rather than the central part of the image. Based on this idea, Wang and Zhang (2001) take into account only peripheral image blocks. The majority of authors prefer using the LUV colour space instead of RGB; however, in Baluja and Rowley (2005), several different colour spaces are used (RGB, YIQ).

First, we extract the luminance and chrominance features. An image of size M by N pixels is converted to YCrCb colour space and then divided by S horizontal blocks, such that the width of the block is equal to the width of the image M and the height constitutes a fraction N/S of the image height (Tolstaya 2007). As noted, with increases in the number of image fragments, classification accuracy increases, but complexity learning and classification processes increase significantly (Fig. 12.1).

In this way, for every colour component, a feature is represented as a vector, whose components are the chrominance or luminance characteristics $\{x_{1,\ldots,} x_{S*K}\}$, where K is the number of characteristics. The following characteristics of each fragment were computed:

Fig. 12.1 Scheme of image blocks. The width of the block is equal to the image width; the height constitutes a fraction N/S of the image height

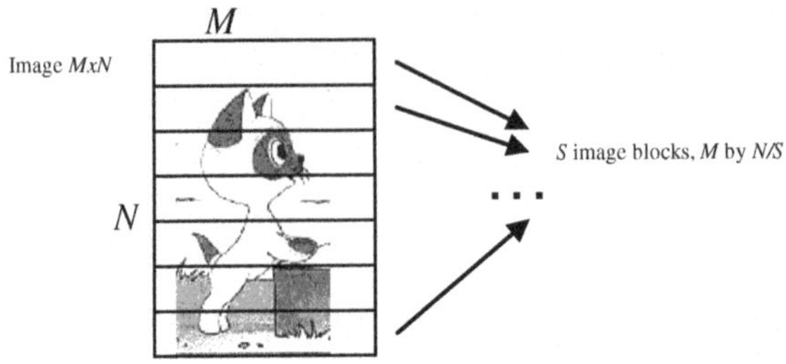

$$\left.\begin{aligned} x_{(p-1)K+1} &= \frac{S}{MN}\sum I(i,j), \\ x_{(p-1)K+2} &= \sqrt{\frac{S}{MN}\sum \left[I(i,j) - x_{(p-1)K+1}\right]^2}, \end{aligned}\right\},$$

where $i \in [1,\ldots, M]$, $j \in [(p-1)N/S,\ldots, pN/S]$, $p = [1\ldots S]$. The feature vector is formed by three vectors of each colour component. This vector is invariant to image flipping relative to the vertical axes. Then the image is divided into S *vertical* blocks, such that the height of the block is equal to the image height N and the width constitutes a fraction M/S of the image width, and the whole procedure of computing features is repeated.

12.3.2 Texture Feature

A texture feature is computed using RGB colour space. First, as previously, the image is divided by S horizontal blocks, such that the width of the block is equal to the width of the image M, and the height constitutes an N/S fraction of the image height. S is significantly smaller for a texture feature than for a colour feature. The mean values of R, G, and B channels are computed and stored in R^m, G^m, and B^m correspondingly. Then for each pixel inside the image block, an angle is computed according to the following formula:

$$A(i,j) = \left| \cos^{-1} \frac{R_{ij}R^m + G_{ij}G^m + B_{ij}B^m}{\sqrt{R_{ij}^2 + G_{ij}^2 + B_{ij}^2}\sqrt{R^{m2} + G^{m2} + B^{m2}}} \right|.$$

A histogram of the angle distribution is constructed. All angles are clipped to [0, π/4], because $R, G, B \geq 0$ and the absolute values of the majority of angles are smaller than π/4. Figure 12.2 provides an illustration of the described texture feature.

An *Angle Histogram* (AH) is generated for each of the S regions to characterize the image's structural and texture information. This choice is based on the

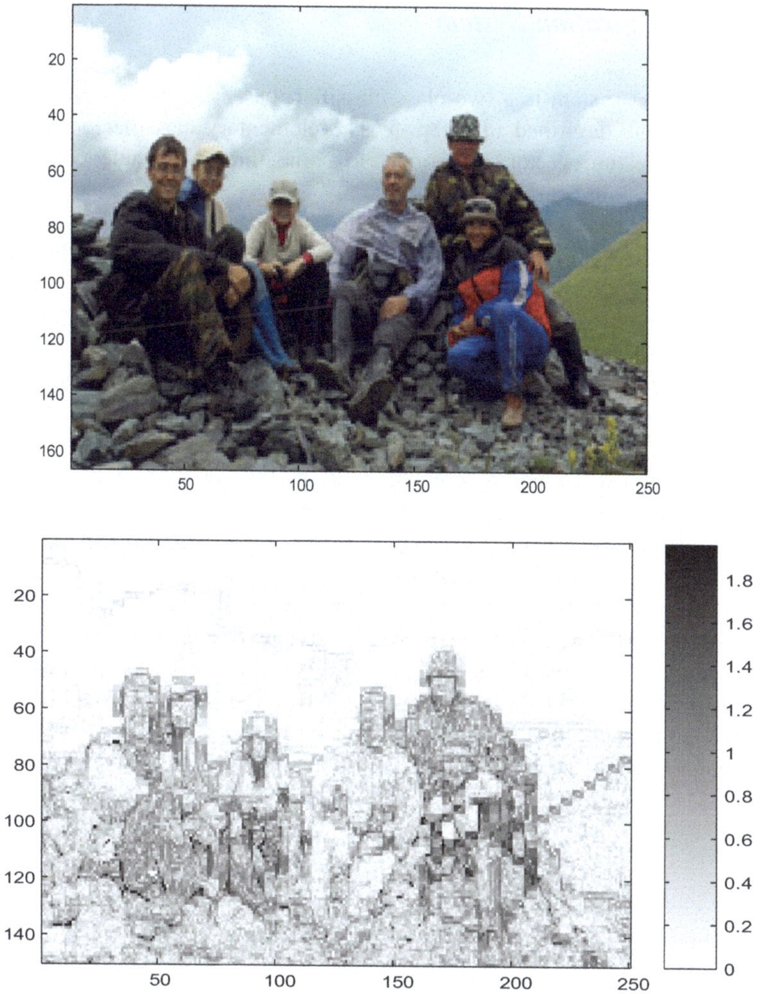

Fig. 12.2 Photo (upper panel) and image of angles (the photo is divided into 50 horizontal blocks; each pixel of the left image is computed according to the formula above) in logarithmic scale

observation that generally more texture is present in the lower part of the image than in the upper part.

This feature vector is invariant to image flipping relative to the vertical axes. The array of edge directions is divided into S vertical blocks such that the height of each block is equal to the image height N and the width constitutes a fraction M/S of the image width, and the whole of the procedure is repeated.

12.3.3 Classification System

The classification system is a two-class classifier, based on Real AdaBoost, very similar to the one described in Chap. 5. The elementary classifier compares the corresponding feature vector component to some threshold value computed in advance at the training stage.

A set of 800 images with known portrait and landscape orientations was selected for the training stage. Images from various sources were selected, including photos from different parts of the world captured by professional and amateur photographers, to prevent the classification model from concentrating on some particular colour combinations. Images were resampled to 200 × 200 pixels by nearest neighbour interpolation. Features were extracted from all images in the set. The number of images within each class was equal to 400.

A set of 861 images, again from various sources, was chosen for testing. Each image was assigned a random orientation of 0°, 90°, or 270° and then rotated accordingly; after that, the image features were extracted. The partitioning of portrait and landscape oriented images was kept at about half and half. Table 12.1 describes the distribution of scene types.

Every image from the testing set was processed according to the following procedure: first, the image was classified as landscape or portrait-oriented. If it was portrait-oriented, first it was classified according to whether it was to be rotated by 90° or 270°, and the probability p_1 was computed. The 90°/270° rotation scheme was improved by additionally rotating the images by 180° and subsequent 90°/270° classification and p_2 evaluation. The final probability is found as:

$$p = \begin{cases} p_1, & \text{if } |p_1| \geq |p_2| \\ -p_2, & \text{if } |p_1| < |p_2| \end{cases}$$

This improved classification scheme provided a significant gain in accuracy. Note that 180° rotations can be simply implemented as feature vector flipping.

Table 12.1 Distribution of scene types S in the dataset

Clear blue sky	87
Cloudy blue sky	122
Overcast sky	200
Sunset, sunrise	36
Night scenes	23
No sky or a small portion of sky	393

12.3.4 Rejection Scheme

Rejection is widely employed by many authors. It means that when the orientation recognition confidence is low, the image is marked as "not detected". Low confidence is typical for close-up views, uniformly textured images, and nearly diagonal rotations. Usually it is defined using small absolute values of SVM output probabilities and confirmed by setting some threshold. If the probability is smaller than the threshold, the orientation is not detected. In Wang and Zhang (2001, 2004), the image is rotated by 180° for testing the confidence, and if the detected orientation of the rotated image differs from the orientation detected earlier, the image is rejected (see Fig. 12.3). The authors also mention that complex rejection schemes do not provide a significant gain in classification accuracy, the gain being within 1%.

The probability distribution, which is the output of classifiers, measures the level of confidence in the detected orientation. A classifier without rejection will assign some orientation even if it is not confident in its choice, that is, when the probability is close to zero. But the majority of misclassifications occur exactly in these cases. So, it seems to be reasonable not to assign orientation when confidence is low and therefore to reject an image from classification. In our work, the rejection scheme is based on a preselected threshold T: if $P <= T$, the image is portrait oriented, and if $P > 1-T$, it is landscape oriented. The rejection rate shows how many images have been rejected relative to the image test set.

12.3.5 Results

The classification results on the training set were 94% for the first classifier (portrait/landscape) and 87% for the second classifier (90°/270° rotation). The overall figures for the classification scheme are presented in Table 12.2 (the number of images in the test set is equal to 861). Here, the classification accuracy is computed as the number of correctly oriented images that were not rejected divided by the total number of images that were not rejected (Table 12.3).

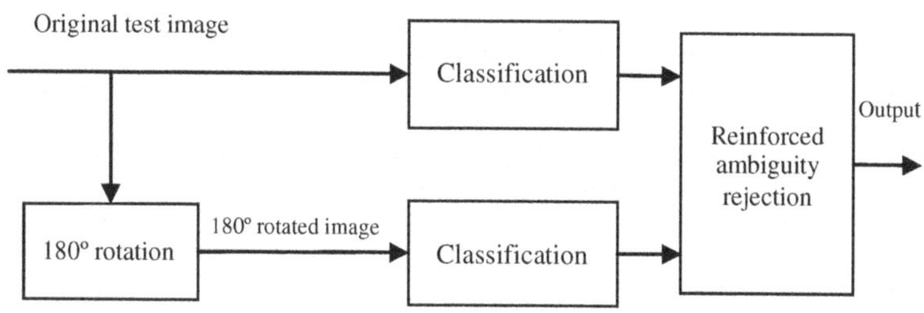

Fig. 12.3 Reinforced ambiguity rejection from Wang and Zhang (2001)

Table 12.2 Classification results for portrait or landscape oriented images (the actual number of images is given in brackets)

All (861 images)			
Rejection rate	0.4% (4)	13.7% (118)	40% (338)
Accuracy	87% (749)	88.5% (657)	90% (473)

Table 12.3 Classification results for 90° or 270° rotation (the actual number of images is given in brackets)

Rotated (586 images)			
Rejection rate	0.3% (2)	15% (92)	47% (275)
Accuracy	77% (451)	79.5% (392)	83% (259)

Table 12.4 Classification results reported by other authors

(Wang et al. 2003)	84–94%, depending on cue, and 94% overall
(Luo and Boutell 2005)	70–82%, without rejection
(Wang and Zhang 2001)	78 and 96%, depending on the classifier architecture (single or double layer), with a rejection scheme
(Wang and Zhang 2004)	78% (with 0% rejection) to 96.5% (with 50% rejection)
(Baluja and Rowley 2005)	27–96%, depending on image type, with the worst cases being backgrounds and close-ups

On comparing the results with other works, we can see that the proposed method delivers quite acceptable results (see Table 12.4).

References

Baluja, S., Rowley, H.A.: Large scale performance measurement of content-based automated image-orientation detection. IEEE Int. Conf. Image Process. **2**, 514–517 (2005)

Egorova, M.A., Safonov, I.V.: Selection of Photos for Album Building Applications, 20th IEEE International Conference on Pattern Recognition (ICPR), pp. 3256–3259 (2010)

Fischer, P., Dosovitskiy, A., Brox, T.: Image orientation estimation with convolutional networks. In German Conference on Pattern Recognition, Springer International Publishing, pp. 368–378 (2015)

Friedman, J., Hastie, T., Tibshirani, R.: Additive logistic regression: a statistical view of boosting. Ann. Stat. **38**(2), 337–374 (2000)

Loui, A.C., Wood, M.D.: A software system for automatic albuming of consumer pictures. In: Proceedings of the 7th annual ACM International Conference on Multimedia (Part 2), pp. 159–162 (1999)

Luo, J., Crandall, D., Singhal, A., Boutell, M., Gray, R.T.: Psychophysical study of image orientation perception. Spat. Vis. **16**(5), 429–457 (2003)

Luo, J., Boutell, M.: Automatic image orientation detection via confidence-based integration of low-level and semantic cues. IEEE Trans. Pat. Anal. Mach. Int. **27**(5) (2005)

Rother, C., Bordeaux, L., Hamadi, Y., Blake, A.: AutoCollage, ACM Transactions on Graphics (SIGGRAPH) (2006)

Szummer, M., Picard, R.: Indoor-Outdoor Image Classification. IEEE International Workshop in Content-Based Access to Image and Video Databases, Bombay, India (1998)

Tolstaya, E.V.: Content-based image orientation recognition. In: Proceedings of 17th International Conference on Computer Graphics. GraphiCon, pp. 158–161 (2007)

Vailaya A., Zhang, H., Yang, C., Liu, F., Jain, A.K.: Automatic image orientation detection, IEEE Trans. Image Process. **11**(7) (2002)

Wang, L., Liu, X., Xia, L., Xu, G., Bruckstein, A.: Image orientation detection with integrated human perception cues (or which way is up). Proc. Int. Conf. Image Process. **3**, 539–542 (2003)

Wang, Y., Zhang, H.: Detecting image orientation based on low-level visual content. Computer Vision and Image Understanding, pp. 328–346 (2004)

Wang, Y., Zhang, H.: Content-based image orientation detection with support vector machines. In: IEEE Workshop on Content-based Access of Image and Video Libraries, pp. 17–23 (2001)

Zhang, L, Li, M., Zhang, H.: Boosting image orientation detection with indoor vs. outdoor classification, Workshop on Applications of Computer Vision (2002)

Chapter 13
Anaglyph Printing

13.1 Introduction

At the present time, there are many sources of stereo images: 3D cameras, 3D movies, and stereo-pairs created from two frames captured by a conventional 2D camera or rendered by catching different views of a 3D object, for example, Google Earth and parallax effects in HTML 5. Examples of stereo-image sources are shown in Fig. 13.1.

Representation of 3D visual information by anaglyphs is the simplest and most economic method. However, this approach has some disadvantages such as loss of colour conformance and discomfort during continuous viewing. In spite of the drawbacks, printed anaglyphs have gained popularity during the last few decades. For example, in 2012, the famous magazine *SLINK* launched the experimental *SLINK 3D* edition, in which all photos were printed as anaglyph images. On the Internet, there are various user groups where the process of 3D colour anaglyph synthesis is discussed (Lo et al. 2010; Sanders and McAllister 2003).

There are several desktop and mobile software applications for the generation of anaglyphs, for example StereoPhoto Maker, Anaglyph Maker, Anaglyph, Anaglyph Workshop, and Z-Anaglyph. They have some shortcomings: the settings for reduction of the ghosting effect do not consider the size of the printed anaglyph and there is no adaptation for given glasses and colours of printer ink/toner. Usually it is assumed that viewing an anaglyph on a display and adjusting the printer colour profile is enough to obtain a printed anaglyph that looks similar. However, it is impossible to obtain similar colour perception while looking at a display and a colour hardcopy produced by a laser or inkjet printer. This can be explained by the difference between the gamuts of the display and printing devices (Foley et al. 1990). That is why printing device manufacturers are interested in raising the printing quality of anaglyph images (Zeng and Zeng 2011).

Measurements of the transmission coefficient for several stereo-glasses [$\tau_r(\lambda)$ for the right filter and $\tau_l(\lambda)$ for the left filter] as well as the reflection spectrum of

I.V. Safonov et al., *Adaptive Image Processing Algorithms for Printing*,
Signals and Communication Technology,
https://doi.org/10.1007/978-981-10-6931-4_13

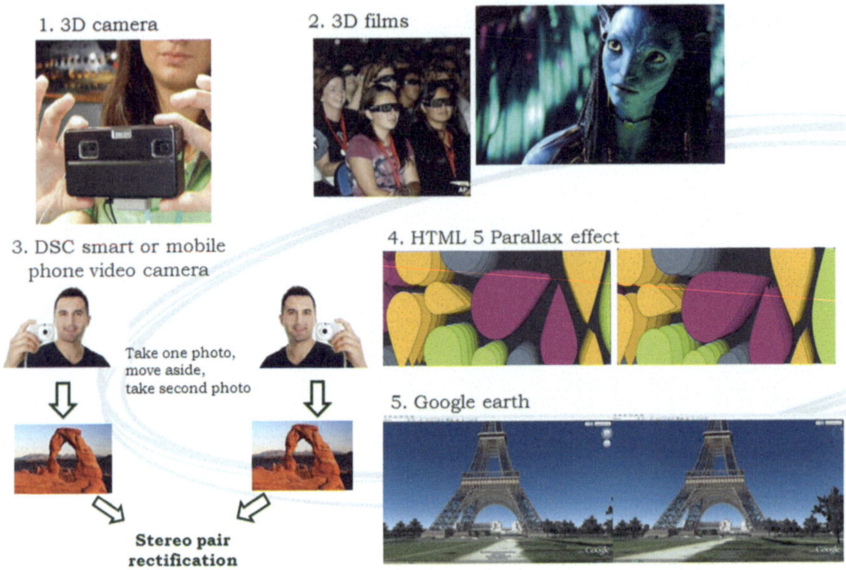

Fig. 13.1 Examples of stereo-image sources

printed colours, that is, magenta $g_M(\lambda)$ and cyan $g_C(\lambda)$, by spectrometer and analysis of the estimated reflection and transmission coefficients make it possible to conclude that full suppression of the cross-talk effect is impossible. Nevertheless, it is possible to reduce cross-talk interference by correct colour setting of a printed anaglyph. As proven experimentally, primary printed colours (cyan and magenta) are not transmitted well, but estimations showed that colour selection according to given glasses makes it possible to decrease ghosting artefacts significantly. In this chapter, we describe anaglyph generation methods and a printing algorithm for adaptation to the transmission characteristics of glasses and printed colours.

13.2 Anaglyph Generation Methods

Anaglyph generation is a difficult problem. Obviously, it is not sufficient to put one colour channel in the left-hand image and another one in a right-hand image. Contradictory problems must be solved during the anaglyph generation process. It is necessary to represent two images in a single view by means of coding of appropriate colour planes for creation of the stereo effect with colour reproduction that is close to natural. Due to stereo and colour conflicts, it appears to be impossible to develop an algorithm for anaglyph creation that always produces good colour representation and sharp details and is permanently free from typical artefacts such as ghosting and region merging.

Dubois (2001), Ideses and Yaroslavsky (2005), McAllister et al. (2010) describe several well-grounded techniques for colour anaglyph generation intended for displays. Sanders and McAllister (2003) discuss three approaches: a Photoshop (PS) algorithm and its variations, a least squares (LS) algorithm, and a midpoint (MID) algorithm that minimizes the sum of the distances between the anaglyph colour and the left and right eye colours in Lab colour space. Usually, the MID method produces excellent colours and details, but it may be subject to significant ghosting effects. Anaglyphs produced by the LS method are darker and contain poor details, but they have no significant ghosting. The PS method is easy to implement and works well for greyscale images but may also suffer from ghosting and poor colour representation.

Ideses and Yaroslavsky (2005) describe several methods for anaglyph enhancement. The enhancements substantially reduce unwanted ghosting artefacts, improve the visual quality of the images, and allow comfortable viewing in 3D as well as 2D mode.

McAllister et al. (2010) present a method for computing pixel colours in anaglyph images depending on knowing the RGB spectral distributions of the display device and the transmission functions of the filters in glasses. It requires the solution of a nonlinear least-squares problem for each pixel in a stereo pair and is based on minimizing the distances in Lab colour space.

Dubois (2001) describes a method that is adapted to the spectral absorption curves of the left and right filters of the anaglyph glasses. A projection technique is used to compute the anaglyph image that yields an image pair (after the glasses) as close as possible to the desired stereo pair. To generate an anaglyph it is necessary to move into the XYZ colour space using a transition matrix:

$$[C]_{kj} = c_{kj} = \int \bar{p}_k(\lambda) d_j(\lambda) \mathrm{d}\lambda,$$

where $\bar{p}(\lambda)$ is a colour-matching function, $d(\lambda)$ is the spectrum of a standard illuminant, $k = 1, 2, 3$ is an index of a colour channel (red, green, blue), $j = 1, 2, 3$ is an index of a colour channel, and λ is the wavelength.

Reflected light from an image passes through the colour filters of the anaglyph glasses and is transformed by two transition matrixes:

$$[A_l]_{kj} = a_{lkj} = \int \bar{p}_k(\lambda) d_j(\lambda) f_l(\lambda) \mathrm{d}\lambda,$$

$$[A_r]_{kj} = a_{rkj} = \int \bar{p}_k(\lambda) d_j(\lambda) f_r(\lambda) \mathrm{d}\lambda,$$

where $f_l(\lambda)$ and $f_r(\lambda)$ are the transmission functions of the left and right filters.

An anaglyph is generated by the following statements:

$$\widehat{V}_{an}(x) = N(R^T WR)^{-1} R^T WC_2 V(x),$$

$$V(x) = [V_{l1}(x) V_{l2}(x) V_{l3}(x) V_{r1}(x) V_{r2}(x) V_{r3}(x)]^T,$$

$$V_l(x) = [V_{l1}(x) V_{l2}(x) V_{l3}(x)]^T,$$

$$V_r(x) = [V_{r1}(x) V_{r2}(x) V_{r3}(x)]^T,$$

where N is the normalizing matrixes for the condition $\hat{V}_{aj} \in [0, 1]$, $N = diag(V'_{aj}/\hat{E}_{aj})$, V'_a is an anaglyph vector with maximum values, which is needed for normalizing $\hat{V}_a$, $V'_a = [111]^T$, $\hat{E}_a = (R^T WR)^{-1} R^T \mathrm{W} C_2 E$, W is a weighted matrix that allows the Y component to be weighted more heavily than X and Z to favour reproduction of the correct luminance, and $R = \begin{bmatrix} A_l \\ A_r \end{bmatrix}$, $C_2 = \begin{bmatrix} C & 0 \\ 0 & C \end{bmatrix}$.

13.3 Adaptive Anaglyph Generation

13.3.1 General Workflow

Our approach is based on the method described in Dubois (2001). We propose a method for adaptation to the spectral characteristics of stereo glasses and the colours of the ink/toner as well as several procedures to reduce the ghosting effect. A general scheme of adaptive anaglyph generation is presented in Fig. 13.2. Before anaglyph generation, a process of transmission function estimation is performed. For this purpose, we first print the colour pattern (see Fig. 13.2) on a target printer and then we estimate the transmission coefficients of given glasses. Then it is necessary to prepare a stereo pair. The preparation includes performing geometrical alignment, colour correction, and enhancement. Stereo pair colour correction is performed by the well-known method of histogram matching (Lo et al. 2010). After that, the stereo pair is enhanced to reduce unwanted artefacts.

There are two main concerns in anaglyph enhancement: decreasing the disparity range of a stereo pair and defocusing the colour components. Then the anaglyph generation is performed with the adapted transmission functions.

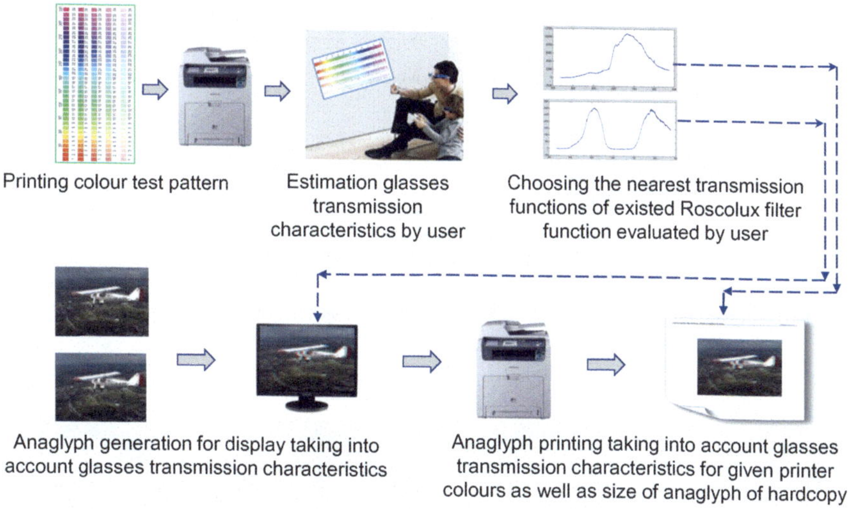

Fig. 13.2 General scheme of an adaptive anaglyph generation

13.3.2 Anaglyph Enhancement

Decreasing the disparity range of a stereo pair includes estimation of the average disparity value of the stereo pair by the following method:

$$\underset{d_i}{\operatorname{argmin}} \sum_{(x,y)\in W} |I(x,y) - I(x+d_i,y)|,$$

where d_i is the disparity that corresponds to the minimal values of SAD (Sum of Absolute Differences) in some region of the left and right images. We compute the average disparity $\bar{d}$ after estimation of the disparity map. The disparity decrease is produced by horizontal shifting of the stereo pair proportionally to $\bar{d}$. If the average disparity is less than four pixels, the stereo pair is not changed.

The main idea is that disparity should not be too large relative to the image size. Therefore, the horizontal shift is computed as $\mathrm{Shift} = \mathrm{Q}\,\bar{\mathrm{d}}$, where Q is a coefficient depending on the ratio of average disparity and image width W, as presented in Fig. 13.3. If the value of the ratio $\bar{\mathrm{d}}/\mathrm{W}$ is greater than 0.03 it might indicate erroneous average disparity estimation and a stereo pair should be shifted on a smaller value. All of the constants of the empirical approach were found during many visual experiments.

Figure 13.4a shows the anaglyph without disparity correction. The image on hardcopy has a ghosting effect. Figure 13.4b demonstrates the anaglyph with decreased disparity. For its generation, the stereo pair was shifted by several pixels. The ghosting effect for this anaglyph hardcopy is almost invisible.

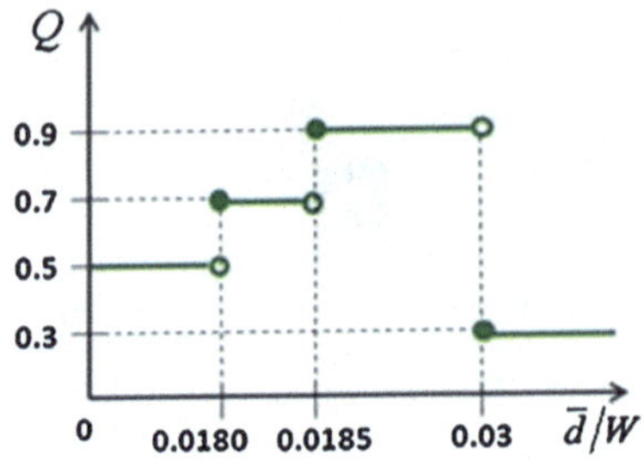

Fig. 13.3 Dependence of coefficient Q on the size and disparity of a stereo pair

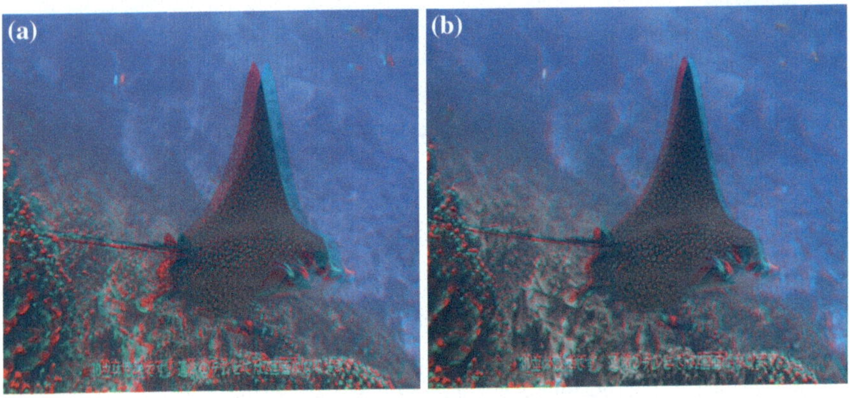

Fig. 13.4 Effect of disparity correction for anaglyphs: **a** without disparity correction, **b** with corrected disparity

A simple and effective way of decreasing crosstalk noise is by defocusing one colour channel for both images of a stereo-pair. We perform smoothing of the red or blue channel or both by means of a low-pass box-filter:

$$\tilde{I}(x,y) = \frac{1}{(2N_x+1)(2N_y+1)} \sum_{n_x=-N_x}^{N_x} \sum_{n_y=-N_y}^{N_y} I(x-n_x, y-n_y),$$

where $2N_x + 1, 2N_y + 1$ is the kernel size. An effective algorithm for box filter calculation is based on the summed area table (also well-known as the *integral image*) (Crow 1984). The value at any point (x, y) in the integral image *I* is just the sum of all the pixels of image *i* above and to the left of (x, y) inclusively:

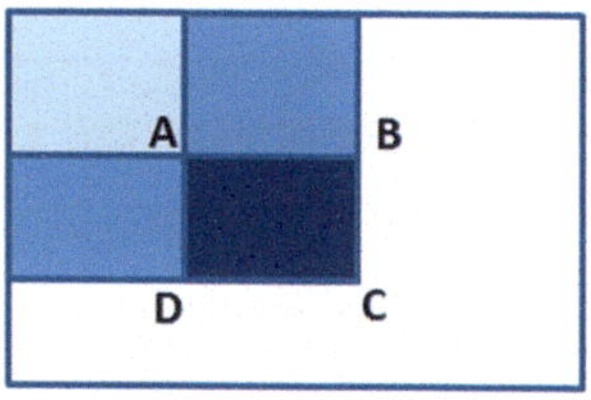

Fig. 13.5 Illustration of box-filter calculation via integral image

$$I(x,y) = \sum_{\substack{x' \leq x \\ y' \leq y}} i(x',y').$$

The integral image I can be computed efficiently in a single pass over the image i using the fact that the value in the summed area table at (x, y) is:

$$I(x,y) = i(x,y) + I(x-1,y) + I(x,y-1) - I(x-1,y-1).$$

The sum of the darkest box shown in Fig. 13.5 can be calculated from the integral image as:

$$\sum_{\substack{A(x) < x' \leq C(x) \\ A(y) < y' \leq C(y)}} i(x',y') = I(A) + I(C) - I(B) - I(D).$$

This method provides an identical processing time for any box size.

13.3.3 Estimation of Transmission Functions by User

To achieve good correspondence between glasses and printed colours, we propose to estimate the transmission functions $f_l(\lambda)$ and $f_r(\lambda)$ for given glasses and printer colours. This makes it possible to reduce the ghosting effect significantly and to improve the quality of printed 3D colour anaglyph images. In this case, we need to know the dependence of the digital colour components and reflection spectrum of these colours, that is, the dependence of the three colour components RGB and the wavelength λ.

For the evaluation of $f_l(\lambda)$ and $f_r(\lambda)$, we print a colour pattern (Fig. 13.6) on the target printer. This pattern is a colour table including all printable colours in HSL (Hue, Saturation and Lightness) colour space. This pattern reveals the dependence of the digital colour components and the reflection spectrum of these colours. *Hue* corresponds to wavelength, *lightness* is the average of the maximum and minimum values of the spectrum, and *saturation* is the distance from the maximum (or minimum) value to the lightness. The rows of the table are colours with various saturations (the step size is 0.2), and the columns are colours with different hues (the step size is 10°).

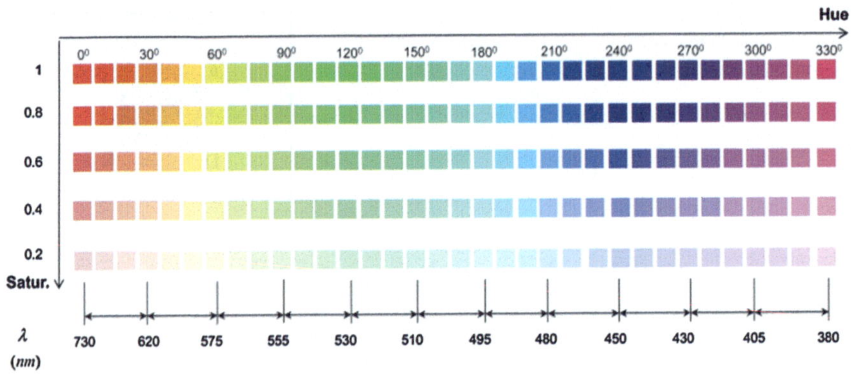

Fig. 13.6 Colour pattern

For the estimation of the transmission function of colour filters of glasses, the user should examine the pattern through the left and right filters of the anaglyph glasses separately. From the visibility level of the colour sample through the filter, the transmission function is evaluated and then applied for anaglyph generation. If the colour sample is invisible, it fully passes through the filter and the maximum of the transmission function is at that location.

Let us assume that the left filter is red and the right filter is cyan. First, the user should examine the first row of the pattern with maximal saturation through the red filter. We suppose that a colour with hue = $10° \pm 10°$ corresponds to red colour with a wavelength range of 700 ± 27.50 nm. If any colour from the band $10° \pm 10°$ is invisible, the maximum of the transmission function is in the range of 700 ± 27.50 nm. If the colour is visible, the user should choose the row with less saturation (one of the lower rows) and look at the corresponding colour sample again. If the colour is still visible, the user should choose the row with the least saturation for analysis. The value of the maximum of the transmission function depends on the saturation that can be extracted from Table 13.1.

Calculation of the transmission function for a cyan filter is similar. We use a cyan filter of glasses and suppose that cyan colour has hue = $180° \pm 10°$, which corresponds to a range of 495 ± 3.75 nm. The user should examine the test pattern and define the visibility level of each colour by the grade (0 … 5) which corresponds to transmission. For each clearly visible point we give a grade of 0. Values

Table 13.1 Dependence of maximum of transmission value on saturation

Row of colour pattern	Saturation	Maximum of transmission (MaxTrans)
1	1	0.90 (90%)
2	0.8	0.85 (85%)
3	0.6	0.65 (65%)
4	0.4	0.50 (50%)
5	0.2	0.45 (45%)

Table 13.2 Dependence of maximum of transmission value on saturation

Transmission	Grade
MaxTrans × 1	5
MaxTrans × 0.85	4
MaxTrans × 0.60	3
MaxTrans × 0.45	2
MaxTrans × 0.20	1
MaxTrans × 0	0

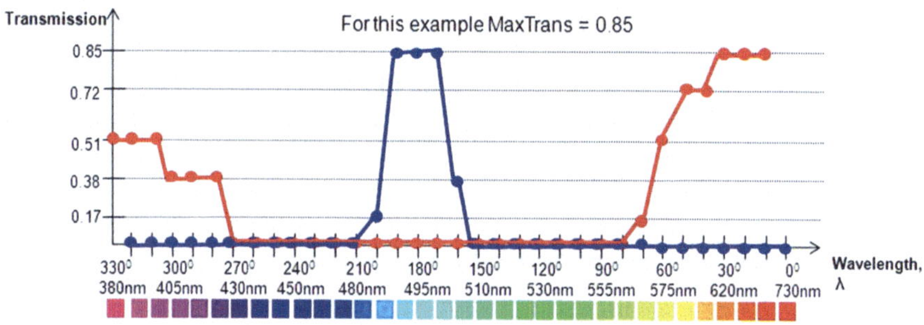

Fig. 13.7 Example of transmission function evaluated by user

of the transmission function depend on the maximum of transmission associated with the row of the colour pattern and the grade given by the observer (see Table 13.2). The user should mark only invisible or semi-transparent colours. In this way, we obtain a rough approximation of the transmission functions for particular glasses. One can see examples of such functions in Fig. 13.7.

The approach to transmission estimation can be used for any type of glass: green, orange, and so on. We suppose that orange colour has a hue = 40° ± 10°, which corresponds to 609 ± 11.25 nm, while green colour has a hue = 120° ± 10°, which corresponds to 530 ± 5.0 nm.

13.3.4 *Algorithm for Selection of Transmission Function of Existing Filter*

Usage of the transmission functions in such a "pure" form leads to an anaglyph image with the wrong colours. We propose to use known transmission functions of Roscolux filters and to calculate the functions of given glasses by selecting the nearest function by rough approximation. Roscolux filters are used in professional photography. We chose 10 filters that have spectra similar to the spectrum of commercial anaglyph glasses and transmission values of glasses computed by spectrometer.

To choose the appropriate filter or glasses with the transmission function F_i that is most similar to our computed transmission function, we note the position of the maximum of the function and its value. Then, we choose F_i as the estimation position of the maximum:

$$\left\|\lambda_{\max(F_i)} - \lambda_{\max(f)}\right\|_i \rightarrow min$$

Then, for the maxima with equal wavelength, we apply the following condition:

$$\|\max(F_i) - \max(f)\|_i \rightarrow min$$

For example, let us compare f (transmission functions for the red filter evaluated by the user) in Figs. 13.8 and F_i (real transmission functions of Roscolux filters) in Fig. 13.9. On the top of the figure, the transmission functions are presented; at the bottom, the enlarged part of the graph with the transmission functions are shown. By analysing the wavelengths that correspond to the maxima of the transmission functions, we can select one of the most appropriate filters.

$\lambda_{\max(f)} = 620$; $\lambda_{\max(F_1)} = 730$; $\lambda_{\max(F_2)} = 660$; $\lambda_{\max(F_3)} = 680$; $\lambda_{\max(F_4)} = 700$; $\lambda_{\max(F_5)} = 730$; $\lambda_{\max(F_6)} = 660$. We choose $F_2 = 0.88$, $F_6 = 0.89$, and $\lambda_{\max(F_2)} = \lambda_{\max(F_6)} = 660$ nm. Then for maxima with equal wavelength, we apply the condition of the nearest transmission and choose the Roscolux filter "Fire" with $F_2 = 0.88$.

The anaglyph in Fig. 13.10a is generated with transmission functions of some common stereo glasses. The hardcopy printed on a Samsung CLP-6240 printer has a perceptible ghosting effect for given glasses. The anaglyph in Fig. 13.10b is generated by transmission functions adapted for given glasses and printer colours. The ghosting effect on this anaglyph hardcopy is reduced significantly.

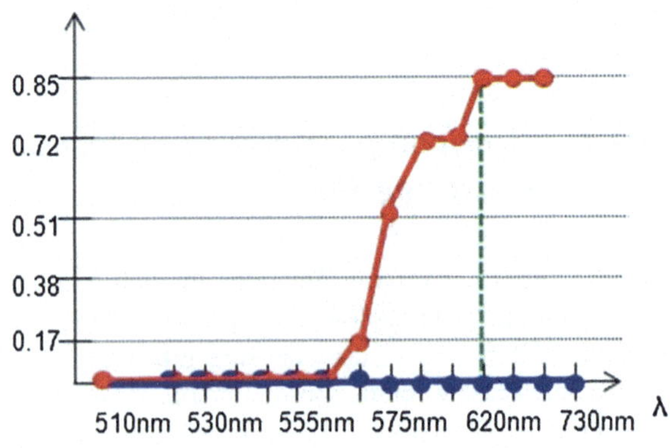

Fig. 13.8 Example of transmission functions f evaluated by user

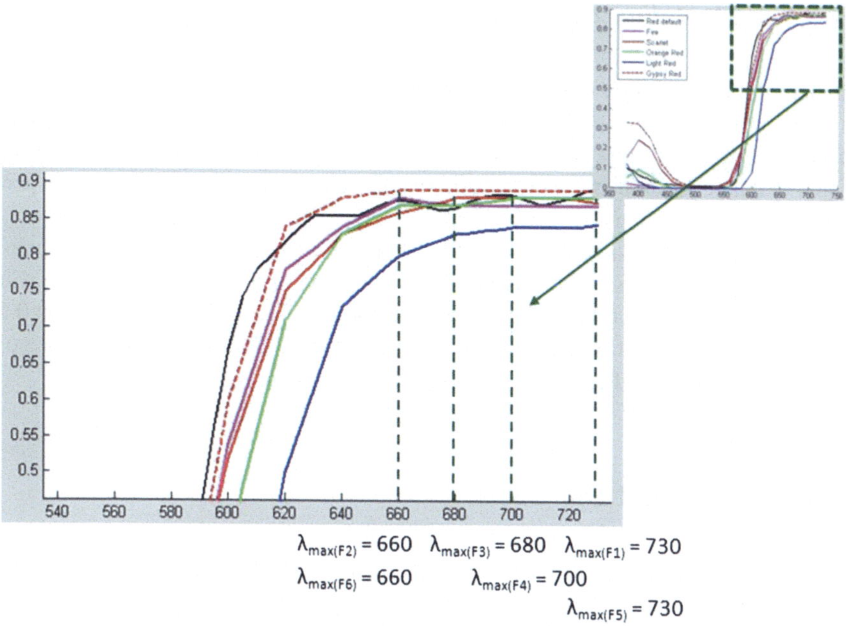

Fig. 13.9 Choosing the transmission function of Roscolux filters

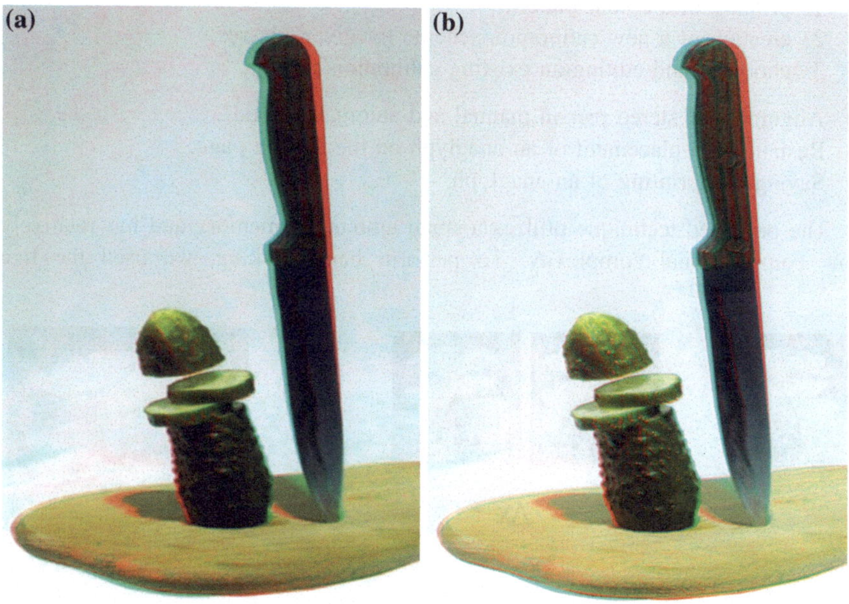

Fig. 13.10 **a** Anaglyph example for transmission functions of some common stereo glasses; **b** anaglyph example with adaptation for given glasses

13.3.5 Adaptation to Size of Hardcopy

Upsizing of an anaglyph image while printing leads to a large increase in disparity and the emergence of the ghosting effect. Downsizing leads to a large decrease in disparity and the disappearance of the stereo effect. We propose to decrease the disparity value by shifting stereo pair images relative to each other before performing the enlargement.

In the case of upsizing $\Delta x = -0.3\bar{d}$, where $\bar{d}$ is average disparity value computed for whole images of stereo pair, Δx is the value of the shift of the stereo pair images relative to each other to decrease the disparity (Fig. 13.11). The constant 0.3 is chosen as a result of a large number of experiments. Conversely, in the case of downsizing, $\Delta x = 0.3\bar{d}$ is the shift for the disparity increase (Fig. 13.12).

13.4 Results and Discussion

All of the mentioned algorithms were implemented in a software application. The user interface of the application is shown in Figs. 13.13 and 13.14. The software application allows the following actions to be performed.

- Calibration:
  1. printing of a colour pattern;
  2. creation of a new calibration scheme for given glasses;
  3. choosing and editing an existing calibration scheme.
- Alignment of stereo pair in manual and automatic modes.
- Resizing and placement of an anaglyph on the printed page.
- Saving and printing of an anaglyph.

The proposed technique utilizes a small amount of memory and has relatively low computational complexity. To perform benchmarking, we used the five

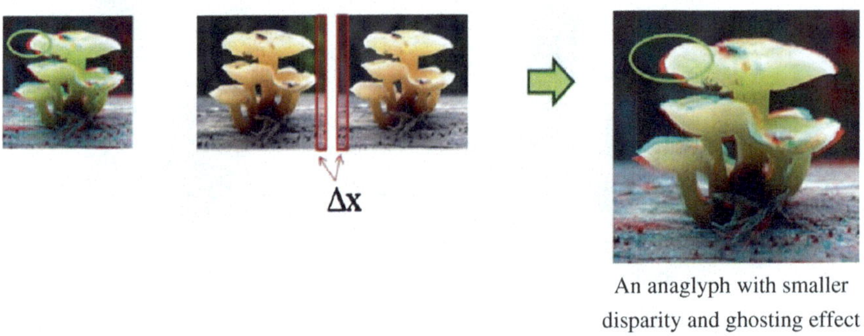

Fig. 13.11 Disparity modification in the case of upsizing

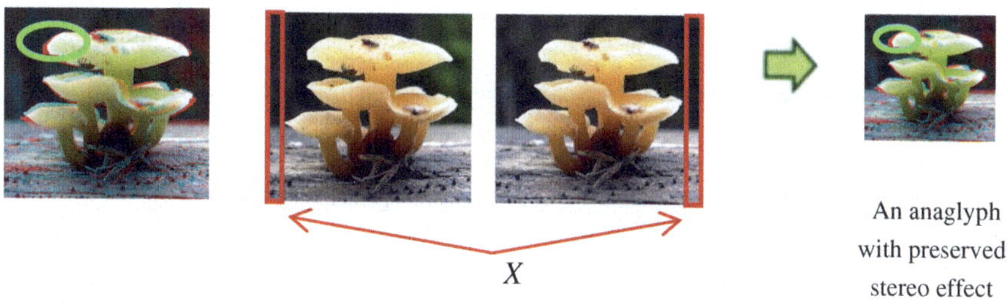

Fig. 13.12 Disparity modification in the case of downsizing

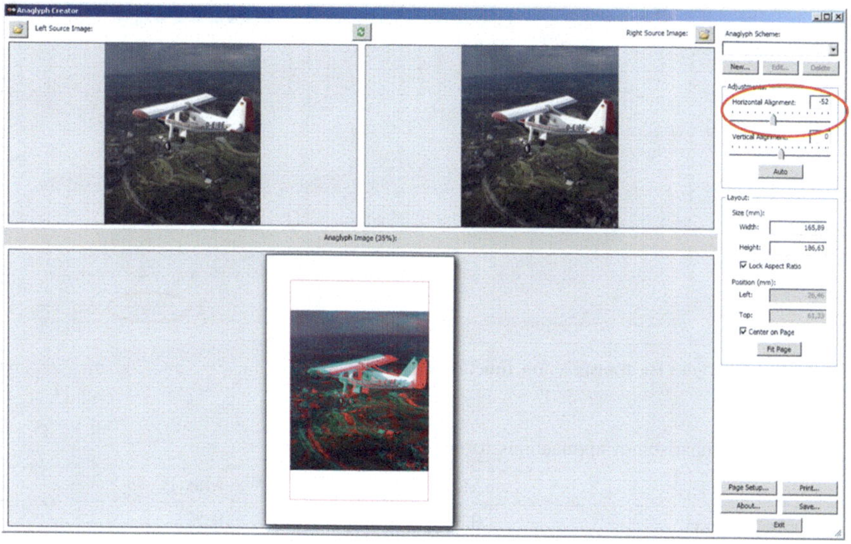

Fig. 13.13 Main window of the application for anaglyph generation and printing

abovementioned solutions for anaglyph generation and for our software application. Evaluation was done by 14 observers with the same viewing conditions and glasses. We screened the subjects for normal colour vision using Ishihara test plates. In our survey, we used a test set of six stereo pairs. The generated anaglyphs were printed on a Samsung CLP-6240 printer. We propose to calculate the quality factor of anaglyph generation as the weighted sum of two subjective estimations:

$$E = \alpha_1 V_1 + \alpha_2 V_2,$$

where V_1 is the subjective quality of 3D visualization, V_2 is the subjective level of colour naturalness, $\alpha_1 = 0.8$, and $\alpha_2 = 0.2$. V_1 and V_2 are changed from 0 to 1 with a step size of 0.2; lower values of V_1 and V_2 are better. We prioritized weights α_1 and α_2 by using the Analytic Hierarchy Process (Saaty 2001). Table 13.3 contains a comparison of the solutions for anaglyph generation analysed in our survey.

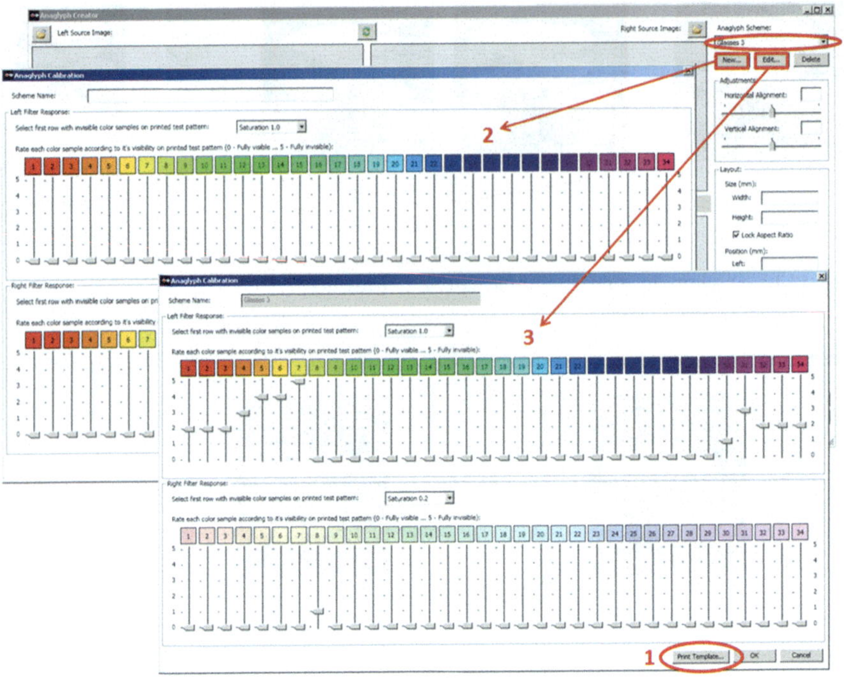

Fig. 13.14 Dialogue for transmission function estimation

Table 13.3 Comparison of applications for anaglyph generation

	Mean E	Max E
StereoPhoto maker	0.33	0.52
Anaglyph maker	0.57	0.72
Anaglyph	0.45	0.56
Anaglyph workshop	0.58	0.76
Z-Anaglyph	0.55	0.74
Proposed technique	0.30	0.58

In general, our algorithm outperforms all tested solutions. The anaglyph generated by StereoPhoto Maker is presented in Fig. 13.15a. The anaglyph on hardcopy has very noticeable cross-talk noise. The anaglyph generated by our method is presented in Fig. 13.15b. The picture looks better.

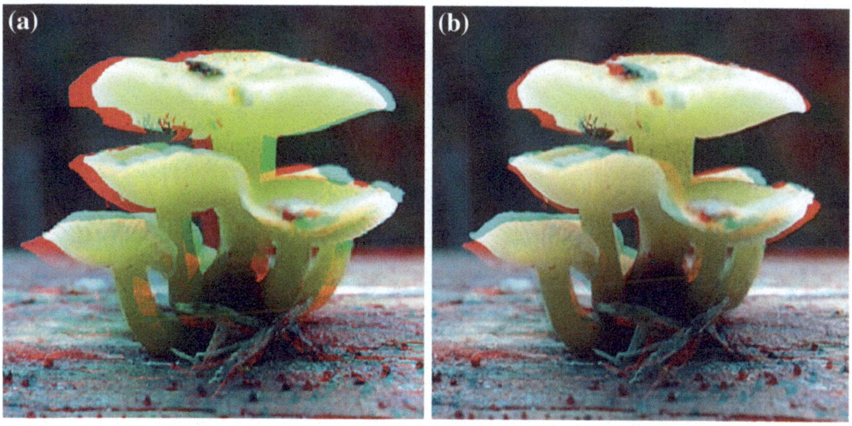

Fig. 13.15 **a** Anaglyph generated by StereoPhoto Maker; **b** anaglyph generated by our method

References

Crow, F.: Summed-area tables for texture mapping, SIGGRAPH '84: Proceedings of the conference on Computer graphics, pp. 207–212 (1984)

Dubois, E.: A projection method to generate anaglyph stereo images. Proceedings of ICASSP, IEEE International Conference on "Acoustic, Speech, and Signal Processing", (Vol. 3, pp. 1661–1664). IEEE (2001)

Foley, J.D., van Dam, A., Feiner, S.K., Hughes, J.F.: Computer Graphics. Principles and Practice, 2nd edn. (1990)

Ideses, I., Yaroslavsky, L.: Three methods that improve the visual quality of colour anaglyphs. J. Opt. A: Pure Appl. Opt. **7**, 755–762 (2005)

Lo, W.-Y., van Baar, J., Knaus, C., Zwicker, M., Gross, M.: Stereoscopic 3D Copy & Paste. Proc. of ACM SIGGRAPH Asia **29**(6) (2010)

McAllister, D.F., Zhoub, Y., Sullivanc, S.: Methods for computing color anaglyphs. In Proceedings of Stereoscopic Displays and Applications XXI, Proceedings of IS&T/SPIE Electronic Imaging, Vol. 7524 (2010)

Saaty, T.L.: Decision Making for Leaders: The Analytic Hierarchy Process for Decisions in a Complex World, New Edition, Analytic Hierarchy Process Series, Vol. 2 (2001)

Sanders, W., McAllister, D.F.: Producing anaglyphs from synthetic images. Proceedings of Electronic Imaging, pp. 348–358 (2003)

Zeng, R., Zeng, H.: Printing Anaglyph Maps Optimized for Display. Proceedings of Color Imaging XVI: Displaying, Processing, Hardcopy, and Applications, vol. 7866 (2011)

Chapter 14
Automatic Generation of Collage

14.1 Introduction

A photo collage is a set of pictures that are put together to make a single picture. It is a nice way to combine several images. Many users like creation and printing of photo collages, but they do not want waste time on manual selection and editing of photos. Automation of such processes is a topical theme. Algorithms for automatic photo-collage generation should include a photo selection process, which should be appeal visually to a user, and which should be constructed from photos that characterize a set of photos in the best way. It is assumed that a user makes only a preliminary choice of photos. Photos could be captured by different cameras, be of various sizes, have different quality and orientation.

There are many types of collages. Most frequently we see collages of the following three types:

- tiling or mosaic (see Fig. 14.1) (Chen et al. 2007);
- photos heap up on table (see Fig. 14.2) (Wang et al. 2006; Battiato et al. 2007), and;
- smooth transition between adjusted photos (see Fig. 14.3) (Rother et al. 2006; Wang et al. 2007).

To determine the most pleasant looking type of collage we conducted a survey. Survey participants were offered 5 sets of images with three different collages as in Figs. 14.1, 14.2 and 14.3, as well as with varying numbers of images in the collages: 4, 6, 8, 15 and 21. Thirty participants were asked two questions.

1. What collage is the most pleasant?
2. What collage contains an excessive number of photos?

Table 14.1 contains the response to the first question. The numbers in the cells of the table contain how many people chose the corresponding collage as the most pleasant. Most of the observers prefer the collage with a smooth transition between

I.V. Safonov et al., *Adaptive Image Processing Algorithms for Printing*,
Signals and Communication Technology,
https://doi.org/10.1007/978-981-10-6931-4_14

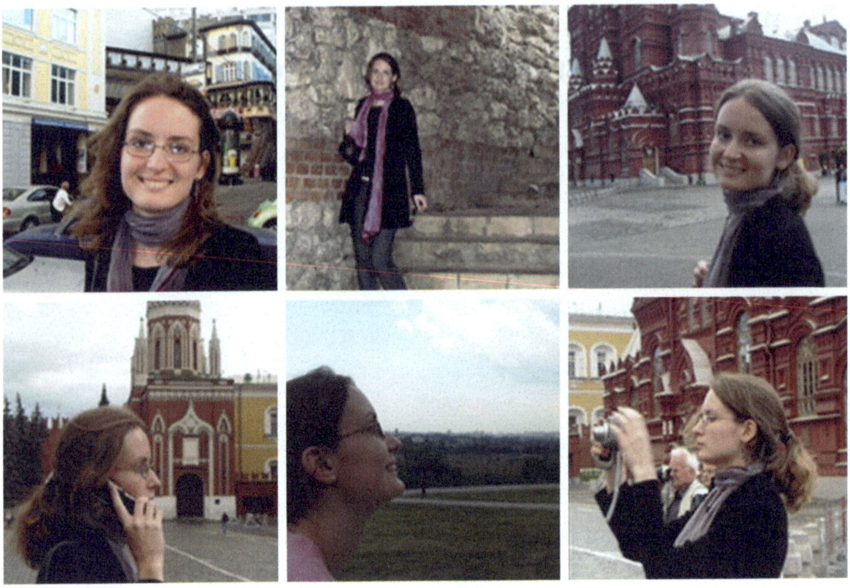

Fig. 14.1 Tiling collage

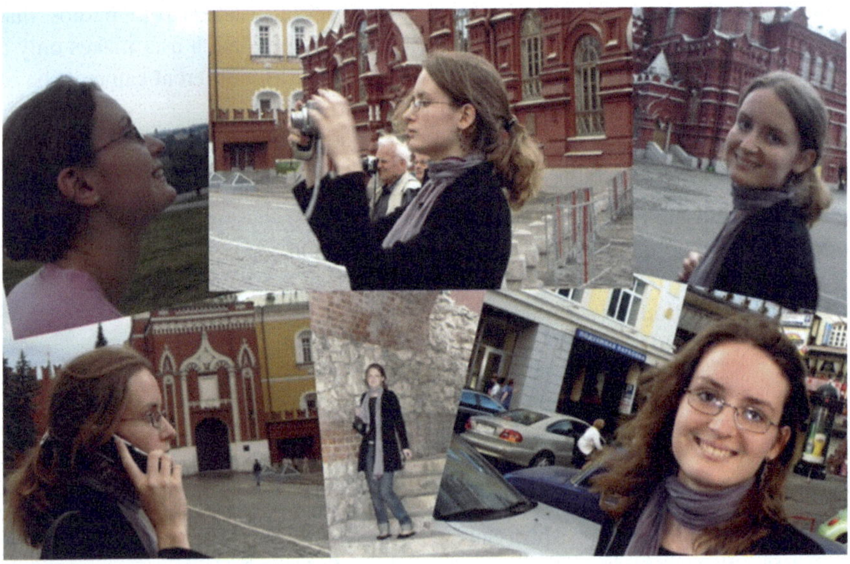

Fig. 14.2 Picture collage

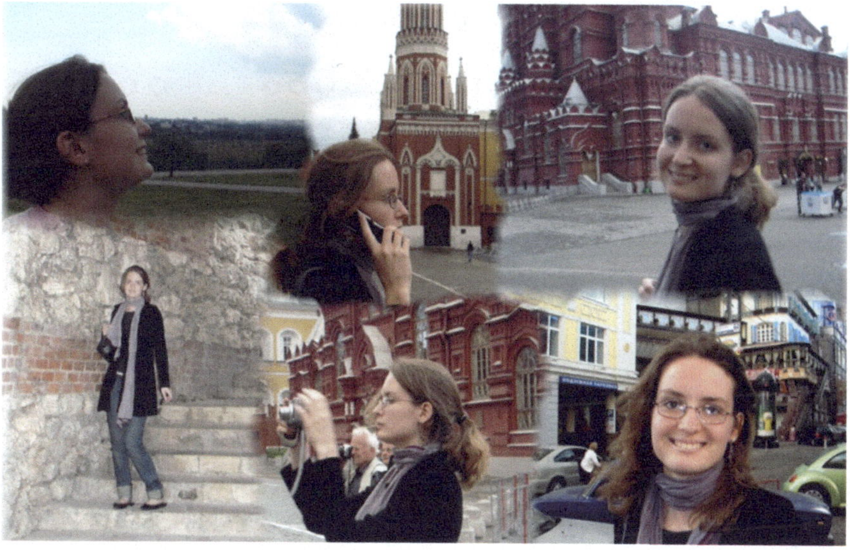

Fig. 14.3 Collage with blending

Table 14.1 Results of survey about collage pleasantness: how many people chose the corresponding collage as the most pleasant

Collage type	1	2	3	4	5	Total
Figure 14.1	5	3	12	12	9	41
Figure 14.2	7	7	5	2	1	22
Figure 14.3	18	20	13	16	20	87

photos, although the content of photos strongly effects on answers. The majority thinks that using more than 8 photos for a collage is excessive. Considering this, we decided to create a collage containing *M* photographs, where *M* varies from 3 to 8 photos.

14.2 Representative Image Selection

The selection of representative photos from a given collection is an important part of an automatic collage creation. Rother et al. (2006) describes an approach for the selection of images in three ways: textually *interesting*, mutually different and with the presence of faces in the image. *Interestingness* of the image is assessed applying entropy of histograms in *ab,* in colour space *Lab*. The mutual difference in distances between the histograms *ab*. Images with the greatest entropy are considered as an interesting. In our opinion, this is not the right way to consider interestingness from

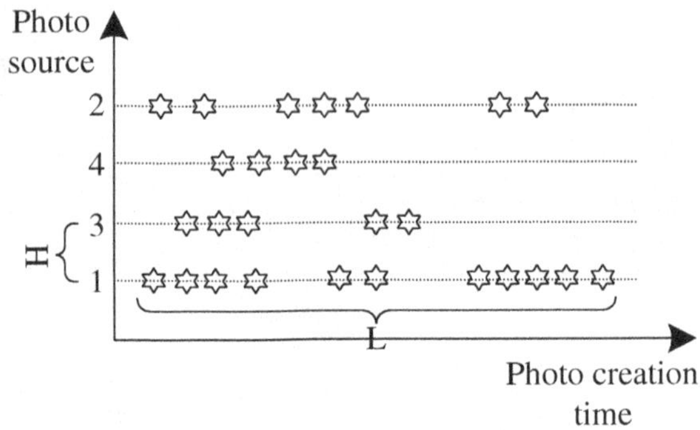

Fig. 14.4 Time-source coordinate system

the point of view of the information theory. For example, it is well-known that as image entropy increases the image noise also increases. Thus, images with high levels of noise are selected for a collage. The assessment of similarity using distances between histograms of images can lead to images of different content being identified as similar.

Chen et al. (2007) exclude from collection blurred, underexposed and overexposed photos, time-based clustering is also used to select photos for a slideshow. Algorithms for photo browsing apply time-based clustering as well (Platt et al. 2003; Huynh et al. 2005). Potapova et al. (2009) describes a procedure for photo selection comprising of image quality assessment and clustering in the *time-camera* coordinate system. Egorova and Safonov (2010) describe classifiers for estimation of contrast (*C*), sharpness (*S*) and compression artefacts (*A*) quality metrics. We borrow the measures for selection of photos with higher quality.

If all photos from a collection were captured by one camera, then time-based clustering provides reasonable selection outcomes. When we have a photo from different cameras, this approach is not as effective. We propose clustering in 2D Cartesian coordinates, where the first axis is *time* and the second axis is the *source* of the photo. The source is camera name from EXIF information. If EXIF is absent, then all images for every present file format are merged into a separate source. Figure 14.4 demonstrates the *Time-Source* coordinate plane. Sources are sorted in ascending order by the number of photos. *L* is the time between the least and the most time for the largest source. Yp_i coordinate on the *Source* axis is calculated as:

$$Yp_i = \begin{cases} H \times (i+1)/2 : & i\, is\, odd \\ H \times (Nps + 1 - i/2) : & i\, is\, even \end{cases},$$

where i is index of source, Nps is the number of sources, $H = L/M$.

The next stage is clustering into M classes by applying k-means techniques. From each class, a photo with maximum integral quality IQ_i is selected for collage:

$$IQ_i = C_i^{w_1} \times S_i^{w_2} \times A_i^{w_3},$$

where w_i are in the range [0, 1], and they are used to control of relative weights of the C, S and A quality measures.

If several photos have the same IQ, then the photo with maximal saturation is selected.

14.3 Photo Arrangement

Initial images can have either portrait or landscape orientation. We determine orientation using information from EXIF and turn the photo, if it is necessary. Unfortunately, part of the photo does not contain EXIF info or orientation tag. In this case, it is wise to use content-based algorithms of orientation detection (see Chap. 12).

Our approach for photo arrangement allows not only the optimal layout of photos on a canvas, but also the determination of scaling factors for images. To arrange the photos, we compose configurations of possible layouts in advance. For example, four possible configurations to lay out three images are shown in Fig. 14.5. Then, we construct a *layout matrix* for each configuration. Let us say w is a width of a minimal cell of a configuration, and h is the height of a minimal cell. The width of the layout matrix equals M. The height of the layout matrix equals the sum of the maximum number of cells in rows of given configurations and the maximum number of cells in columns. One in layout matrix indicates that the image occupies corresponding cell. Figure 14.6 demonstrates the layout matrix for configurations shown in the left-top corner in Fig. 14.5. Data in first column of layout

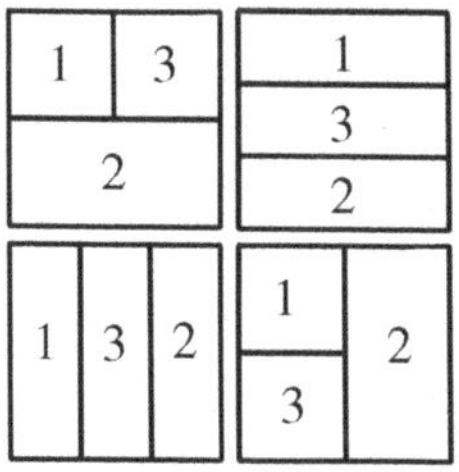

Fig. 14.5 Four possible configurations to lay out 3 images

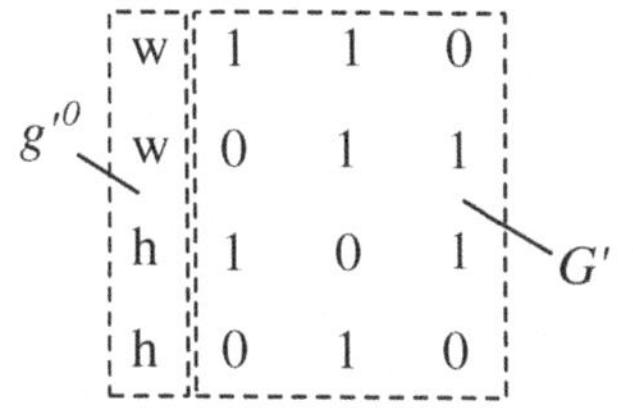

Fig. 14.6 Layout matrix for configuration shown in left-top corner of Fig. 14.5

matrix relates to image #1 of the configuration. The image occupies a smallest cell with w and h dimensions in the first row and the first column of the configuration. The second column relates to image #2. The image has width equal to $2w$, that correspond two columns of the configuration, and height equal h, that corresponds to the second row of the configuration. The third column of the layout matrix contains data for image #3. The image has a width equal to w, that corresponds to the second columns of the configuration, and height equal to h, that corresponds to the first row (Egorova et al. 2008).

To determine the scaling factors, we use the least-squares method:

$$\overline{\alpha} = (G^T G)^{-1} G^T g^0,$$

where $\overline{\alpha}$ is the vector of scaling factors for images forming a collage;

$$G_{ij} = \begin{cases} G'_{ij} w_i, & if\ g'^0_i = w \\ G'_{ij} h_i, & if\ g'^0_i = h \end{cases},$$

where w_i, h_i are width and height of ith image, g'^0 and G' are created for each configuration;

$$g^0 = \begin{cases} w_c, & if\quad g'^0_i = w \\ h_c, & if\quad g'^0_i = h \end{cases},$$

where w_c, h_c are canvas width and height correspondingly.

For example, canvas width is 800 and height is 600; selected images have sizes: (300, 400), (50, 150), (350, 150). For G' in Fig. 14.6 G, g_0 and $\overline{\alpha}$ equal:

$$G = \begin{bmatrix} 300 & 50 & 0 \\ 0 & 50 & 350 \\ 400 & 0 & 350 \\ 0 & 150 & 0 \end{bmatrix}, \quad g^0 = \begin{bmatrix} 800 \\ 800 \\ 600 \\ 600 \end{bmatrix}, \quad \overline{\alpha} = \begin{bmatrix} 1.29 \\ 4.54 \\ 1.47 \end{bmatrix}.$$

We compute a discrepancy as:

$$\left\| w(g_i) - g^0 \right\|,$$

where $w(g_i)$ is resulting canvas size, obtained from new sizes of images for a given scheme.

Using the discrepancy, the best layout scheme in terms of maximization of canvas area coverage and minimization of images overlap is determined. Then, we change scaling factors by 15% to be able to provide a seamless smooth transition between the adjacent pictures.

14.4 Seamless Blending

Figure 14.7 demonstrates the scheme of our seamless blending algorithms. The approach includes elements of randomness. Images I_1 and I_2 are overlapped partially. Fragments IF_1 and IF_2 are overlapping parts of the images. Figure 14.8 shows the system of coordinates for overlapping regions. The origin is left bottom angle of the regions OF; $x_{\max}$ is the width and $y_{\max}$ is the height of the region. Our aim is to create a seamless transition in the region OF. Figure 14.9 illustrates stages of blending for two images.

The first step is blurring of overlapped parts of the images by applying of finite impulse response (FIR) low-pass filter:

$$B_1 = h * IF_1,$$

$$B_2 = h * sIF_2.$$

Then, we make blending of the blurred images with respective parts of the initial images:

$$G_1(x, y) = \alpha_G(x)B_1(x, y) + (1 - \alpha_G(x))IF_1,$$

$$G_2(x, y) = (1 - \alpha_G(x))B_2(x, y) + \alpha_G(x)IF_2,$$

where α_G is smooth alpha-function with random slope:

$$\alpha_G(x) = (2^{x^{(1+rand(25)/10)}} - 1)/x_{\max},$$

where function *rand*(*x*) returns uniformly distributed pseudo-random value in the range [1, x].

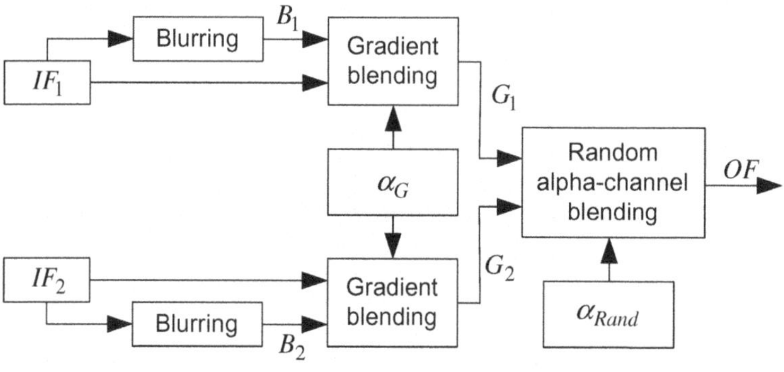

Fig. 14.7 Scheme of seamless blending with elements of randomness

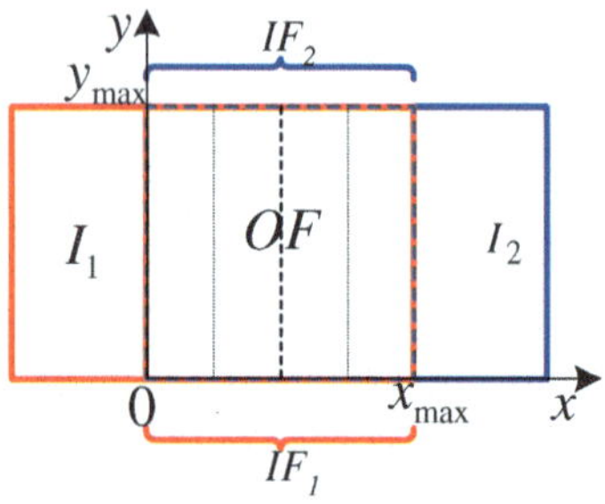

Fig. 14.8 Designations for overlapping regions

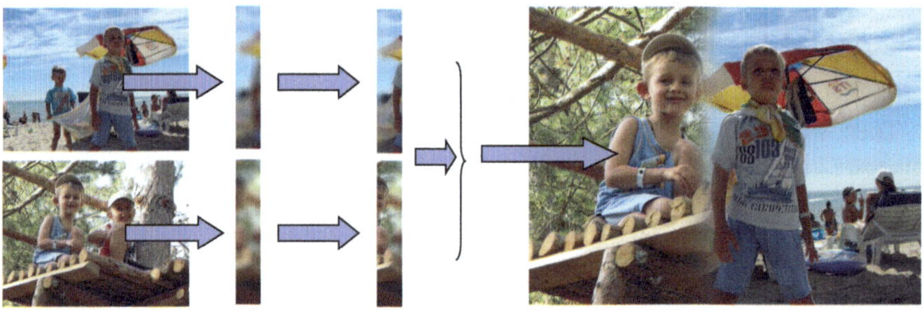

Fig. 14.9 Illustration of seamless blending for two images

The final step is the blending of the two photos via random alpha-channel α_{Rand}:

$$OF(x, y) = \alpha_{Rand}(x, y)G_1(x, y) + (1 - \alpha_{Rand}(x, y))G_2(x, y),$$

$$\alpha_{Rand}(x, y) = \begin{cases} 1, & x \le T(x, y) \\ 0, & x > T(x, y) \end{cases},$$

where $T(x, y)$ is the composite Catmull-Rom spline with the following control points P_i ($i = 0 \ldots N$):

$$P_0 = (0, 0),$$

$$P_1 = (x_{max}/2, 0),$$

$$P_{N-1} = (x_{max}/2, y_{max}),$$

$$P_N = (0, y_{max}),$$

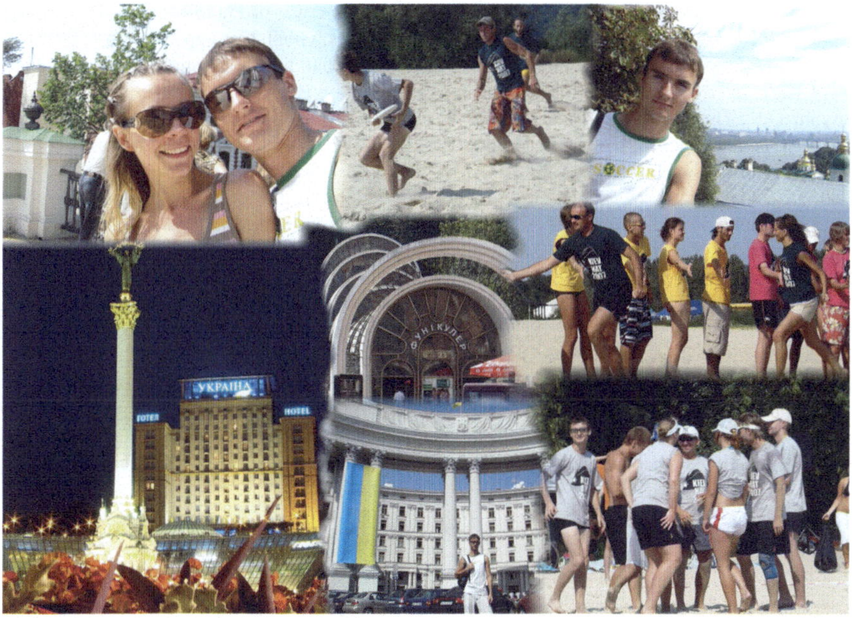

Fig. 14.10 Example of collage by our method

$$P_{ix} = x_{\max}/4 + rand(x_{\max}/2),$$

$$P_{iy} = P_{(i-1)y} + y_{\max}/30 + rand(y_{\max}/30).$$

Composite Catmull-Rom spline in parametric form is:

$$P(t) = \frac{1}{2}\begin{bmatrix} t^3 & t^2 & t & 1 \end{bmatrix} \cdot \begin{bmatrix} -1 & 3 & -3 & 1 \\ 2 & -5 & 4 & -1 \\ -1 & 0 & 1 & 0 \\ 0 & 2 & 0 & 0 \end{bmatrix} \begin{bmatrix} P_{i-2} \\ P_{i-1} \\ P_i \\ P_{i+1} \end{bmatrix},$$

where parameter t grows from 0 to 1.

Figure 14.10 shows collage automatically produced by our technique. According to survey, most observers were pleased with the collage.

References

Battiato, S., Ciocca, G., Gasparini, F.: Smart photo sticking. In: Proceedings of the 5th Workshop on Adaptive Multimedia Retrieval (2007)

Chen, J., Chu, W., Kuo, J., Weng, C., Wu, J.: Tiling Slidesho: an audiovisual presentation method for consumer photos. In: Proceedings of ACM Multimedia Conference, pp. 36–45 (2007)

Egorova, M.A., Safonov, I.V., Korobkov, N.V.: Collage for cover of PhotoBook. In: Proceedings of 18th International Conference on Computer Graphics. GraphiCon, pp. 160–163 (2008)

Egorova, M.A., Safonov, I.V.: Selection photos for album building applications. In: Proceedings of IEEE International Conference on Pattern Recognition (ICPR), pp. 3256–3259 (2010)

Huynh, D.F., Drucker, S.M., Baudisch, P., Wong, C.: Time quilt: scaling up zoomable photo browsers for large. In: Unstructured Photo Collections, CHI'05 Extended Abstracts on Human Factors in Computing Systems, pp. 1937–1940. ACM (2005)

Platt, J.C., Czerwinski, M., Field, B.A.: PhotoTOC: automatic clustering for browsing personal photographs. In: Proceedings of IEEE Pacific Rim conference on Multimedia, pp. 6–10 (2003)

Potapova, E.V., Egorova, M.A., Safonov, I.V.: Automatic photo selection for media and entertainment applications. In: Proceedings of 19th International Conference on Computer Graphics. GraphiCon, pp. 117–124 (2009)

Rother, C., Bordeaux, L., Hamadi, Y., Blake, A.: Autocollage. ACM Trans. Graph. **25**, 847–852 (2006)

Wang, J., Sun, J., Quan, L., Tang, X., Shum, H.Y.: Picture Collage. In: Proceedings of Computer Vision and Pattern Recognition Conference (CVPR), pp. 347–354 (2006)

Wang, T., Mei, T., Hua, X.S., Liu, X., Zhou, H.Q.: Video collage: a novel presentation of video sequence. In: Proceedings of IEEE International Conference on Multimedia and Expo, pp. 1479–1482 (2007)

The manufacturer's authorised representative in the EU is Springer Nature Customer Service Centre GmbH, Europaplatz 3, 69115 Heidelberg, Germany. If you have any concerns regarding our products, please contact ProductSafety@springernature.com

Printed and bound by CPI Group (UK) Ltd, Croydon, CR0 4YY
25/09/2026
02222892-0011